An Illustrated Guide to

COMPANION

Animal Breeds

By Teresa F. Sonsthagen
North Dakota State University

cognella® | ACADEMIC PUBLISHING

Bassim Hamadeh, CEO and Publisher
Kassie Graves, Director of Acquisitions and Sales
Jamie Giganti, Senior Managing Editor
Miguel Macias, Senior Graphic Designer
Mark Combes, Senior Field Acquisitions Editor
Sean Adams, Project Editor
Luiz Ferreira, Senior Licensing Specialist
Abbey Hastings, Associate Production Editor
Kat Ragudos, Interior Designer

ISBN: 978-1-5165-5340-2

cognella® | ACADEMIC PUBLISHING

Contents

AVIANS

Dedication

I would like to dedicate this book to my dad, who arranged a ride-along with our veterinarian who told me about the veterinary technician program starting at NDSU. If not for Dad and Dr. Malmedahl, I would have been a forest ranger!

Acknowledgements

I want to thank all of the students who over the years have kept me "honest" by asking, "what about (insert any breed or species of companion animal you can think of)?" as we went through our companion-animal breeds course. It kept me on my toes and opened my eyes to include other critters that people enjoy having as pets that I always thought of as "uncompanionable" (insert mouse)! They were right—a little copper-colored mouse or a big slobbery Newfoundland does have its own unique charm, so there is truly a pet for everyone in this big, diverse world!

I also want to thank my husband and cousins LJ and Gary for their encouragement and support while writing this book. I appreciated the short breaks we took together over the long winter to keep my perspective fresh.

Introduction

This book is designed to assist veterinary-medicine or veterinary-technology students studying identification of common breeds of domestic animals. The goal is to provide the student with basic information about each breed. It is not intended to be a diagnostic tool or be a definitive answer to potential problems.

The book is laid out utilizing the American Kennel Club (AKC) grouping of dogs. They divided 189 breeds of dogs into seven groups. These groups are generally based on what the breed was developed to do for man. The eight groups are Herding, Hound, Non-Sporting, Sporting, Terrier, Toy, and Working. The Cat Fanciers' Association recognizes 41 pedigree breeds for showing and are grouped as Longhair Specialty Breeds or Shorthair Specialty Breeds. Birds are grouped by passerine or parrot species and in alphabetical order. Pocket pets follow in alphabetical order.

Every breed or species of animal seems to have a group of people who are dedicated to the preservation and continuation of the breed or species. Information about breed characteristics, behaviors, and diseases was gathered from these and other sites and then compiled in a consistent format so that the student can find information quickly. Each animal will have a short paragraph on its origin or domestication, followed by general characteristics, and finally a list of often-seen genetic diseases or conditions.

Animals come in a variety of colors and color patterns which are often used to identify a specific breed or species. Each animal will have a Quick Facts Table that will provide the life span, height (if available), weight, coat color(s), coat type(s), and for the dogs general head, ear, and tail shape. There is also a Comparison Chart at the beginning of each chapter that will indicate which breeds or species are further discussed in the following chapter.

There are a number of terms that are also species specific. Each chapter will contain a nomenclature chart that will include terms used to identify genus species, male, female, offspring, group terms, and parturition terms. In addition, other terms will be in **bold** type throughout the chapters with the definitions available in the glossary.

Finally this work is not all-inclusive of every known breed or species of dog, cat, avian, or pocket pet out in the world! It would take several volumes to accomplish that job. What is in this body of work are those species that are fairly common or are regionally common in the United States that you may see in your career in veterinary medicine.

CHAPTER 1

Canines

- Herding Group
- Hound Group
- Non-Sporting Group
- Sporting Group
- Working Group
- Terrier Group
- Toy Group

Domesticated dogs can be traced back to 4000 BC. Where and how they became "man's best friend" is shrouded in mystery, but there is some conjecture. One popular thought is that wolf pups were taken in and raised by people. The pack nature of wolves kept them with their "pack," and their natural instinct for hunting as a pack and protecting territories were of benefit to the people. Another thought is that wolves are scavengers as well as hunters, and as people congregated together, their scraps were utilized by wolves that either developed or naturally had a small fight-or-flight radius. This means that they tolerated people coming in close, and eventually that radius became smaller and smaller until the "wolf" eventually moved into camp. The natural instinct to hunt as a pack and protect their territory again extended to their adopted pack. The people realized that this addition to their camp helped them with hunting and security. Either way it is known that all domestic dogs have 98% of the same genes as wolves and other wild dogs!

There are over 700 known breeds of dogs in the world. Where did they all come from? Genetic testing shows that the earliest dogs can be linked to northern Eurasia. It is thought that as man moved out from those areas, they needed different types of dogs to help them in different ways. Dogs were selected and bred for behavior, size, hair coat, and color. In a relatively short period of time, as measured in evolutionary time, dogs became specialized for hunting, herding, guarding, protecting, and companionship. People picked dogs that were smaller or larger, long haired or short haired, loyal or aloof, friendly or fearsome and bred them to each other and, when available, to other breeds to develop the type of dog they wanted or needed to help them in their daily lives. In return, dogs were fed, housed, and protected, thus allowing them to thrive. It is hard to imagine what man would be or would have been without dogs. The Table below indicates the nomenclature for dogs.

TABLE 1.00 Dog Nomenclature

Genus species	Intact Male	Intact Female	Young	Group	Parturition
Canis lupus familiaris	Dog	Bitch	Puppy	Pack Brace = 2	Whelping

Purebred dogs are those breeds that are recognized by a group or organization as "breeding true" to form and function. Specific height, weight, conformation, coat type, and color are set as a breed standard. All dogs of a particular breed must conform to this standard, and when bred to each other, their offspring also conforms to this standard. Developing a purebred dog takes an immense amount of time and diligence in selecting appropriate partners to breed true to the desired form and function. To get the desired look, size, or behavior, parent-offspring matings were used which often resulted in genetic diseases and conditions to develop along with the breed. Good breeders are constantly working towards eliminating these issues and will cull a bloodline to prevent the continuation of sometimes horrible diseases or conditions. Unfortunately, some breeders care only for the profit that can be made by the sale of puppies without regard to health. If selecting a purebred dog, you should ask that both parents be seen, and you should ask questions about health and behavior issues before selecting a puppy.

Cross-bred dogs are the offspring of the mating of two purebred dogs. In recent years this practice has produced a variety of "designer" dogs with cute names. For example, Labradoodles are crosses involving Labradors and Poodles, and Puggles are a Pug and Beagle cross. Often these cross-breeds seem to have the best characteristics of the parents although they often inherit the genetic conditions and diseases from both sides as well. The other downside of this is that people are willing to pay exorbitant prices for these cross-breed dogs that are often more than what the purebred parent dog costs. The same advice for selecting a purebred puppy goes for selecting a "designer" puppy.

Mutts, mongrels, or mixed-breed dogs are those whose parentage is in question. They may resemble a purebred dog, so you will hear them described as a "Shepherd mix" or a "Poodle cross" or a "Terrier type." It is often a guess as to what a puppy with mixed parents will turn out like. It is difficult to determine how big they will get or how they will behave. A general rule of thumb for what their adult size may be is to look at their feet. A puppy with large feet and long limbs may grow to be quite large as an adult. But, again, unless you have seen the parents, it is a guess only as to how big or small your dog will be when it reaches adulthood. The figure below demonstrates the various sizes of dogs as measured at the withers, which is where the shoulder and the spine meet on the back of the dog.

FIGURE 1.00 Dog Sizes

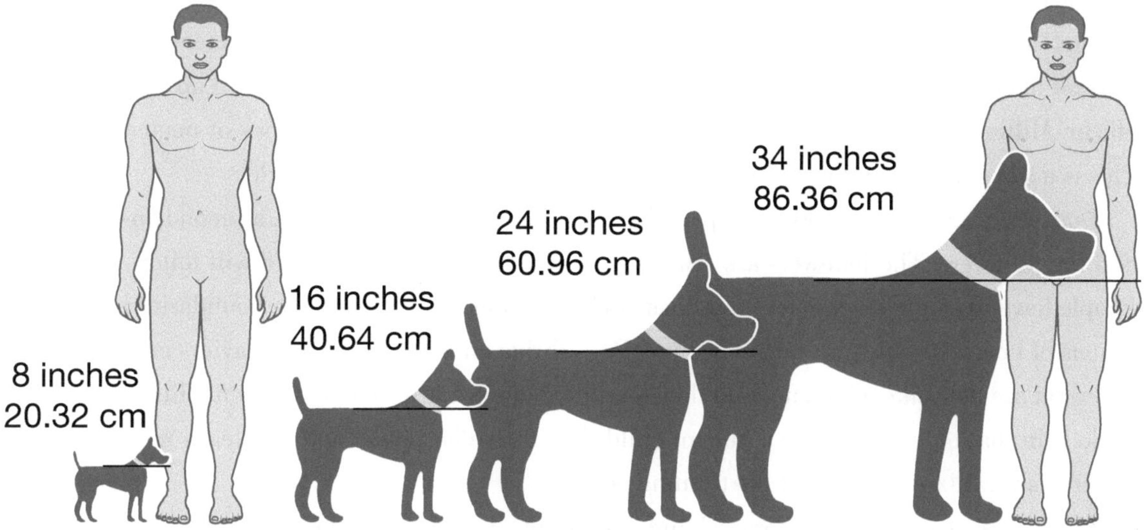

Dog estrus cycles vary greatly in length and occur every six to nine months. The cycle is divided into four periods; Proestrus, Estrus, Anestrus, and Diestrus. Anestrus is a resting phase where the female does not allow breeding to occur. Proestrus is sometimes called the readying phase. This is when the vulva swells, and a bloody discharge is seen. It can last for five-to-nine days, and the males will be interested, but the females will be non-receptive. The Estrus phase is when the female will accept the male for breeding. The vulva remains swollen, but the discharge turns into a light straw color. The male will mount the female and when fully penetrated the vagina closes around the glans penis, and the two will "tie." This ensures that the sperm will enter the uterus and make their way to the oviduct to fertilize the eggs. The Estrus phase can last for five-to-nine days, and during that time the pair may tie a number of times. If the female is bred Diestrus follows, and the embryos attach to the wall of the uterus. The gestation period for dogs is typically 58–68 days. If the female is not bred, the Diestrus phase can last for 80 days before the female moves into Anestrus. The Anestrus phase is called the resting phase, and the female is not receptive to males.

Puppies are generally weaned at six-to-eight weeks of age and are ready for new homes at 10–12 weeks. Male and female dogs are usually neutered or spayed at six months of age. This is usually right before sexual maturity.

Whatever the breed of dog, if it is not to be used for breeding, it is important to have it neutered or spayed to prevent unwanted litters. There are multiple benefits for both dog and owner when this has been done. Intact males are at a higher risk of testicular cancer than neutered males, and spayed females do not develop urine leakage and mammary cancer, thus both living longer lives. Neutered males do not go looking for females, being content to stay home, versus intact males running around looking for love and potentially being hit by a car. Spayed females cannot develop a pyometra, which is an infection of the uterus, if it has been removed. This also translates into saving money in veterinary bills for treating often catastrophic illnesses, injuries or fines if said male is picked up by the humane officer. Millions of dogs are euthanized every year because owners do not spay or neuter their pets. This is a sad situation for these dogs that didn't ask to be brought into the world.

Dogs are pack animals and every pack has a hierarchy; there is a leader, a second-in-command, and the followers. The followers are always subservient to the leader(s) and will follow their lead. People have to be the leaders for their dogs. To be a leader, you must set the boundaries and expectations of how you want your dog to behave. Dogs thrive on knowing what behaviors are acceptable and what are not. Like a human child they learn by being shown what to do, then telling them when to do it in that order. Also like a human child they thrive on praise and will learn more readily if rewards are utilized throughout the training period.

If the rules are not straightforward and consistent, the dog will take over and become difficult to handle, including biting if the owner should press any issue. Some examples of setting the rules: When going out the door, you go first, and then invite the dog out. This lets the dog know that you are in control, and he or she must do what you say. While out on a walk, the dog must be at heel or a step behind you. At heel a strong-willed dog may look at you as equal, so this dog must be shown that you are the leader, and he or she must be a step behind you. Leaders eat before followers; never feed your dog from the dining room table. Doing so invites the dog to be underfoot and to beg while you are preparing meals or trying to eat. This can be very embarrassing when entertaining if while you are dining, the dog is whining or howling for food. Feeding dogs from the table also gives them permission to get onto the table to help themselves because they know eventually you will feed them from there, so why not just take care of business now! Dogs should be told to go to their kennel or bed and to not come around until called. Another thing to remember is that all dogs require daily exercise and mental stimulation. They can become unruly, anxiety ridden, or neurotic if this doesn't happen. Luckily there are different breeds of dogs with different levels of energy, so if you know that you have time for only one walk a day, then pick a dog that has a lower energy level.

Training dogs to be good citizens is the responsibility of every owner. There are multiple ways to go about training a dog, but the most straightforward way is to use positive or negative reinforcement. Positive reinforcement is "catching" the dog doing what you want it to do! It uses verbal praise, vigorous petting, food treats, or access to toys as a reward for the dog whenever it does something you want it to do. This reward must be given **immediately** after the behavior, so the dog connects the two together. "Oh! If I put my butt on the ground when master says SIT, I get a biscuit!" "Gee, If I potty outside, my master tells me I'm a good boy!" "If I run through this tunnel, I get to play with a squeaker toy!" Dogs thrive on heaps of praise, treats, and pets which is all it takes to reward a dog for learning what you want it to do!

Negative reinforcement is NOT beating or striking the dog. It is however as simple as ignoring a behavior or as harsh as pinning the dog to the ground. An example of ignoring a dog is when it is looking for attention, but you are not willing to give it to the dog at the present time. If you ignore the dog's attempts to interact with you, usually lifting your hand or arm with its nose, eventually it will go looking for something else to keep it occupied. If you pet the dog that has nosed your arm, you are teaching him or her; "in order to get attention I simply have to lift my owner's arm or hand up with my nose!" This is fine until the dog lifts your arm with its nose just as you are about to take a sip of hot coffee. Pet the dog when YOU want to pet it, NOT when it wants to be petted! The next step up from ignoring the dog is a quick reaction to something the dog is doing that you don't care for, such as jumping on you. If you grab the dog's paws and/or push the dog away, it will be interpreted as play time! Dogs love to bounce off of each other, and they think this is a great game to play with their humans as well! It may be fun out in the backyard with old jeans on, but what about when you are dressed in that brand new outfit or suit? To teach a dog to not jump on you ever, you simply lift your knee, and let the dog run into it as he or she is coming at you. This doesn't have to be brutal; it just has to be every time. The dog will quickly learn that "if I jump on master, I run into that hard thing, and it kind of hurts! I guess I'll quit jumping on master!" This is an important thing to teach all dogs but especially the large breed dogs that can injure a child by jumping on the child. Another thing to keep in mind when bringing up puppy is to never let a dog put his or her mouth on you. If you do, it increases the likelihood of the dog thinking it is ok to bite if it doesn't want to do something you are asking it to do! If a dog puts its mouth on you, your reaction needs to be swift. Quickly and firmly grasp its muzzle with one hand, and place your other on the back of its neck, and in a very deep voice say, "No biting!" Release the dog and ignore it; do not resume the activity leading up to the offense. This is how mother dogs teach their puppies to not bite them. They place their mouth over the puppy's muzzle, a paw is placed on the back of the neck, and a warning growl is issued. If the puppy doesn't quit, the next step the mother takes is to pin the puppy to the ground, this time with her mouth over the puppy's neck and her paw on its body to keep it pinned until it stops struggling. Then it is released and ignored. When you do this, you are being the "mother," and the message is loud and clear: "Don't bite mother." Everyone in the family should have "mother" status!

There are other ways to train your dog that if used correctly teaches the dog what is and isn't acceptable. It is highly recommended that owners take their dog to obedience training starting with puppy socialization classes or if adopting an adult as soon as possible. This is a great opportunity for the owner to learn how to tell the dog what he or she wants it to do and for the dog to learn basic commands like "Sit," "Down," and "Stay." If there are other members in the family, everyone should attend the training sessions and take a turn at working with the dog. This teaches the dog its place in the family hierarchy. You and the dog will also have the added benefit of socializing with other dog owners and dogs! While training, owners need to be authoritative but fair by showing the dog what is wanted. For example, if you want the dog to sit, place one hand on its rump and the other hand under its jaw. Say in a firm voice "Sit!" pushing down on the rump and pushing up at the neck. Don't ask the dog to "sit," tell the dog to "SIT!" If a dog thinks it is stronger and smarter than you, then you have already lost the dog's respect, and the dog will assume leadership. All commands should be made with an authoritative voice; baby talk will be ignored by some breeds, and for others it indicates that you are weak because you sound like a puppy! If everyone would keep these simple ideas in mind plus providing a daily or twice-daily walk, we wouldn't have so many dogs surrendered or abandoned because of behavior problems.

Socializing your dog is also a must, and there is a bit more to it than just going to obedience classes. It is important to expose dogs to many types of situations to help them realize that everything in the world is not out to kill them! Ball parks and soccer fields are great places to expose puppies to all kinds of noises and people. Let everyone that wants to pet the puppy do so, and if it is small enough, hold it as well. Don't allow it to hide behind you or run from or snap at anyone. Take the dog for car rides; walk the dog past noisy areas like car washes or building sites. Anything that is out of the ordinary will work to expose the dog to noises, other people, and other dogs. Your response to the dog's response during this exposure is very important. If the dog is cowering or trying to get away, ignore it. Don't talk to the dog, pet the dog, or try to comfort the dog as this reinforces the behavior ("if I shy away, master pets me"). Remain where the disturbance is, and when the dog calms down, then you praise or pet it. The message here is "if I'm calm, then master will reward me." If your dog is extremely shy or timid, don't rush into an area that will put him or her into panic mode. Stop well away from the noise and just stand there; reward the dog if it stays calm with a pet or word of praise. Move closer and wait; if the dog is calm, reward it again. Remember, too, that dogs are great at reading body language, and if you are nervous, the dog will become nervous, too. Take a deep breath and exude calmness and confidence; this will be picked up by your dog, and he or she will follow your lead by being calm and confident.

Dogs communicate with each other in a relatively easy-to-understand way by using body language. For example, if two dogs are approaching each other, you may see their heads go up, ears perk forward, tail up or straight out and wagging. This often indicates that the dogs are interested or alerted to each other. If the other dog approaches with the same body language and proceeds to sniffing, usually a budding friendship is in the works. However, if one should approach straight on with shoulders squared, head dropped between shoulders, and eyes locked on the others, this is considered an aggressive or dominant approach. If the other dog matches this approach, then a fight to see who is dominant usually ensues. If one steps to the side and looks away, it has conceded that the approaching dog is dominant and usually submits to a thorough sniffing. Occasionally the dominant dog will make a cursory snap or even growl as if to solidify its position as dominant dog. Remember this when approaching a strange dog; turn your body to the side and refrain from locking eyes with the dog. This is a benign approach that indicates you are not a threat or an aggressor. Once contact has been made you need to become the leader by giving short, firm, and consistent commands.

As mentioned earlier, there are over 700 breeds of dogs in the world. The American Kennel Club (AKC) recognizes 189 breeds of dogs with more being recognized every year. The AKC has placed the breeds into seven groups based upon how they are used by man as well as their size, hair coat, or behavior. Also as previously mentioned, each breed has a breed standard that each representative of the breed must consistently show in order to be considered a purebred dog. With that being said, there are certain head, ear, and tail shapes that are seen in various breeds along with coat colors, patterns, and types. The following table shows representative pictures of each of these traits. In the chapters that follow you will find a "Quick Facts Table" that will indicate the shapes of heads, ear and tail shapes, coat color, pattern, and types for each breed. In addition, life expectancy, height, and weight are also included on the table. A size comparison of a 6 foot man is available for each breed in the Size Comparison Charts, located in the Appendix on page 335.

TABLE 1.01 Dog Head, Ear, Tail Shapes, Coat Colors, Color Patterns, and Coat Types

Head Shape	 Figure 1.01 – **Brachycephalic** a shorted muzzle usually resulting in an under bite – example, Boston Terrier	 Figure 1.02 – **Dolichocephalic** – a long narrow, often pointed muzzle – example, Saluki	 Figure 1.03 – **Mesocephalic** – a wedge or "normal" dog shaped head – example, English Pointer
Ear Shape	 Figure 1.04 – **Bat, Blunt, Round Tip** – large pinna that stands erect without surgery, blunt tipped and wide at the base – example, French Bulldog	 Figure 1.05 – **Button** – medium sized pinna with half or more of it folded over obscuring the ear canal – example, Irish Terrier	 Figure 1.06 – **Candle Flame** - variation of the Bat ear in that the base is as wide but the tip is more pointed – example, Basenji
Ear Shape	 Figure 1.07 – **Cropped** – surgically altered ears to make them stand upright in a sharp point – example, Doberman Pinscher	 Figure 1.08 – **Drop, Hanging or Hound** – long, wide pinna that hangs alongside of the face – example, Beagle	 Figure 1.09 – **Folded** – long, wide pinna that is so wide it folds as it hangs alongside the face – example, Bloodhound

Ear Shape	 Figure 1.10 – **Prick, Erect, or Natural Upright** – stand erect without surgery, sharply pointed – example, Siberian Husky	 Figure 1.11 – **Rose** – pinna folds in half but towards the back of the head, these ears can be held erect when alert – example, Whippet	 Figure 1.12 – **Semi-Prick or Cocked** – top third of an otherwise prick ear bends over, the pinna itself tends to be a bit larger than a prick ear – example, Shetland Sheepdog
Ear Shape	 Figure 1.13 –- **V-Shaped or Pendant** – medium length, triangular pinna that hangs along-side the face – example, Brittany Spaniel		
Tail Shape	 Figure 1.14 – **Bobtail** – naturally docked tail – example, Australian Shepherd	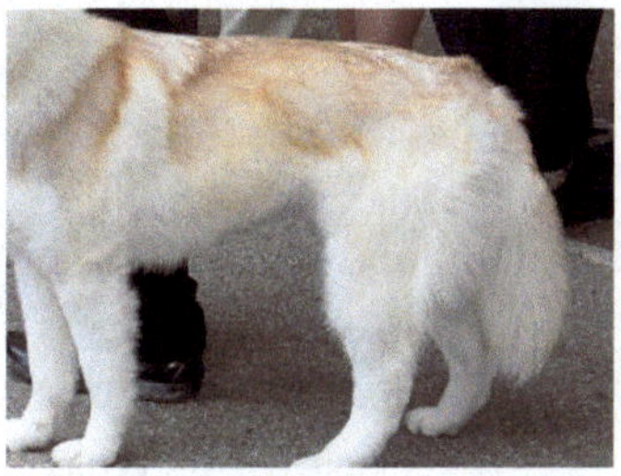 Figure 1.15 – **Brush** – fully furred tail that looks as thick at the base as it does at the end – example, Siberian Husky	 Figure 1.16 – **Corkscrew or Crank** – has more than one curl and is held tightly against the back – example, Bulldog

Tail Shape

Figure 1.17 – **Docked** – end has been surgically removed, usually at 3–4 days of age, length depends on breed standard. Banned in European countries – example, Spaniels

Figure 1.18 – **Gay** – naturally short or docked tail that is carried straight up when alert or in motion – example, Fox Terrier

Figure 1.19 – **Otter** – a medium to long tail, that is thick at the base and tapers slowly to a blunt end, useful for direction control when swimming – example, Labrador Retriever

Tail Shape

Figure 1.20 – **Plume** – is often saber shaped but usually has long hair that hangs from the underside – example, Irish Setter

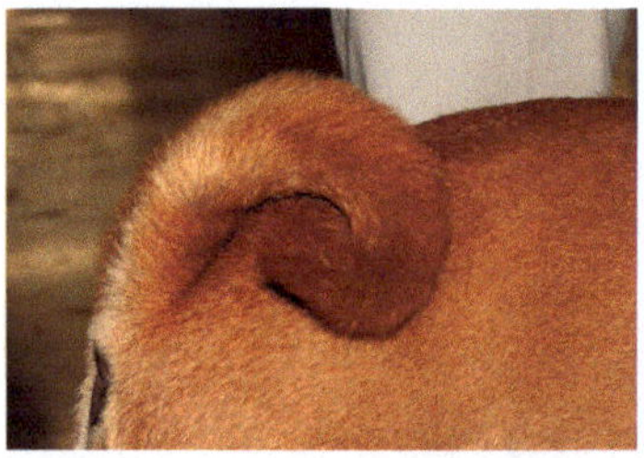

Figure 1.21 – **Ring, Snap, or Squirrel** – either a long or short tail that curls around and is carried naturally over the back. – example, Shar Pei, Alaskan Malamute long hair covers tail

Figure 1.22 – **Saber** – fairly thick at base, tapering to a point, carried with a slight curve, much like the sword – example, German Shepherd

Tail Shape

Figure 1.23 – **Sickle** – long slender tail with an upward curve, it is carried up and away from the body when alert or in motion – example, Saluki

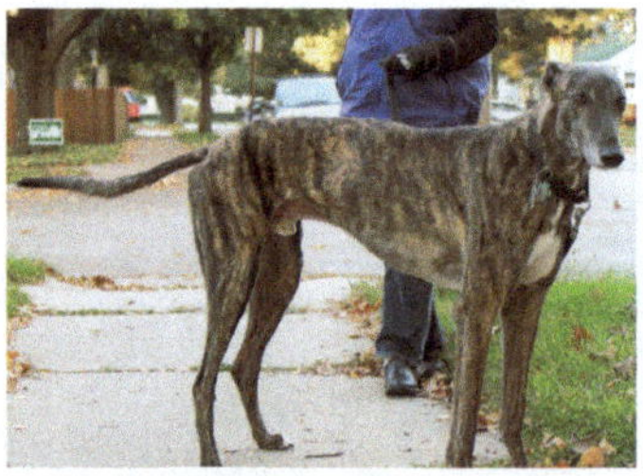

Figure 1.24 – **Whip** – a long slim tail that tapers to a point – example, Greyhound

Coat Colors

Figure 1.25 – **Belton** – white base with colored hair in irregular shaped specks, blue, lemon orange, or liver

Figure 1.26 – **Blue Merle** – mottled or marble coloring – blue-gray/black – illusion of blue coloring

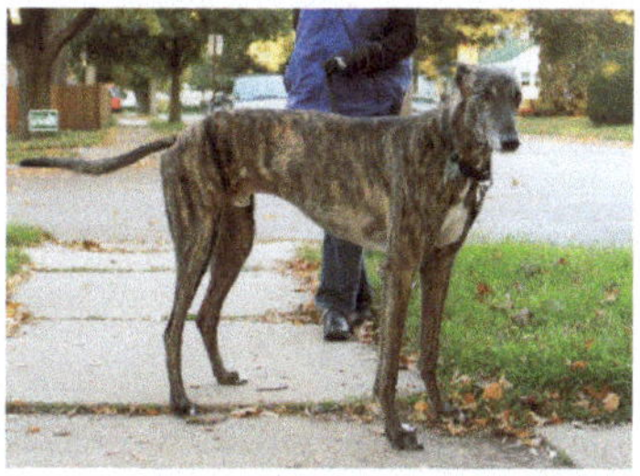

Figure 1.27 – **Brindle** – fine even mixture of black hairs with hairs of a lighter color, usually tan, brown, or gray.

Coat Colors

Figure 1.28 – **Fawn** – solid light to dark tan color over body

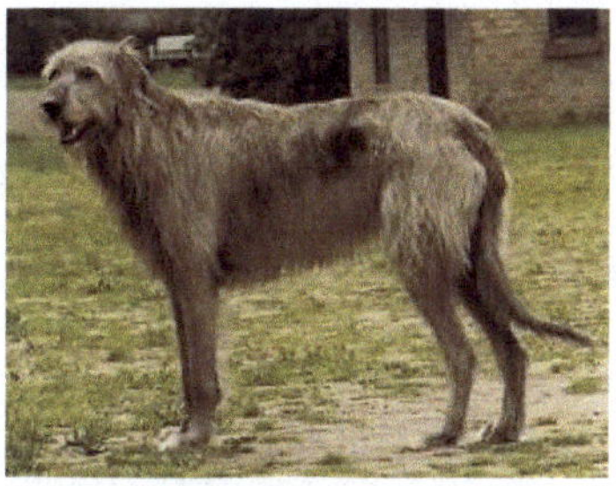

Figure 1.29 – **Grizzled** – bluish-gray colored mixed with lighter and darker shades of the same color

Figure 1.30 – **Harlequin** – patched or pied coloration, usually black on white usually only used in reference to Great Danes

Coat Colors

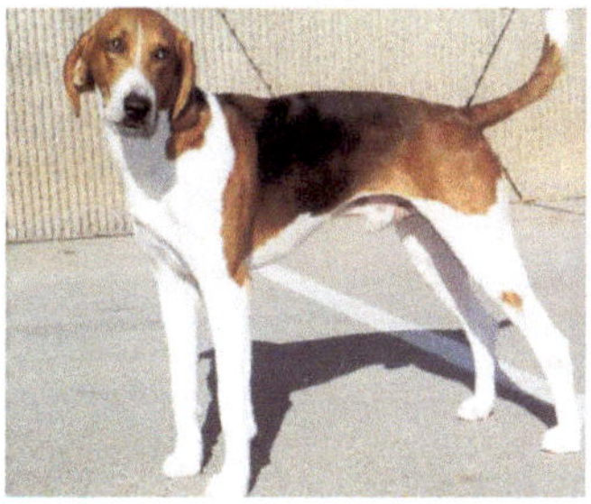

Figure 1.31 – **Hound Colored** – ground color of white with tan and/or black patches on the head, back, legs and tail

Figure 1.32 – **Irish Spotted** – white around neck, on chest, legs, and a white ring around muzzle with or without a Blaze, sometimes referred to as a Tuxedo

Figure 1.33 – **Parti-colored** – variegated patches of color usually only used in spaniels

Coat Colors

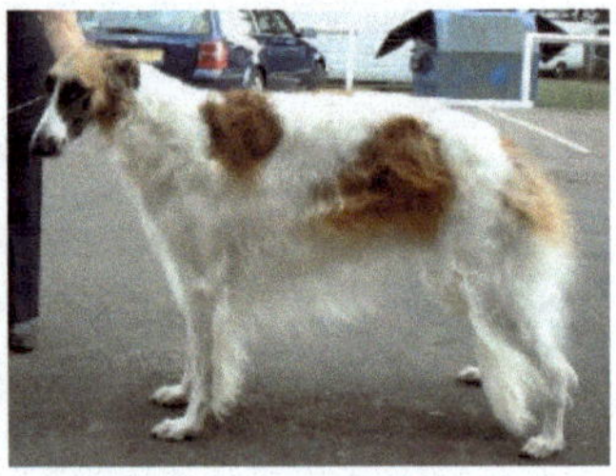

Figure 1.34 – **Piebald** – comparatively large patches of one or more colors on a white body

Figure 1.35 – **Roan** – fine mixture of colored hairs with white hairs; blue, orange, lemon – this example is a patched roan

Figure 1.36 – **Sable** – varying shades of light gold to dark mahogany with black tipped hair

Coat Colors

Figure 1.37 – **Salt & Pepper** – mixture of black and white hairs. Looks like the pepper was used more than the salt!

Figure 1.38 – **Speckled** – flecks of colored hair interspersed with white, blue, or red

Figure 1.39 – **Ticked** – small, isolated areas of colored hairs thickly dispersed on white ground

Coat Colors

Figure 1.40 – **Tri-colored** – includes all three colors: black, white, and tan

Figure 1.41 – **Wheaten** – pale yellow or fawn

Color Marking

Figure 1.42 – **Black Mask** – dark hair covers from eyes to muzzle, and ears

Figure 1.43 – **Blaze** – a white stripe running up the center of the face usually between the eyes

Figure 1.44 – **Butterfly nose** – a parti-colored – spotted with flesh color (the left side of nose has a small pink spot)

Color Marking

Figure 1.45 – **Dudley nose** – flesh colored nose

Figure 1.46 – **Feathering** – long strands of hair on the back of legs, abdomen, chest, and tail

Figure 1.47 – **Frill** – longer hair below the neck on the chest (Apron)

Color Marking

Figure 1.48 – **Kiss marks** – tan spots on the cheeks and over the eyes

Figure 1.49 – **Saddle** – black marking over the back, like a saddle or blanket

Figure 1.50 – **Spectacles** – shadings or dark markings over or around the eyes or from eyes to ears

Color Marking

Figure 1.51 – **Trace** – a darker colored hair in a stripe down the back

Coat Type

Figure 1.52 – **Double Coated** – soft under coat that is water resistant, long guard hairs make up the outer coat and usually grow through the undercoat

Figure 1.53 – **Corded** – hair grows in length and twirls around the undercoat creating the cords

Figure 1.54 – **Curly** – coats also grow continuously in length and can be made to go into cords

Coat Type

Figure 1.55 – **Straight Long hair** – many breeds have hair that grows in length, some will have a single layer and some will have an undercoat

Figure 1.56 – **Straight Short hair** – hair fibers are short and usually lay flat against the skin, some will have a single layer and some will have an undercoat

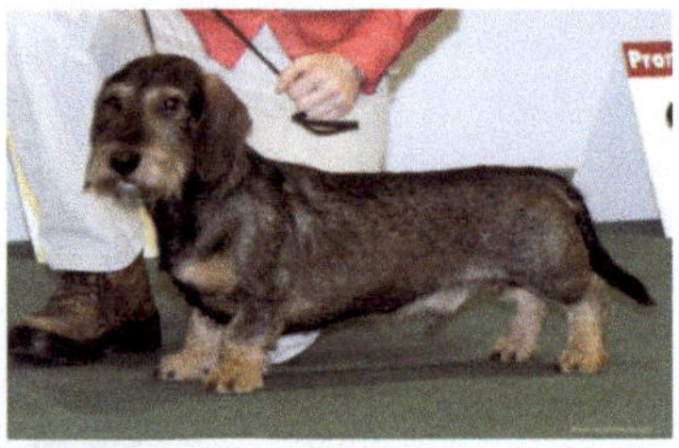

Figure 1.57 – **Wirehaired** – fibers come out of the follicle in more than one strand making the fur feel harsh or wiry to the touch

Herding Group

HERDERS AND HEELERS

The Herding Group was split from the American Kennel Club (AKC) Working Group in 1983. These dogs all have an innate ability to herd or move animals from one place to another. In general, these dogs are also quick to learn basic commands to control flocks of sheep and herds of cattle from a few head to hundreds. They are devoted to their work and to their people. They make good pets if they have something to do such as agility trials, playing, or exercising with their people every day. They will sound the alarm if danger is in the area but quickly recognize friends and welcome them home. However, they have been known to herd people into the house, so beware the backwards step as a friendly nip on the rear or ankle will keep you moving in the right direction, not because they are aggressive but because of their instinct to keep "animals" moving forward! You will see these dogs trying to herd everything from kids to turtles.

These beautiful dogs can be divided further into two groups: herders or heelers. To make livestock move, the herders use stealth and the pressure points all livestock respond to whether it comes from a human or a dog. Herders will provide the occasional nip to recalcitrant animals or to get the herd to move faster. Many of these herding dogs have found a niche in law enforcement and in the armed services. The heelers were developed to bark and nip at the heel or head to make the animals move forward, back up, and/or turn around. They tend to be a bit more independent and perhaps not as willing to please their owners as the herders.

With urbanization there are fewer chances for these dogs to do this sort of work. Fortunately, this group of dogs likes people and most will settle nicely into a "town" home. However, they are high-energy dogs requiring daily exercise and mental stimulation. Without this they can become bored, anxious, and develop behavior problems such as constant barking, destructiveness, or mental instability. They excel at agility, obedience trials, fly ball, tracking, herding trials, and other activities that make them think and move.

Australian Cattle Dog

FIGURE 2.00 Australian Cattle Dogs, Blue and Red

Origins: Australian cattle producers needed an extremely hardy, athletic, hardworking dog to assist them in driving their often wild cattle across great distances and through towns. The Collies of the times were incapable of this endeavor. A number of people started experimenting crossing Collies with Dingoes, which gave them a hardy, athletic dog that would help herd the cattle but were apt to help themselves to a calf when no one was looking. These dogs were then crossed with Dalmatians and eventually the Black and Tan Kelpie. The Dalmatian added a love of horses and loyalty to owner; however, some of the hardiness was lost, and a lot of barking was added. This cross also changed the hair coat from a merle pattern to the current-day speckled pattern. Like the Dalmatian, Australian Cattle Dog puppies are born white and change to the speckled pattern as they age. The addition of the Kelpie added the working ability of a sheepdog, thus bringing back the hardiness and athletic ability while quieting the dog to barking only as necessary to move a stubborn animal. The breed was recognized by the AKC in 1980.

Characteristics: As the name implies the dog's prime function, and one in which it has no equal, is the control and movement of cattle in both wide open and confined areas. Alert, extremely intelligent, watchful, courageous, and trustworthy, with an unwavering devotion to duty are the hallmarks of this breed. Loyalty and protective instincts make it a self-appointed guardian to the herdsman, his herd, and his property. Naturally suspicious of strangers, they should be watched carefully until it is made clear that the stranger is a friend or foe. They need intense socialization if they are not going to be used as a guard dog as they can be dog aggressive. Firm but fair training techniques work well with this dog, and they can be trained to a high level due to their intelligence. If in a town home, it is important to exercise this dog twice a day and offer it toys to make it think. “Heelers” can become very unhappy if left alone and unstimulated for long periods of time. Their short hair coat is easy to take care of, although they do shed on a seasonal basis.

FIGURE 2.01 Australian Cattle Dog with New Puppy

QUICK FACTS 2.00 Australian Cattle Dog

Quick Facts	Height	Weight	Coat	Features
Life Span: 11–13 years	**Males:** 18–20 inches 46–51 centimeters **Females:** 17–19 inches 43–48 centimeters	**Males:** 50–65 pounds 25–29 kilograms **Females:** 40–55 pounds 18–25 kilograms	**Colors:** Blue, blue-mottled or blue speckled with or without black, blue, or tan markings. Red speckle with or without darker red markings. **Type:** Double coat with short, dense undercoat and an outer coat that is short, dense, and flat making it rain resistant.	**Head:** Blunt muzzle, scissor bite. **Ears:** Prick. **Tail:** Brush.

Genetic Diseases or Conditions: Eczema, unilateral or bilateral deafness, hip and elbow dysplasia, and Progressive Retinal Atrophy.

Australian Shepherd

Origins: There are many theories about the origin of the Australian Shepherd. The breed probably originated in the Basque region of the Pyrenees Mountains between Spain and France. It was dubbed the Australian Shepherd because of its association with Basque shepherds who came to

FIGURE 2.02 Australian Shepherd

FIGURE 2.03 Australian Shepherd, Color Pallet

the United States from Australia in the 1800s. The breed as we know it today was developed exclusively in the United States. It became popular after WWI and has been enjoyed by many families since. It was recognized by the AKC in 1991.

Characteristics: This extremely intelligent working dog has a strong herding and guarding instinct with the stamina to work all day. They make loyal companions, being attentive and animated with even dispositions. These are problem-solving, high-energy dogs that require 2–3 hours of daily exercise. Owners must provide a daily job, teach them to do tricks, or enter them into competitions such as agility, fly ball, Frisbee, or herding competitions, all with their people. If they don't get the exercise, mental stimulation, and attention they crave, they can resort to destructive behaviors and barking. Luckily they are easy to train as they love to please their people and have a high level of intelligence. "Aussies" are slightly longer than tall, lithe and agile, and solidly muscled. They come in a wide variety of coat colors. Their coats are medium length and do require weekly brushing.

QUICK FACTS 2.01 Australian Shepherd

Quick Facts	Height	Weight	Coat	Features
Life Span: 12–15 years	**Males:** 20–23 inches 50–58 centimeters **Females:** 18–21 inches 45–53 centimeters	**Males:** 50–65 pounds 25–29 kilograms **Females:** 40–55 pounds 18–25 kilograms	**Colors:** Blue or red merle, black, or red, all with or without white markings and/or tan points. **Type:** Undercoat varies in thickness; outer coat is medium length, straight to wavy, feathered legs.	**Head:** Blunted muzzle, scissor bite. **Eye color:** Brown, blue, or amber. **Ears:** V-shaped or rose ear. **Tail:** Docked or bobtail.

Genetic Conditions and Diseases: Malocclusions (undershot, overshot, wry mouth, Polydontia/ Oligodontia, dropped center incisors), Patent Ductus Arteriosus, hip dysplasia, eye issues (Cataracts, Collie Eye Anomaly, Iris Coloboma, Persistent Pupillary Membrane, and Distichia), epilepsy, and deafness associated with merle-to-merle matings.

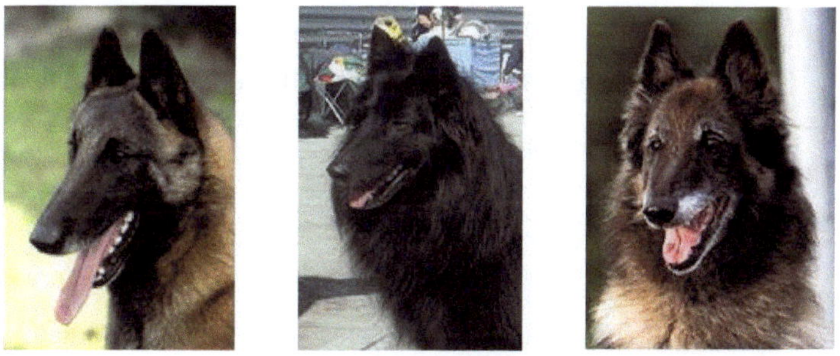

FIGURE 2.04 Belgian Malinois, Shepherd, Tervuren, Heads

Belgian Malinois – Belgian Tervuren – Belgian Sheepdog

Origins: The Belgian Malinois, Tervuren, Sheepdog, and Laekenois (not shown) share common ancestors and the same breed standards in every country but the United States. These Belgium farm dogs were depended upon to assist with herding flocks and guarding property. They were recognized by the AKC in 1959.

FIGURE 2.05 Belgian Malinois

Characteristics: All three have high puppy-like energy levels, some well into their fifth year of age. They are protective of home and people, alert, and hard-working. Many have an exceedingly high "prey drive," which is an instinctive desire to find, pursue, and capture prey. This can result in chasing after rabbits, kids, or strangers. Mental stimulation and exercise are a must for the Belgians. If skipped, many develop destructiveness and neurotic behaviors. They are very easy to train but need experienced, firm, and consistent training. They are intelligent and readily respond to rewards such as treats, praise, or toys.

FIGURE 2.06 Belgian Tervuren

They form strong bonds with their humans, and many of them, especially the Malinois and Tervuren, are used as police or armed forces dogs. "Belgians" are also excellent search-and-rescue dogs, bomb- and narcotic-sniffing dogs, and they excel at sports like fly ball, agility trials, obedience, tracking, and herding events.

They tend to be a bit shyer around strangers and will need some extra time with socialization. This dog is really not for the average dog owner as the dog's energy level and tendency to become the leader in the family can make the dog very difficult to control. The Tervuren and

FIGURE 2.07 Belgian Shepherd

Shepherd will need weekly brushing to keep their coats neat. The Malinois will also require brushing but less often, and all three will shed seasonally.

QUICK FACTS 2.02 Belgians Malinois, Tervuren, and Shepherd

Quick Facts	Height	Weight	Coat	Features
Life Span: 12–14 years	**Males:** 24–26.5 inches 61–66 centimeters **Females:** 22–24.5 inches 56–61 centimeters	**Males:** 64–75 pounds 29–34 kilograms **Females:** 55–66 pounds 25–30 kilograms	**Colors:** Malinois and Tervurens are fawn to mahogany, and Shepherds are black. **Type:** Tervurens and Shepherds have long guard hairs and feathered legs. Malinois have a short, dense coat.	**Head:** Long, moderately pointed muzzle, scissor or level bite. **Ears:** Erect. **Tail:** Saber.

Genetic Diseases or Conditions: Cataracts, epilepsy, thyroid disease, Progressive Retinal Atrophy, hip dysplasia, bloat, torsions, and Pannus. Selective breeding has minimized many of these problems.

Bearded Collie

FIGURE 2.08 Bearded Collie, Black

Origins: As one of Britain's oldest breeds, they are thought to be descended from the Magyar Komondor of Central Europe. The breed wasn't established in England until the 1950s and wasn't eligible to show in the U.S. until 1976 when the AKC placed them in the miscellaneous class and then later moved them to the Herding group.

Characteristics: Unlike other collies, these have long facial hair with great mustaches and beards, thus the nickname "Beardies." Beardies are high-energy dogs that require early obedience training with calm, firm, and consistent handling. They require regular daily exercise and toys that keep them busy while you are gone. The females tend to be independent thinkers and can be a handful if they don't respect people as pack leaders. They mature slowly and so stay "puppy like" longer, which can cause their people to lose hope of ever having a calm, adult dog. Once past the puppy phase, they are loyal and loving, funny and charming. A herding characteristic of this breed is the "Beardie bounce." They will bounce on their front legs and bark at a stubborn animal until it moves. The bounce is not reserved only for four-legged animals, so be ready for a bounce up to lick your nose or a bounce on your back side to get you moving! They have a double coat with the outer coat being harsh, flat, and shaggy. Their coat requires regular brushing to avoid mats.

FIGURE 2.09 Bearded Collie, Fawn

QUICK FACTS 2.03 Bearded Collie

Quick Facts	Height	Weight	Coat	Features
Life Span: 12–14 years	**Males:** 21–22 inches 53–56 centimeters **Females:** 20–21 inches 51–53 centimeters	40–60 pounds 18–27 kilograms	**Colors:** Black, blue, brown, or fawn with or without white markings. **Type:** Double coat with long guard hairs, long hair on cheeks and chin create mustache and beard.	**Head:** Blunt muzzle, scissor bite. **Ears:** Hanging covered with long hair. **Tail:** Sickle, covered with long hair.

Genetic Conditions and Diseases: Cancer, cerebrovascular disease, chronic kidney failure, Addison's disease, hypothyroidism, arthritis, cruciate ligament rupture, colitis and diarrhea, and urologic diseases.

Border Collie

Origins: Border Collies (first classified as the "Scotch Sheep Dog") are the world's premier sheep-herding dog. From the highlands of Scotland and England, these dogs were bred for brains, work ethic, and then beauty. AKC recognized them in 1995.

Characteristics: The Border Collie as we know it today has "stock sense" and the ability to work long days over all types of terrain. Their stock sense is a unique style of gathering and fetching stock with wide sweeping outruns, controlling the flock with an intense gaze known as "eye" which is coupled with a stalking style of movement. Border Collies are intense, energetic,

FIGURE 2.10 Border Collie

FIGURE 2.11 Border Collie, Working

and highly trainable with an extraordinary instinct for herding and an uncanny ability to reason. One of its greatest assets is the ability to work out of sight of its master without commands. This workaholic of the dog world is extremely energetic and requires exercise beyond a walk around the block or a romp in the yard. They are psychologically happy when they have a job to do and are interacting with people and other animals. They are affectionate towards family and friends but are reserved towards strangers. These dogs are not happy if left for long periods of time without companionship or something to do. They can become neurotic and develop anxiety problems if not exercised and socialized with every day. When approached, they should stand their ground but stand down when it is a known friend. They have medium long hair shedding seasonally and require regular brushing.

QUICK FACTS 2.04 Border Collie

Quick Facts	Height	Weight	Coat	Features
Life Span: 10–14 years	18–22 inches 46–55 centimeters	**Males:** 50–65 pounds 25–29 kilograms **Females:** 40–55 pounds 18–25 kilograms	**Colors:** Any color in bi-color, tri-color, merle, sable, or solid patterns. **Type:** Rough or smooth.	**Head:** Relatively flat, blunted muzzle, scissor bite. **Ears:** Erect and/or semi-erect. **Tail:** Brush.

Genetic Diseases/Conditions: Hip dysplasia, Osteochondritis Dissecans (OCD), "Border Collie Collapse," retinal and other eye diseases, Patent Ductus Arteriosus (PDA).

Bouvier des Flandres

FIGURE 2.12 Bouvier des Flandres

Origins: Southwest Flanders and Northern France is where the Bouvier was first found as a cattle dog. It wasn't until the early 1900s that the breed was refined and developed by monks that used the Irish Wolfhound and Scottish Deerhounds to breed them up in size. They were not only herding dogs but were guard and draft animals as well. The breed was almost wiped out in World War I, but a few men, including Germans, succeeded in keeping their dogs alive. A Belgian dog by the name of Ch. Nic de Sottegem was a "trench dog" and after the war was shown and is the foundation dog with many descendants that can be found on pedigrees today. The AKC recognized them in 1931.

FIGURE 2.13 Bouvier des Flandres

Characteristics: Pronounced "Boov-yay day Flahn-druh," this powerful dog is compact in the body with a large head that sports a heavy beard and mustache. Ears were cropped and tails docked until 2006 when Europe outlawed it and currently in the U.S. the American Veterinary Medical Association opposes this practice. They are known to be good family pets because of their loyalty and protective natures. They should be socialized early in life to keep them from becoming overly protective or dog-aggressive. The "Bouvier" is a high-energy dog and will require long walks or even better runs twice a day. Obedience training is important, and an experienced dog owner is best for a Bouvier as the dog can become dominant over a more timid human. They are fast learners and do better with short training bouts as they can get bored which can be mistaken as being naughty. Their hair coat requires weekly brushing to prevent mats from forming. Bouviers are used in police work as tracking and drug-detection dogs.

QUICK FACTS 2.05 Bouvier des Flandres

Quick Facts	Height	Weight	Coat	Features
Life Span: 10–12 Years 2–3 years to mature.	**Males:** 23–28 inches 58–71 centimeters **Females:** 22–27 inches 56–69 centimeters	**Males:** 80–120 pounds 36–54 kilograms **Females:** 60–80 pounds 27–36 kilograms	**Colors:** Black, gray fawn, gray brindle, pepper and salt. **Type:** Double coat.	**Head:** Wide and oval shaped, scissor bite. **Ears:** Rounded V-shape or cropped. **Tail:** Docked.

Genetic Conditions and Diseases: Bloat and hip dysplasia.

FIGURE 2.14 Briard

FIGURE 2.15 Briard – Black and Tawny

Briard

Origins: Pronounced "Bree-ard," records in France show that early on they were used as guard dogs against wolves and poachers, but after the revolution they were used as herding dogs and guarding property. The Marquis de Lafayette and Thomas Jefferson are credited with introducing the breed to the U.S. They were recognized by the AKC in 1928.

Characteristics: This versatile dog is certainly a herding dog but can also be used for tracking and hunting, for military purposes, or as a draft dog. Their training needs to be started early in life, including socialization with exposure to multiple situations. Proper training for leaving and returning is also a must for owners to understand in order to keep their Briard from developing separation anxiety. The Briard develops a very strong attachment to people and will set up a fuss that can go on for a very long time after the owner leaves. They have really good memories and a lesson learned good or bad is remembered for a long time. This was an important trait when taking care of flocks and herds when society was more agricultural. In urban life this might be construed as stubborn or dumb. The Briard is far from dumb. As with most herding dogs, they need regular exercise and/or a job to do to stay healthy and well balanced mentally. This is not the dog for people with little time; they need to be with their people and doing something with their people. They excel at agility, obedience trials, fly ball, tracking, and herding events. Their hair coat is long and will require regular brushing to avoid mats; double dewclaws on the rear legs are a breed characteristic and are not to be removed.

QUICK FACTS 2.06 Briard

Quick Facts	Height	Weight	Coat	Features
Life Span: 10–12 years	**Males:** 24–27 inches 61–69 centimeters **Females:** 23–26 inches 58–65 centimeters	**Males:** 66–68 pounds 30–40 kilograms **Females:** 55–77 pounds 25–35 kilograms	**Colors:** Tawny, black, gray. **Type:** Double coat with long guard hairs that lay flat. Hair covers face.	**Head:** Scissor bite **Ears:** V-shaped, covered with long hair or cropped. **Tail:** Sickle, end has a hook called a "crochet."

Genetic Conditions and Disease: Stationary Night Blindness and Progressive retinal atrophy.

Collie

Origins: Though its origins are not known, it is thought that the Collie was developed in Scotland and then further refined in England. They were first shown in England in 1860, and a registry was started. They were imported to the U.S. in 1879. The AKC recognized them in 1885.

FIGURE 2.16 Rough and Smooth Coated Collies

Characteristics: Rough and Smooth Coated collies are basically the same dog with different hair coats. The Rough Collie has a dense undercoat with long guard hairs that grow through the undercoat. The Smooth Collie has a short undercoat and short hair over its entire body. This consistently popular dog is intelligent and loyal to the entire family. They are sensitive and should be trained in a gentle and consistent manner. They require regular exercise or a job to do on a daily basis. Long walks, playing with children or other dogs, and retrieving are great everyday activities. Obedience,

FIGURE 2.17 Collie, Sable

agility, and herding trials are also suited for Collies. They thrive on human contact, and if they don't get enough, they will resort to entertaining themselves, which often translates into destructive behavior and lots of barking. The Rough Coated Collie will require weekly brushings to keep the coat and skin in good shape.

QUICK FACTS 2.07 Collie

Quick Facts	Height	Weight	Coat	Features
Life Span: 8–12 years	**Males:** 24–26 inches 61–66 centimeters **Females:** 22–24 inches 56–61 centimeters	50–65 pounds 25–30 kilograms	**Colors:** Sable, Tri-color, Blue Merle, White. White markings. **Type:** Rough Collies are double coated, with abundant mane and feathering on legs. Smooth Collies are short-haired all over.	**Head:** Long, pointed muzzle, scissor bite. **Ears:** Semi-Prick. **Tail:** Plume on Rough-coated dogs, Saber on Smooth.

Genetic Diseases or Conditions: Bloat, interception, gastric dilatation and volvulus, hip dysplasia, Dermatomyositis, "Collie Eye Anomaly" (CEA), Choroidal Hypoplasia, Chorioretinal change, Staphyloma, vascular disease, Progressive retinal atrophy, retinal detachment, corneal dystrophy and Malignant Fibrous Histiocytoma, and Nodular fasciitis. Seventy percent of Collies have the mutant MDR1 gene which predisposes them to multiple drug sensitivities. Adverse drug reactions to over a dozen different drugs have been documented. There are tests for the mutant MDR1 gene, Progressive retinal atrophy, Canine Renal Dysplasia, and Canine Cyclic Neutropenia (Gray Collie).

FIGURE 2.18 German Shepherd

German Shepherd

Origins: Derived from old breeds of herding and farm dogs, in 1899 the Society for German Shepherd Dogs developed the breed as we know it today. This dog is of high intelligence making it quick to learn, and its agility is renowned. The AKC recognized this dog in 1908.

Characteristic: Over the years the "Shepherd" has served as a companion, guardian, police or military dog, a guide for the blind, and a helper for the handicapped, and its keen sense of smell makes it an excellent search-and-rescue, cadaver search, narcotics, explosives, accelerant, and land-mine detection dog. They should be self-confident and not shy or fearful. Unfortunately some aggression has been seen because of indiscriminate breeding. Selection of puppies should be made if both parents can be evaluated for aggression, shyness, or fearfulness. They exude a direct and fearless personality, maintaining a certain aloofness around strangers. They don't make immediate or indiscriminate friendships, so a slow approach with these dogs is appropriate. Shepherds are extremely loyal to their family and partners, many having laid down their lives for their caretakers. They are very quick to learn and require a firm, fair, and consistent approach to training. Owners cannot be timid or meek as this dog will take over the leadership role and can become unstable around other people and dogs. This dog is a high-energy dog and will require long, twice-daily walks and a couple of hours romping in the dog park. Their hair coat requires occasional brushing, more often during season changes.

FIGURE 2.19 German Shepherd, White

QUICK FACTS 2.08 German Shepherd

Quick Facts	Height	Weight	Coat	Features
Life Span: 12 years	**Males:** 24–26 inches 60–65 centimeters **Females:** 22–24 inches 55–60 centimeters	**Males:** 66–88 pounds 30–40 kilograms **Females:** 49–71 pounds 22–32 kilograms	**Colors:** Black and tan, black, white, grey. **Type:** Double coat with medium-length guard hairs.	**Head:** Long wedge, scissor bite. **Ears:** Prick. **Tail:** Saber.

Genetic Diseases or Conditions: Gastric torsion, hip dysplasia, elbow dysplasia, Von Willebrand's Disease (vWD), and moist dermatitis.

Old English Sheepdog

Origins: These dogs were developed in the western part of England. Their history is a bit clouded as to what breeds were used, with some feeling that the Scotch Bearded Collie and a Russian Owtchar had a hand in its development. It was recognized by the AKC in 1888.

FIGURE 2.20 Old English Sheepdog

FIGURE 2.21 Old English Sheepdog, Pup

Characteristics: The adjectives that describe an Old English Sheepdog are home-loving, lover, not a fighter, agile, intelligent, and calm. They require obedience training that is firm, fair, and consistent. This dog can be quite the clown with extra energy, although they can also be "couch potatoes," adapting well to any living situation as long as they are exercised regularly or have a job to do. Their herding instincts will kick in around children, but a gentle head bump verses a nip is their mode of moving "livestock." They have a unique shuffling gait that is similar to a bear, and with their fuzzy coat it can be quite a shock when seen for the first time. If not born as a natural bobtail, the tail is docked at a 2–3 days of age, but people are moving away from this practice and are leaving the tails long. Their hair coats if left long do require thrice-weekly brushing. Owners often have a professional groomer trim the hair coat about inch in length with a clipper. It is important to leave the hair long over the eyes as the white color reflects sunlight into the eye and can cause damage.

QUICK FACTS 2.09 Old English Sheepdog

Quick Facts	Height	Weight	Coat	Features
Life Span: 7–11 years	**Males:** 21–22 inches 53–56 centimeters – females are slightly smaller	Up to 101 pounds 46 kilograms	**Colors:** Any shade of gray, grizzle, blue, or blue merle with or without white markings. **Type:** Profuse, shaggy not curly, hard texture.	**Head:** Level or tight, scissor bite. **Ears:** V-shaped, carried flat. **Tail:** Docked or bobtailed.

Genetic Diseases or Conditions: Hip dysplasia, cataracts, glaucoma, entropion, thyroid problems, deafness, diabetes, allergies, and skin problems. Heat stroke is a serious concern for full-coated dogs, and cancer is the number-one killer of Old English Sheepdogs.

Puli

FIGURE 2.22 Puli, Black

Origins: The Puli is an ancient sheepdog of Hungary. This breed was amongst the Komondor and Kuvasz sheepdogs, but it was almost lost in the 16th century when Hungary was invaded. People from Western Europe resettled in Hungary and brought with them their sheepdogs, and the resident Puli interbred with these dogs. Initially there were two types of hair coats, shaggy and curly, and four sizes—a large for police work, a medium for working herds, a small, and dwarf for pets. The AKC recognized the Puli in 1936.

FIGURE 2.23 Puli, White

Characteristics: The most outstanding feature of the Puli is its hair coat. At six months of age the long soft undercoat tangles with the long top coat, creating long cords or "dreadlocks." As the cords form, large lumps need to be separated into smaller cords by hand. This coat must be maintained with careful grooming and drying to prevent mildew occurring in the cords. An occasional bath may be required to remove excess mud or dirt; drying should be done with a drier since otherwise it may take a few days, and if it is hot outside, dermatitis could develop. This coat is a natural protection against the hazards associated with herding and protecting sheep on the range. The Puli has a springy bounce to its step and is extremely agile. It will stay puppy playful its entire life, loving a romp outdoors or on the farm with a job. They do require regular vigorous exercise and will do well in agility and herding trials. They are very intelligent and empathetic towards their owners and need to be trained firmly yet gently as they are sensitive. It is naturally protective of its people and place as they do not tolerate strangers near their property. They are sensitive and will hold a grudge if treated roughly.

QUICK FACTS: 2.10 Puli

Quick Facts	Height	Weight	Coat	Features
Life Span: 12–14 years	**Males:** 17 inches 43 centimeters **Females:** 16.5 inches 42 centimeters	23–25 pounds 10–11 kilograms	**Colors:** Black, grey, white, cream. **Type:** Double coat that forms into cords.	**Head:** Rounded head, scissor bite. **Ears:** Hanging, covered with long hair. **Tail:** Ring.

Genetic Conditions and Diseases: Contact dermatitis if the hair coat is allowed to stay wet but otherwise the Puli is a pretty hardy breed. Reputable breeders will check for hip dysplasia, osteochondritis of elbows, patellar luxations, congenital cardiac and eye diseases, degenerative myelopathy, thyroid, and hearing tests.

Shetland Sheepdog (Sheltie)

FIGURE 2.24 Shetland Sheepdog and Rough Coated Collie

Origins: The Shetland Islands are remote, rough, and sparsely inhabited. The animals on these islands are small, hardy, and tough. This includes the ancestors of the Sheltie which were used to drive the often half-wild sheep into pens, away from gardens, and from one grazing area to another. They also lived with the herds on other islands, chasing birds of prey away from the lambs by barking. In the early 1900s tourists took a liking to these little dogs and started taking them home. The English Shetland Collie Club was established in 1914; however, the Collie breeders objected to these "nondescript" little dogs being called Collies, and the name was changed to Shetland Sheepdogs. The English breeders started crossing the Shelties with Collies to try to make miniature collies. In the 1920s and 30s when the dogs were exported to America, they were still quite large and were listed as Collie crosses. It took careful breeding to get the smaller "apartment-sized Collies," and to this day they are still bigger and have more white on them than the English version of the Shetland Sheepdog. AKC recognized the "Sheltie" in 1911.

Characteristics: It is the Sheltie's nature to obey willingly and naturally with few or no lessons needed. They instinctively guard property and give a watchdog warning which translates into loud, continuous barking. This makes them good farm and home protectors, although this in turn results in a dog that is wary of strangers. If you don't wish this, they must be socialized early and to many situations to develop a confident and friendly dog. They also must be trained to stop barking when commanded; otherwise you will have a noisy friend. Shelties are devoted to their families, usually docile around children, and have a keen intelligence. Their training is accomplished by even novice owners, so they are good starter dogs. Shelties are very agile and of strong constitutions and so hold their own during agility trials, flyball competitions and, of course, obedience trials. Their coats require twice weekly brushing to keep from matting, and they do shed heavily on a seasonal basis.

FIGURE 2.25 Shetland Sheepdog, Blue Merle

QUICK FACTS 2.11 Shetland Sheepdog

Quick Facts	Height	Weight	Coat	Features
Life Span: 14–15 years	13–16 inches 33–40.6 centimeters	25–35 pounds 11.3–15.9 kilograms	**Colors:** Sable, blue merle, tri-color, bi-black, and bi-blue (absence of any brown). White markings. **Type:** Double coated, abundant mane, feathered legs.	**Head:** Pointed muzzle, scissor bite. **Ears:** Semi-prick. **Tail:** Plume.

Genetic Diseases or Conditions: Although not common in the breed, the following problems have been seen: hip dysplasia, thyroid disease, eye diseases, Dermatomyositis (Sheltie Skin Syndrome), Von Willebrands disease, and epilepsy are some of the known health problems. "Border Collie Collapse Syndrome" has also been seen in Shelties.

FIGURE 2.26 Welsh Corgi – Cardigan

Welsh Corgi – Cardigan

Origins: The Celts brought the Cardigan from central Europe to Wales in about 1200 BC. The dog was a member of the same family that produced the Dachshund. Later when the land was owned by the crown but tenants were allowed to graze the land, there were no fences, and so the Cardigan was utilized to chase off the neighbors cattle or sheep. The Cardigan was recognized by the AKC in 1935.

QUICK FACTS 2.12 Welsh Corgi – Cardigan

Quick Facts	Height	Weight	Coat	Features
Life Span: 12–15 years	10.5–12.5 inches 26.6–31.75 centimeters	**Males:** 30–38 pounds 14–17 kilograms **Females:** 25–34 pounds 11–15 kilograms	**Colors:** White points on neck, chest legs, muzzle, underbelly, and tip of tail. Body can be shades of red, sable, brindle, black or blue merle with or without tan or brindle points. **Type:** Double coat with guard hairs that lay flat.	**Head:** Wide and flat between ears, scissor bite. **Ears:** Candle flame, large, rounded tips. **Tail:** Brush, set low and carried parallel to ground.

Welsh Corgi – Pembroke

Origins: Brought to Wales by Flemish weavers in 1107, of the same family that includes the Schipperkes, Keeshond, Pomeranian, Samoyed, Chow Chow, Norwegian Elkhound, and the Finnish Spitz. The Pembroke and Cardigan were crossed before the middle of the 19th century, and for this reason we see the similarities. In 1934 the two breeds were recognized as individuals, and the cross breeding stopped. The AKC recognized the Pembroke in 1934.

Characteristics: Both Corgis are very loyal to their families and will be a bit standoffish to strangers. The Pembroke is shorter in body length with straighter and lighter-boned legs; the coat texture

FIGURE 2.27 Welsh Corgi – Pembroke

is finer, the ears are pointed, and, of course, the tail is a natural dock or docked very short. The Cardigan is a heavier dog with large, rounded ears and, of course, a long tail. Both Corgis are good house dogs with affectionate natures, neither will force its attentions on you, they are intelligent, and they are very alert. Both require obedience training and some sort of job to do to keep them physically and emotionally fit. Because of their instinct to drive "animals" away, they sometimes nip at an owner's ankles or at little children's bottoms. Obedience training will help to "nip that behavior" by calling them off. Corgis tend to bark and will often bark at their owners if they believe they have the leadership role. Firm and consistent training and good leadership skills are essential for the owners to learn and implement; otherwise the Corgi can become incorrigible. The Pembroke will tend towards heaviness if given too much food. Both have short double coats that are easily kept up with an occasional brushing and are seasonal shedders.

QUICK FACTS 2.13 Welsh Corgi – Pembroke

Quick Facts	Height	Weight	Coat	Features
Life Span: 12–15 years	10–12 inches 25–30 centimeters	**Males:** 25–30 pounds 11–13.6 kilograms **Females:** 23–28 pounds 10–13 kilograms	**Colors:** White points on neck, chest legs, muzzle, underbelly, and tip of tail. Body can be shades of red, sable, fawn, or black. Head can be tri-color, and brindle, black or blue merle with or without tan or brindle points. **Type:** Double coat with longer hair on chest, neck, and rear hindquarters.	**Head:** Foxy face, scissor bite. **Ears:** Candle flame, slightly pointed. **Tail:** Bobtail, or docked.

Genetic Diseases or Conditions: Cervical disc syndrome, epileptic, urolithiasis, dystocia, degenerative myelopathy, hip dysplasia, and VonWillebrand's disease.

Hounds

Hounds are amazing dogs that were bred to chase and catch prey by sight or by scent. The "sight" hounds were developed to spot prey with sharp eyesight and then rely on speed to chase it down. These hounds were used to catch deer, antelope, wolves, bears, and lions for food and for sport. The "scent" hounds were developed to search out prey by using their keen sense of smell. Once a scent is picked up, they follow it to the animal and give chase until it climbs a tree or goes underground. At this point the hounds will start to "bell" or bark in such a way that signals to their owners that the prey has been found and is being held in place. These hounds are used to hunt foxes, badgers rabbits, and raccoons.

Because of these differences in hunting techniques, the hound group also has a wide variety of body styles. The sight hounds have large, deep chests for heart and lung power and long legs for speed. The scent hounds have large noses to pull in the scent and range from long legs to short legs. The hounds also have the distinction of having the tallest dog, the Irish Wolfhound, and the shortest, the Dachshund, giving us a choice of size to fit most lifestyles. Even though they were bred to hunt, many of the hounds can be couch potatoes and become overweight if allowed to eat unlimited amounts of food. Regular exercise and measured amounts of food will keep hounds in shape. Another thing to remember about the scent hounds is that they often were worked in packs and so are very sociable. This can cause some behavior issues if their owners do not remember this important part of their psychological wellbeing. They do not do well if left to their own devices day in and day out without being included in the family's activities. If they are not invited in or along, you

FIGURE 3.00 Afghan Hound

may come home to a yard full of holes and angry neighbors because of the barking.

A note about feeding Hounds: The sight hounds can vary in the amounts required to maintain their body mass. For example the Borzoi will require about the same calories as a medium-sized dog. The daily portion for the sight hounds should be divided into three meals and elevated to reduce the risk of bloat or gastric torsion. The scent hounds are "easy keepers" meaning that they will tend toward obesity if allowed to eat freely. Their meals need careful measuring and consistency in order to keep them fit. It is a good idea to make sure the kitchen garbage can is protected as they, like most dogs, are likely to see if there are any tidbits available!

SIGHT HOUNDS

Afghan Hound

Origins: The Afghan Hound was developed in present-day Afghanistan, but how it came to Afghanistan is a mystery, although plenty of research has been done to try to verify its origins. The Afghani people made the greatest contributions to the breed as we know it today by combining the northern compact and heavy-coated version with the gangly, sparsely coated version from the south. The AKC recognized them in 1926.

FIGURE 3.01 Afghan Hound

Characteristics: Because of its long, aristocrat nose and monkey-like movement facilitated by hipbones that are higher than their withers and set farther apart than other dogs, the Afghan is called the "monkey-faced hound." The hips allow them to traverse hilly, rugged, uneven ground at greater speeds (monkey-like) without slipping or stumbling. The Afghan is an aloof breed with a regal carriage and a "don't mess with me" attitude, which makes sense since it was hunted in either a male and female pairing or in packs for jackals, wolves, leopards, hares, and marmots. Owners should always be asked if the dog can be touched before doing so. When touching the dog, it should be in full view of the owner, and he or she should be in contact with it. Afghans are sensitive dogs and do require gentle and consistent handling in order for them to respond well. Afghan hounds should always be approached slowly and touched without being spoken to; they should never be cornered, as it will elicit a strong fight-or-flight response. They do require exercise on a daily basis and will need to be able to run free to maintain a healthy mind. This must be in a well-fenced yard or dog park to keep them safe. Their hair coats are fine and require weekly bathing to keep the coat long and shiny. Many owners opt to have a professional groomer trim the coat short to reduce the amount of time spent maintaining the coat.

QUICK FACTS 3.00 Afghan Hound

Quick Facts	Height	Weight	Coat	Features
Life Span: 12–14 years	25–27 inches 63.5–68.5 centimeters	50–60 pounds 23–27 kilograms	**Colors:** Any solid color. **Type:** Thick, long, silky, very fine texture.	**Head:** Dolichocephalic, with a slight Roman nose, level bite. **Ears:** Hanging and feathered. **Tail:** Sickle tail, set low, carried high.

Genetic Diseases or Conditions: Bloat, hypothyroidism, hip dysplasia, juvenile cataract, corneal dystrophy, cataract, necrotic myelopathy, and allergic reactions to anesthesia, grass, flea collars, and meat with lots of fat.

Basenji

Origins: The Basenji come to us from Central Africa where they were treasured for their intelligence, adaptability, and courage. They were used for pointing, retrieving, or driving game for hunters. They were first brought to America in 1941 and recognized by the AKC in 1944.

FIGURE 3.02 Basenji

Characteristics: Known as the "barkless dog," they actually are not silent. They do manage a yodel-like sound, something like a "barooo," as well as howls and other vocalizations. Another unique feature is that females only cycle into estrus once a year versus the two to three times per year other dogs do. The Basenji is aloof, curious, and fastidious, often cleaning themselves like a cat. They will climb a chain-link fence

FIGURE 3.03 Basenji, Tricolor and Red

like stairs, and if they see something they are curious about, they will stand on their back legs even leaning against a wall to stay upright. They have a strong prey drive and must be introduced to cats at an early age if expected to live with them peaceably. The Basenji responds to training that is calm, firm, consistent, and positive, with a firm set of rules and plenty of food treats. They will become resentful and unresponsive if punished. Basenjis are one-person dogs and will tolerate others in the family group. They are active dogs that require vigorous exercise to maintain healthy bodies and minds. They have a short hair coat that requires very little grooming and provides protection from cold weather.

QUICK FACTS 3.01 Basenji

Quick Facts	Height	Weight	Coat	Features
Life Span: 10–12 years	16–18 inches 41–46 centimeters	20–26 pounds 9–11.8 kilograms	**Colors:** Red, black, tri-colored, brindle, with white feet, chest, and tail tip. **Type:** Short and smooth.	**Head:** Mesoce-phalic. **Ears:** Candle flame. **Tail:** Curled.

Genetic Diseases and Conditions: Fanconi syndrome, hypothyroidism, progressive retinal atrophy, immunoproliferative systemic intestinal disease, hemolytic anemia, and sensitivity to environmental and household chemicals which can cause liver damage.

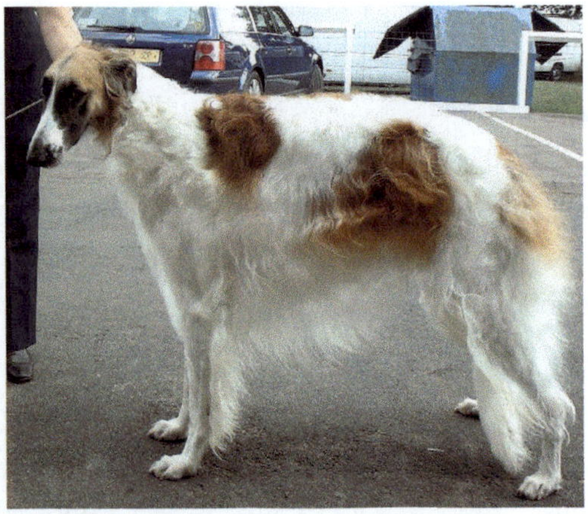

FIGURE 3.04 Borzoi

Borzoi (Russian Wolfhound)

Origins: The Borzoi was developed in Russia by the aristocracy to chase after hare and wolf. Large groups would go out on horseback, and when a wolf was spotted, a trio of dogs would be sent to chase it down and hold it until the hunters caught up. The first Borzoi was brought to the U.S. in 1889 and was recognized by the AKC in 1891. They are still used in parts of the U.S. for coyote control.

Characteristics: For a "hunting" dog, these great beasts are gentle, affectionate, quiet, charming, and unassuming. They rarely bark and are quite content to be couch potatoes. However, they will need daily exercise and should be released only in well-fenced yards as they will chase after small game. This strong instinct to chase down, catch, and hold fast-moving critters can land them into trouble as the critter may be the house cat or other dogs when playing. Do not mistake this as aggression as it is what they were developed to do! Also, they are very fast, and if they are allowed to run in an unfenced area, it is not uncommon for them to cover many miles from the point of origin in a short period of time. Fences do need to be over six feet in height as the Borzoi can clear anything lower than that with very little effort. As with most hounds they are bored by repetitive, seemingly pointless commands, which the uninformed often mistakes as the dog being dumb or stubborn. It is important to find their motivation for obeying commands, whether it be food or toys. The Borzoi is slow to mature, usually taking three years to reach its full height and weight. Although considered a large breed, they actually eat amounts of food that are equivalent to what a medium-sized dog would consider enough. Regular brushing will keep their coats in good shape, and owners should expect heavy seasonal shedding.

FIGURE 3.05 Borzoi

QUICK FACTS 3.02 Borzoi

Quick Facts	Height	Weight	Coat	Features
Life Span: average 4–6 years may live up to 12–15 years	26–28 inches 66–110 centimeters	**Males:** 75–105 pounds 34–47 kilograms **Females:** 60–85 pounds 27–38 kilograms	**Colors:** Any. **Type:** Medium-long, silky, either flat, wavy, or rather curly.	**Head:** Dolichocephalic, should not have a Roman nose. **Ears:** Rose. **Tail:** Sickle Tail.

Genetic Diseases or Conditions: Bloat, hygromas over elbows or other bony prominences, osteochondritis dissecan, and progressive renal atrophy. The Borzoi are allergic to cholinesterase inhibitors (flea collars) and barbiturates of any type, and it can be very chancy to anesthetize these dogs.

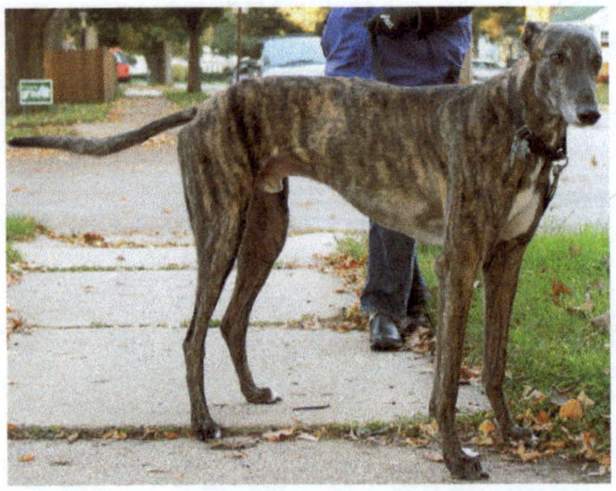

FIGURE 3.06 Greyhound, Brindle

Greyhound

Origins: Some references suggest that the Greyhound descended from the coursing dogs found in Egypt as early as 2000–2500 BC as tombs bear pictures of dogs resembling the Greyhound. However, recent genetic testing suggests that they are closely related to herding dogs that were thought to be brought with the Celts from Eastern Europe. They were utilized for hunting all types of game, although the hare is their preferred prey. The AKC recognized these dogs in 1885.

Characteristics: The Greyhound is the fastest dog alive with sustained average speeds in excess of 40 miles per hour (64 kilometers) and sprinting speeds of over 43 mph. Their deep chests and muscular hind quarters aid in this remarkable speed as do their red blood cells. They have more red blood cells on average than other breeds, which in turn carries more oxygen to muscle tissues. Because of this excess of red blood cells, the Greyhound is often used as a universal blood donor in veterinary medicine. However, the very thing that gives them remarkable speed also can hurt them. The deep chest is one of the hallmarks for a propensity to bloat or develop gastric torsion, and their lean muscle mass makes it very difficult to anesthetize them safely. Greyhounds are very lovable, intelligent, loyal, sweet, unaggressive, and tractable (even racers) dogs, making them ideal companions. Gentle and consistent training is required, and socialization to smaller pets at a young age is a must, but even then they should not be trusted with these animals because of their strong prey drive. They adapt well to apartment living as they can be true couch potatoes, but they will need regular exercise on leash or in a fenced yard as they cannot be held accountable if a rabbit should happen along. Their single-layer, short-hair coats require little grooming which decreases allergens, so they are often touted as non-allergenic. Because of their short coats, they do require housing indoors to protect against cold weather.

QUICK FACTS 3.03 Greyhound

Quick Facts	Height	Weight	Coat	Features
Life Span: 10–12 years	27–30 inches 68–76 centimeters	**Males:** 65–70 pounds 29–32 kilograms **Females:** 60–65 pounds 27–29 kilograms	**Colors:** Any. **Type:** Short, smooth, and firm.	**Head:** Dolichocephalic. **Ears:** Rose. **Tail:** Whip tail.

Genetic Diseases or Conditions: Esophageal achalasia, bloat, decubital ulcers (if allowed to rest without padding), cancers, hypothyroidism, megaesophagus, hemophilia, Von Willebrand's disease, tail irritations in both young and adult, dystocia, and sensitivity to barbiturates. Racing injuries include: luxation, fractures, dropped muscle, fracture to accessory carpal bone, "jack" selling and hematoma or seroma of the medial surface of the hock, and injured digits.

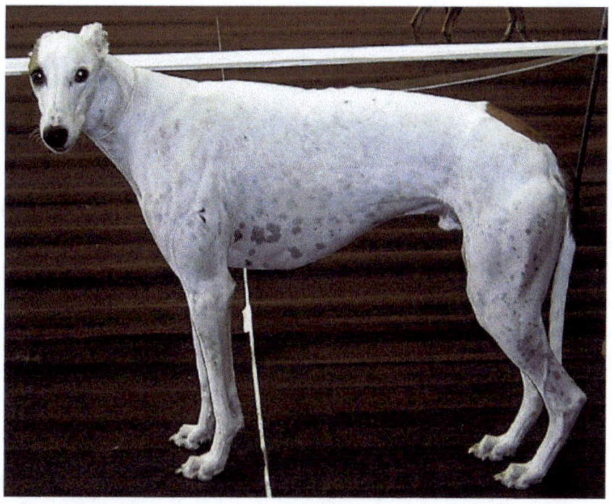

FIGURE 3.07 Greyhound

Ibizan Hounds

Origins: Another ancient breed of dog that closely resembles statues found in tombs of Egyptian kings. At some point the ancestors of these hounds were certainly taken to the islands off the coast of Spain and France to hunt rabbit and other small mammals. The AKC recognized these dogs in 1978.

FIGURE 3.08 Ibizan Hound

Characteristics: Although in the "sight" hound category, this remarkable hound uses both sight and scent to hunt. They are athletic, sensitive, and energetic, often retaining puppy-like energy well into adulthood. They are protective and clean. They do not respond well to harsh training techniques, thinking themselves "on par" with their humans. "Beezers" are outstanding jumpers and require fences that are over six feet tall, which may or may not keep them enclosed as they are known to be escape artists. It is almost as if they can figure out how to unlock gates! They do require early socialization, and it is not recommended that there be a household cat as they have a very strong prey drive, and even their buddy indoors may be fair game if found outside and on the loose. A last tidbit about Ibizan Hounds: they "blush" when excited, and their ears will turn a bright pink. These dogs require a great deal of exercise; twice daily walks and a long romp with children will be about right to help them maintain body and mind. The three types of hair coats are all easy to maintain, and they are average shedders.

QUICK FACTS 3.04 Ibizan Hound

Quick Facts	Height	Weight	Coat	Features
Life Span: 10–12 years	22–29 inches 56–74 centimeters	45–65 pounds 20–29 kilograms	**Colors:** Red or red and white. **Type:** Single-layer, smooth, long, and wire coated.	**Head:** Dolichoce-phalic, scissor bite. **Ears:** Candle flame **Tail:** Sickle, ring, or saber.

Genetic Diseases and Conditions: Allergies, epilepsy, deafness, eye diseases, hip dysplasia, and they too are sensitive to barbiturate anesthesia.

Irish Wolfhound

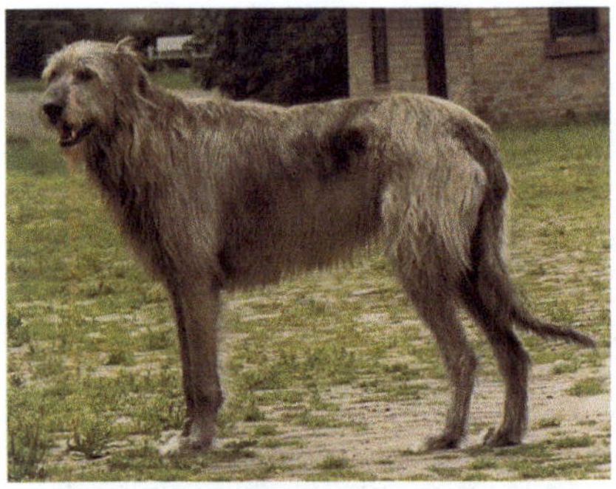

FIGURE 3.09 Irish Wolfhound

FIGURE 3.10 Irish Wolfhound, Head

Origins: Ancient records show the Irish Wolfhound hunting and fighting alongside their people as far back as 390 AD. They were used to hunt wolves and elk, and as war dogs to take people off horses. The breed was almost lost as the last wolves were hunted down in Ireland in the late 1700s. However, an Englishman by the name of George Graham is credited with saving the breed with a careful program of adding Mastiff, Deerhound, Great Dane, and Borzoi to increase the breeding gene pool. The AKC recognized this breed in 1897.

Characteristics: This is a big-hearted dog that would rather curl up and snuggle than fight. They will defend themselves and their people but not property. If provoked, the mode of defense is seizing the antagonist by the back and shaking him to death, which his great strength enables him to do with little or no effort. Fortunately, this dog is mild mannered with a peaceful disposition and calming influence within the family circle because of their immense size. They do require exercise and a means to move them about, so small apartments and cars are not practical for the Irish Wolfhound owner. It takes wolfhounds about two to three years to fully mature, and they must be exercised and fed carefully to prevent rapidly growing joints from damage over the long term. They should be fed three to four small meals daily, and the food should be elevated to prevent bloat or gastric torsion throughout their life span. They are extremely trustworthy around children and will be quite protective of the entire family. They do require regular grooming with a twice-a-year pluck to remove dead hair.

QUICK FACTS 3.05 Irish Wolfhound

Quick Facts	Height	Weight	Coat	Features
Life Span: 6–10 years, 7 being average	30–32 inches 76–81 centimeters	**Males:** 120 pounds 54 kilograms **Females:** 105 pounds 48 kilograms	**Colors:** Gray, grizzled, brindle, red, black, white, fawn. **Type:** Rough and hard on body, legs, and head; wiry and long over eyes and under jaw.	**Head:** Dolichocephalic, scissor bite. **Ears:** Rose, semi-prick when alerted. **Tail:** Long, whip.

Genetic Diseases or Conditions: Cardiomyopathy, bone cancer, bloat, tail injuries, elbow hygroma (thick callus that can break open if required to sleep on hard ground), rhinitis syndrome in puppies, hip dysplasia, and other bone diseases. The Irish Wolfhound is extremely allergic to barbiturates and worm medications because of its relatively low body-fat content.

Pharaoh Hounds

Origins: The national dog of Malta, an island nation in the Mediterranean, the Pharaoh Hound is another "sight" hound claiming ancient lines dating back to the pharaohs and can be confirmed from frescos and statues. They have a similar appearance to the Ibizan Hound and were used in much the same way as a rabbit hunter. The breed was brought to the U.S. in 1967, and it was recognized as a hound breed by the AKC in 1984.

FIGURE 3.11 Pharaoh Hound

Characteristics: The Pharaoh Hound requires early socialization not for aggressiveness but for shyness. They can be very timid and unhappy in unfamiliar situations, so exposing the Pharaoh to parks, playgrounds, ball diamonds, or soccer fields is a must to have a well-adapted dog. They will remain wary of strangers but love their families greatly and will be quite protective. They are fun and funny, exuberant and well behaved if trained consistently and gently. Do not allow Pharaoh Hounds off leash, because if something runs away, it must be chased! All off-leash activities must be in a well-fenced and supervised area. They share two traits with the Ibizan Hound: blushing and being able to clear fences of up to six feet easily. They enjoy playing with children,

FIGURE 3.12 Pharaoh Hound, Head

and that combined with long daily walks will keep the Pharaoh Hound in good shape. Their short-hair coats will require little grooming, and they can be kept indoors.

QUICK FACTS 3.06 Pharaoh Hound

Quick Facts	Height	Weight	Coat	Features
Life Span: 12–14 years	**Males:** 23–25 inches 58–63 centimeters **Females:** 21–24 inches 53–61 centimeters	45–55 pounds 20–25 kilograms	**Colors:** Chestnut (rich red) or tan with white on toes, tail tip, and chest only. **Type:** Short, smooth.	**Head:** Dolichocephalic. **Ears:** Candle flame. **Tail:** Whip.

Genetic Diseases and Conditions: These dogs are generally very healthy, other than some dermatology issues, hip dysplasia, patellar luxations, and eye issues. Being a sight hound, they, too, will negatively react to barbiturates.

FIGURE 3.13 Rhodesian Ridgeback

Rhodesian Ridgeback

Origins: The original dogs were found on the African Cape Peninsula with the Khoikhoi people by the Dutch. These dogs were ferocious guard dogs and had a peculiar "cowlick" of hair running the length of their back. An infusion of Great Dane, Bloodhound, Greyhound, and Terrier furthered the development of the Boer hunting dog. In 1870 the Reverend Charles Helm took two of these Boer dogs with him to South Africa where a big-game hunter by the name of van Rooyen bred them with his dogs to hunt lions, baboons, and wild pigs. In 1950, specimens were imported to the U.S., and the AKC admitted them to the hound group in 1955.

Characteristics: Trustworthy with children, these "people" dogs like to be where their families are, the closer the better, preferably curled up next to you on the couch. They are big, athletic dogs that require daily exercise to stay in shape. They must be socialized but should never be trusted to be alone with small mammals or be left by themselves for long periods of time. They are reserved with strangers and more often than not will ignore them. They are protective of their people and will need to be shown who is a friend as they can grasp who is foe quite readily enough on their own. "Ridgies" are even tempered and sensitive, so training should be consistent and fair.

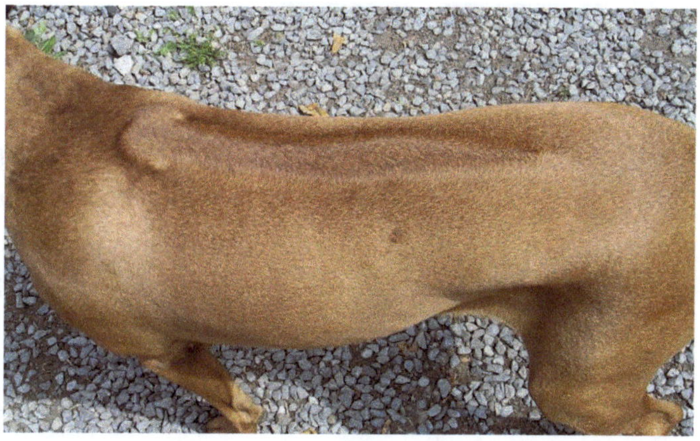

FIGURE 3.14 Rhodesian Ridgeback, Back

The hallmark of this breed is the *ridge* on the back which is formed by the hair growing in the opposite direction to the rest of the coat. This "cowlick" is clearly defined, tapers from wider to narrower starting over the withers and ending at the hip. This is a genetic mutation, and "ridgeless" puppies are seen in litters. They are usually sexually altered and sold as pets. They are almost tireless and require vigorous exercise to maintain a solid mind. Long, twice-daily walks and a good romp in the dog park will be just enough! A smooth short-hair coat is easy to maintain, and they must be kept indoors during cold weather.

QUICK FACTS 3.07 Rhodesian Ridgeback

Quick Facts	Height	Weight	Coat	Features
Life Span:	24 to 27 inches 61–68 centimeters	70–85 pounds 32–38 kilograms	**Colors:** Light wheaten to a red wheaten, small white mark on chest and toes. **Type:** Short, dense, sleek, and glossy.	**Head:** Mesocephalic. **Ears:** Pendent. **Tail:** Saber.

Genetic Disease or Conditions: Dermoid sinus occurs at the neck or rump and is a funnel of outer skin growing inward from the surface down to and frequently connecting to the dura mater of the spinal cord. If left untreated, it can cause an abscess to form which can cause meningitis, often leading to death. Surgical removal of the skin is required, and the dog should be sexually altered. Cervical vertebral deformities, degenerative myelopathy, hypothyroidism, bloat, hip and elbow dysplasia, and deafness are also seen.

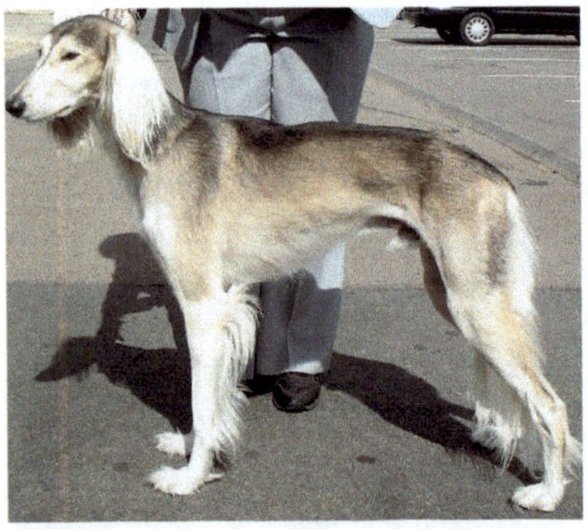

FIGURE 3.15 Saluki

Saluki

Origins: The Saluki is truly an ancient breed, stretching from China to Egypt. Saluki-like dogs are depicted in petroglyphs, paintings, statues, pottery, and other rock arts dating back to 8,000–10,000 BC. It is also said that "whenever the word 'dog' is seen in the Bible, Koran or the Avesta, it is referring to the Saluki." The royal dog of Egypt was recognized by the AKC in 1927.

Characteristics: Don't let this frail-looking dog fool you. They were used to hunt gazelles and so not only had to be fast but also have endurance and agility as well to run over extremely rough ground. They take about four years to mature and require socialization and exposure to a great many different situations to prevent shyness and timidness. The Saluki must be trained using gentle and consistent methods as they will "shut down" if treated harshly. They do grow attached to their owners; however, they are not overly demonstrative with their affection. With that being said, they do respond to affection from their owners and seem to require this connection to maintain their mental health. Extensive exercise will keep the Saluki in balance, and if one is looking for a jogging companion, this is one of the dogs to choose. They love a good run, and, if available, a fenced area is the only place they should be allowed to run free. Salukis are average shedders and will require an occasional brushing to maintain their hair coats. They do get chilled and prefer to be indoors, especially in cold weather.

FIGURE 3.16 Saluki, Head

QUICK FACTS 3.08 Saluki

Quick Facts	Height	Weight	Coat	Features
Life Span:	23–28 inches 58–71 centimeters	29–66 pounds 13–30 kilograms	**Colors:** White, cream, fawn, golden, red, grizzle, tan, tricolor, and black and tan. **Type:** Soft silky texture, smooth, slight feathering on the back of legs.	**Head:** Dolichocephalic. **Ears:** Hanging covered with long silky hair. **Tail:** Sickle.

Genetic Diseases or Conditions: Cancer, heart diseases, dermatitis, Saluki Sudden Death Syndrome, ear mites, or mange. The Saluki reacts negatively to barbiturates of all kinds and is sensitive to cholinesterase inhibitors. Broken bones should be plated.

Scottish Deerhounds

Origins: An ancient breed of the Scots, the Scottish Deerhound has remained true to form even to this day. They had been used for coursing and deer-stalking in the rough terrain of Scotland and England. They were used to hunt wolves and grizzly bears in the U.S. and kangaroo and wild boar in Australia. The AKC recognized them in 1886.

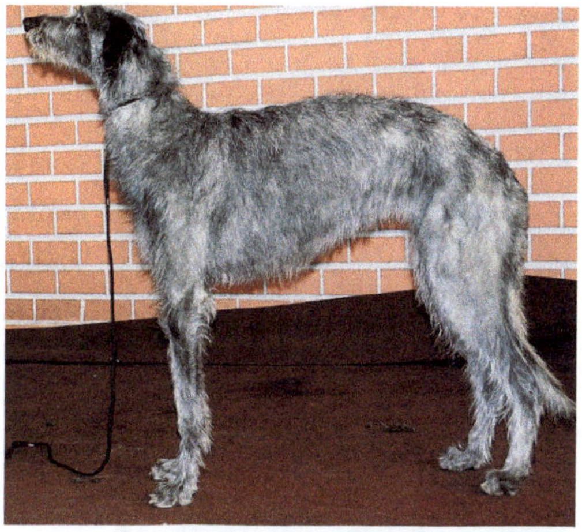

FIGURE 3.17 Scottish Deerhound

Characteristics: Brave, easygoing, sensitive, loyal, and well-mannered are the adjectives used to describe this gentle and friendly beast. They are able to adapt to city life if given the proper amount of daily exercise, which is not a walk around the block. They need to run freely, preferably with another dog, play a rousing game of fetch and/or perform an agility activity. This is not only important for adult dogs but also for puppies' proper formation of muscles and strengthening of joints. They are great around people and other animals, but it is, of course, important to be mindful that they are hounds and will have the "chase it if it moves" instinct. The Scottish Deerhound does resemble a Greyhound in size and shape. However, their hair coat is wiry and harsh to the touch. This shaggy hair covers the face, neck, body, and upper legs. The coat doesn't require much grooming, and they are average shedders.

FIGURE 3.18 Scottish Deerhound, Head

QUICK FACTS 3.09 Scottish Deerhound

Quick Facts	Height	Weight	Coat	Features
Life Span: 8–10 years	28–32 inches 71–81centimeters	**Males:** 85–110 pounds 38–50 kilograms **Females:** 75–95 pounds 34–43 kilograms	**Colors:** Dark blue-gray, grizzle, brindle, yellow, sandy red, red fawn. **Type:** Wiry.	**Head:** Dolichocephalic. **Ears:** Rose. **Tail:** Saber.

Genetic Diseases and Conditions: Cardiomyopathy, allergies, bloat, stomach or splenic torsion, osteosarcoma, kidney disease, Factor VII Deficiency (blood disorder), epilepsy, and Addison's Disease.

FIGURE 3.19 Whippet

FIGURE 3.20 Whippet, Fawn

Whippet

Origins: Developed in England by crossing smaller greyhounds with terriers and later additions of Italian greyhound, Whippets were kept small to hunt hares and rabbit. They were brought to the U.S. by mill operators and were primarily used as racing dogs and so were known as the "poor man's race horse." They were recognized by the AKC in 1888.

Characteristics: The Whippet is known as a "snap-dog," derived from how they "snap up and shake" the rabbit or a lure when they catch it. Whippets are very quiet, with a calm disposition, and they prefer to be neat and clean, thus making them easy to house-train. New owners should be aware that as puppies they are destructive chewers, so early training on what can and cannot be chewed on is a must. Training must be consistent and patient as they are sensitive dogs. The Whippet will require daily exercise and will enjoy a jog as well as a romp in the dog park. Because of their short hair, owners will need to provide sweaters or a blanket, sometimes even in the summer because of air conditioners, or they may find their Whippet in their bed under the covers.

QUICK FACTS 3.10 Whippet

Quick Facts	Height	Weight	Coat	Features
Life Span: 12–16 years	18–22 inches 45–56 centimeters	75–110 pounds 34–50 kilograms	**Colors:** Any. **Type:** Short, smooth, and firm.	**Head:** Dolichocephalic. **Ears:** Rose, semi-prick when alerted. **Tail:** Whip, carried low in a curve.

Genetic Diseases or Conditions: Cryptorchidism, weak digestive tracts (bland food usually works to keep them from having flare-ups), and variable tolerance for anesthetics.

SCENT HOUNDS

Basset Hound

Origins: This French dog was used to trail rabbits and deer. Slower than their sight-hound cousins, they are steadier on the trail as they use their noses rather than their eyesight to follow the scent laid down by the prey. Basset Hounds are second only to the Bloodhound in scenting abilities. In 1884 the first Basset Hound was shown at the Westminster Kennel Club. The Basset was recognized by the AKC in 1885.

Characteristics: Basset Hounds are able to work alone or in packs. They are great family dogs as they will romp with the kids for hours or snuggle on the couch for hours, which they would prefer to do. Walks and exercise are important as Basset Hounds are "easy keepers," as are most scent hounds. This means the Basset will gain weight to obesity levels without careful feeding and exercise. The Basset's long body and short stature can fool you as to how much they weigh, with many in the 60–70 pound range. New owners need to know that if their Basset latches on to a scent, they will follow it regardless of pleas to come back. Therefore, exercise must be on leash or in a fenced yard. Bassets can be hard to house train and need consistent and patient training techniques in order to thrive. The Basset's ears are both his charm and his bane. Those glorious, long, silky ears and excess skin give him that classic "hangdog" look that many an owner falls head over heels for. But because their ears are so heavy, if they should get infected, it is very difficult to clear up because the ear canal provides the perfect environment for bacteria to thrive. To prevent this, when the Basset is being bathed, place cotton balls in the ear canals to keep water out. Bassets have a short-hair coat that constantly sheds but requires little grooming, although they do require an occasional bath to remove doggy odor.

FIGURE 3.21 Basset Hound

FIGURE 3.22 Basset Hound, Puppy

QUICK FACTS 3.11 Basset Hound

Quick Facts	Height	Weight	Coat	Features
Life Span: 8–12 years	14 inches 35 centimeters	40–60 pounds 18–27 kilograms	**Colors:** Hound color (tan, black over back, white points). **Type:** Hard, smooth, short.	**Head:** Mesocephalic, pronounced occipital protuberance, dewlap, and heavy flews. **Ears:** Extremely long, hanging ears. **Tail:** Saber, carried gaily.

Genetic Diseases or Conditions: Front leg injuries from jumping from heights, ear infections, entropion, ectropion, skin-fold dermatitis, hypothyroidism, glaucoma, inter-vertebral disc disease, pyometra, mammary gland tumors, inguinal hernias, platelet disorders, lymphoma, Von Willebrand's disease, epilepsy, and hip and elbow dysplasia.

Beagle

Origins: England developed these lively dogs from the packs of foxhounds that were used to hunt fox and deer. The little Beagles were used to go after hares and rabbits. AKC first recognized them in 1885, and they have been a family favorite ever since.

FIGURE 3.23 Beagle

Characteristics: This happy little dog has won the hearts of many people. They come in two sizes, 13- and 15-inch, each at the withers. They have short, easy-to-care-for hair coats that do shed seasonally. Beagles are very happy living and working in a pack, so when a single dog joins a family, they look at the family members as their pack. It is important that owners remember this for two reasons. First, in a pack there is a hierarchy, and it is important that the Beagle knows who is in charge; otherwise it will be him. Secondly, it is a very sad Beagle that is left in the backyard and not included in the daily goings-on of the family, a.k.a. "his pack." The dog will bark and bay, which is loud for its size, howl at the moon (even if it is not up), dig to China, and chew on the siding off the house. Don't let this happen to your Beagle, yard, or neighborhood! Beagles are also "easy keepers" and must have meals measured out and daily exercise to keep them trim. Beagles are great with kids of all ages and tend to be very social with other dogs and people. They must be walked on a leash because if they find a rabbit trail, the head goes down, the tail comes up, and the ears shut off. Some find their way back home, but, unfortunately, others don't because of car accidents, or they become lost. Today's Beagles are also used as detection dogs on everything from bed bugs to drugs to vegetables.

FIGURE 3.24 Beagle

QUICK FACTS 3.12 Beagle

Quick Facts	Height	Weight	Coat	Features
Life Span: 12–15 years	13 & 15 inches 33 & 38 centimeters	15–30 pounds 6.8–13 kilograms	**Colors:** Any hound color, including reds, lemons, and whites. **Type:** Short, close, hard.	**Head:** Mesocephalic, scissor bite; however, malocclusions are not uncommon. **Ears:** Hanging. **Tail:** Saber, carried gaily, brush on underside.

Genetic Diseases or Conditions: Glaucoma, factor VII and hemophilia A, epilepsy, deafness, demodectic mange, hypothyroidism, "Funny Puppy," disc disease, obesity, arthritis, eye diseases and conditions, and reverse sneezing.

Bloodhound

Origins: The Bloodhound's history is a long one, but most resources agreed that the St. Hubert Monastery in Belgium is where it got its start. Bloodhounds were further developed and refined in the U.S., and the AKC standard was written in 1996. Present-day Bloodhounds will closely resemble their U.S. ancestors as the standards have not changed.

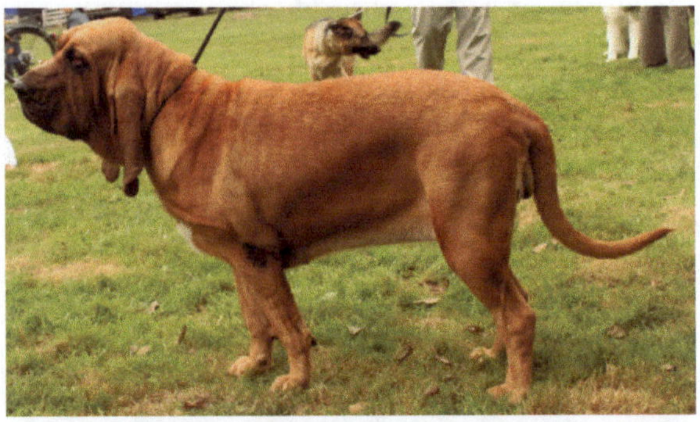

FIGURE 3.25 Bloodhound

FIGURE 3.26 Bloodhound

Characteristics: Bloodhounds are valued for their scenting ability on even days-old trails. In the U.S. they are used by law enforcement agencies to find lost people. The number of olfactory receptor cells in a Bloodhound is 4 billion, compared to a human's puny 5 million and a cat's 200 million. Scents explode inside a Bloodhound's nose. Bloodhounds are gentle giants that will put their owners through multiple tests of patience. They will follow their noses through mud and muck, they are not the most graceful and will knock stuff over when excited, and they have been known to eat things they shouldn't, thus requiring expensive surgeries to retrieve said item, which usually needs to be replaced after journeying through their intestinal system. If ignored they will dig, climb over or jump fences, and bay LOUDLY! Training must be positive, consistent and firm. It should be started at an early age both for obedience and tracking or trailing. Realize that no matter how obedient a Bloodhound is, if it picks up an interesting scent, it is almost impossible to pull them off until they find the source! Mental stimulation is also a must because the instinct to follow a trail or to track is strong and if not utilized may drive the Bloodhound nutty. These gentle giants have been known to allow children to crawl, pull, and pester them without so much as a twitch. Please be an advocate for the Bloodhound and teach children to respect this great dog. They have short hair coats which require little grooming, but careful attention must be paid to the ears as they can develop ear infections quickly that are difficult to clear up because of the heavy ear flap.

QUICK FACTS 3.13 Bloodhound

Quick Facts	Height	Weight	Coat	Features
Life Span: 8–10 years	**Males:** 25–27 inches 63–69 centimeters **Females:** 23–25 inches 58–63 centimeters	**Males:** 90–110 pounds 41–50 kilograms **Females:** 80–100 pounds 36–45 kilograms	**Colors:** Black/tan, liver/tan and red, small amounts of white on chest, feet, and sternum. **Type:** Short hair.	**Head:** Mesocephalic, pronounced occipital peak, scissor or level bite, excessive flews. **Ears:** Folded pendent. **Tail:** Saber, long.

Genetic Diseases or Conditions: Bloat, ectropion, entropion, cancer, ear infections, hematoma of the pinna, moist dermatitis, elbow and hip dysplasia, toxoplasmosis. Bloodhounds require lower anesthetic doses for it size and weight.

Dachshund

Origins: According to 15th Century European historical accounts, this short-legged, long-bodied dog was used to hunt badgers. The Dachshund would follow the badger into its hole and fight to the death. It is thought that there is some Terrier in the Dachshund's makeup, for when they capture their prey, they vigorously shake their heads, thus killing whatever they've caught. The German word for badger is "dachs," and "hund" means dog—thus the name! They were listed in the AKC stud book in 1885.

FIGURE 3.27 Dachshund, Shorthair

Characteristics: The tubular Dachshund is a hardy, vigorous, tireless hunter outdoors. Their short but not crooked legs are situated such that they can creep through burrows and dig in their heels when fighting with their prey. Indoors they are affectionate, responsive companions that are hilarious at play, charming their way into their owner's hearts. These awesome little dogs bring smiles and delight people wherever they go. Most of them are happy-go-lucky characters that delight in meeting various people. Unfortunately, Dachshunds have little-dog syndrome which causes them to pick fights with bigger dogs. It is suggested that they be fitted with a harness so that when they pick a fight, they can be pulled out of harm's way. Dachshunds will announce strangers at the door and will have to be trained to stop barking when asked. Dachshunds will try to train their owners and take over the household. Consistent and firm training that includes food-treat rewards will produce the best results. Dachshunds come in three varieties (smooth, longhaired, and wirehaired), in three sizes (standard, miniature, and toy), and in a multitude of colors. The longhaired "Dachs" will need weekly brushings to keep their fur from matting, and the wirehairs will need a twice-yearly plucking to remove dead hair. The shorthaired variety are often chilled and will require sweaters or a blanket it can get under to stay warm.

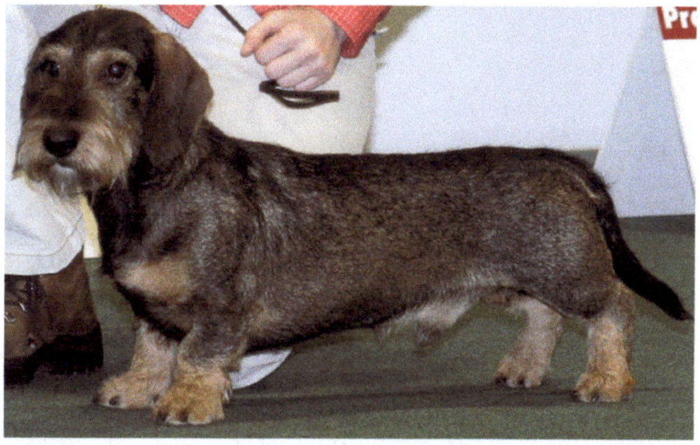

FIGURE 3.28 Dachshund, Wirehair

QUICK FACTS 3.14 Dachshund, Standard, & Miniature

Quick Facts	Height	Weight	Coat	Features
Life Span: 13–15 years	**Toys:** Under 10 inches Under 25 centimeters **Miniatures:** 10 inches 25 centimeters **Standards:** Over 10 inches Over 25 centimeters	**Toys:** 5 pounds 2.2 kilograms **Miniatures:** 6–9 pounds 2.2–4 kilograms **Standards:** 12–20 pounds 5.4–9 kilograms	**Colors:** Solid colors: red, red/yellow, yellow, or brindle. Two-color: deep black, chocolate, gray, and white each with brown points. Dappled colored: brownish or grayish or even white ground with dark irregular patches of dark gray, brown, yellow, or black. **Type:** 3 types: **Smooth** - short, smooth, shiny. **Wirehaired** - short, thick, rough, hard, with a beard and bushy eyebrows. **Longhaired -** soft, sleek, slightly wavy.	**Head:** Dolichocephalic, scissor bite. **Ears:** Pendent. **Tail:** Whip, depending on the hair coat: wirehaired has a bristly underside, longhaired has the longest tail with long hair on the underside, all are held straight out from the body, but the longhaired will hold it upright like a flag.

Genetic Diseases or Conditions: Intervertebral disc disease, obesity, eye disorders, demodex, dermoid cysts, osteoporosis, cardiac diseases, uroliths, diabetes mellitus, epilepsy, pattern baldness, alopecia, thyroid diseases, glaucoma, cataracts, and mast cell tumors.

FIGURE 3.29 Dachshund, Longhair

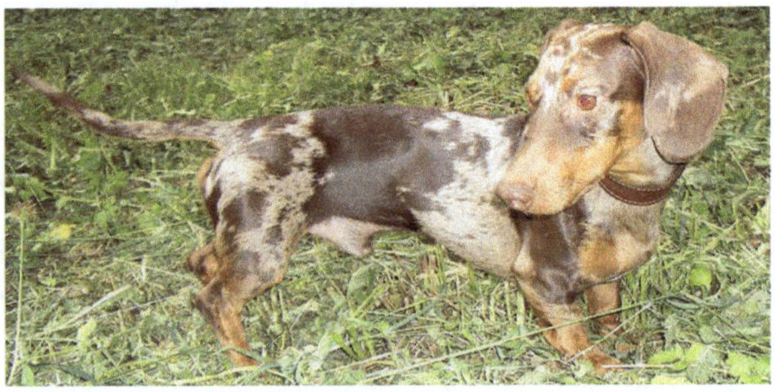

FIGURE 3.30 Dachshund, Shorthair Dapple

American Foxhound

FIGURE 3.31 Foxhound, American

Origins: The American Foxhound, a cousin to the English Foxhound, is one of many strains of Foxhounds in the U.S. As the name suggests, they were bred to hunt fox in packs with the hunters on horseback. They were first recognized by the AKC in 1886.

Characteristics: The American Foxhound is a large dog that looks similar to a Beagle with longer legs. While on the hunt they will "sing" with a musical bark that changes as they close in on their prey. Owners can tell if the prey has been cornered by the change in the sound. For this reason there are few Foxhounds that are kept in urban areas. They tend to be reserved around strangers but are calm and have a sweet disposition which allows them to get along well with children and other dogs. Obedience training, walks on leash, and patience when training is recommended for Foxhounds. Foxhounds are shorthaired and require protection from cold weather.

QUICK FACTS 3.15 American Foxhound

Quick Facts	Height	Weight	Coat	Features
Life Span: 11–13 years	21–25 inches 53–63 centimeters	65–75 pounds 29–34 kilograms	**Colors:** Any hound color. **Type:** Medium-length, close, hard.	**Head:** Mesocephalic. **Ears:** V-shaped. **Tail:** Saber, head erect.

Genetic Diseases and Conditions: These dogs are fairly healthy with few genetic diseases or common health concerns. Obesity is one issue that can be controlled with measured meals and plenty of exercise.

FIGURE 3.32 Norwegian Elkhound

Norwegian Elkhound

Origins: Remains identified as being from an Elkhound-type dog were found in a cave in Western Norway dating back to 4000–5000 BC. A hunter of large game such as moose, elk, and bear, they were also companions and guard dogs of the Vikings. The AKC recognized them in 1913.

Characteristics: The Norwegian Elkhound is devoted to man and is a fine hunting dog. As a companion and guardian, they are great around children and will be a good watch and guard dog for the family. They can be territorial and will sound off if strangers approach. They must be trained to stop barking once the owners have been alerted; otherwise they can become obsessive barkers. Elkhounds are heavily muscled dogs with good stamina. They can track and then hold prey at bay by barking and jumping toward and away from the prey to hold it in place until the hunter arrives. In keeping with the hound personality, they are a bit difficult to train and will need to have firm and consistent handling. They can be a bit aggressive toward other dogs and should be socialized from an early age. They are highly energetic dogs and will require twice daily walks, jogs, or runs alongside a bike, plus a romp in the yard with the children. They have a thick, double coat that sheds seasonally and will require brushing to remove the dead hair. Otherwise it will mat, and the dog can develop skin sores.

QUICK FACTS 3.16 Norwegian Elkhound

Quick Facts	Height	Weight	Coat	Features
Life Span: 12–14 years	19–20 inches 49–52 centimeters	48–55 pounds 21–25 kilograms	**Colors:** Variations of gray, undercoat light silver as are legs, stomach, buttocks, and underside of tail. Muzzle, ears and tail tip are black. **Type:** Double with thick, dense undercoat, outercoat and guard hairs are of medium length.	**Head:** Mesocephalic, scissor bite. **Ears:** Erect. **Tail:** Ring tail.

Genetic Diseases or Conditions: Hot spots, subcutaneous cysts, seborrhea, progressive retinal atrophy, cataracts, glaucoma, some drug sensitivities, hip dysplasia, obesity, and kidney disease in old age.

Otterhounds

FIGURE 3.33 Otterhound

Origins: An English-bred hound that was used to control river otters, the Otterhound was a large shaggy dog that was thought to have some Terrier in its background. It wasn't until hunting otters for sport after WWI that the breed became fixed. Bloodhounds and Griffon Nivernais (a French dog) were crossed with the Otterhounds to establish hunting ability and the breed standard. They were brought to the U.S. in 1907 and were recognized by the AKC in 1909. They are still a relatively rare breed with less than 1,000 worldwide.

Characteristics: The Otterhounds are the opera singers of the hounds. They enjoy throwing back their heads and howling, grunting, sighing, and ruffing along with other hounds, humans, sirens, or just going solo. With their deep voice it can be quite enjoyable if not in the middle of the night. Being a large dog they need multiple small meals of a good quality food. Exercise is a must and, of course, must be on leash or in a fenced-in yard. They love to swim and will do so for hours even diving to retrieve items. Their coats do require weekly grooming and if softer two to three times a week. This should be started as a puppy so that it is conditioned to accept this care. The long hair on their ears and beard will hold water. Many Otterhounds love to drink at the bottom of the water dish and eat to the bottom of the food dish, so special attention should be paid to that area of the body.

QUICK FACTS 3.17 Otterhound

Quick Facts	Height	Weight	Coat	Features
Life Span: 12 years	24–27 inches 61–68 centimeters	80–115 pounds 36–52 kilograms	**Colors:** Grizzle, red, wheaten, sandy, blue, black and tan, blue and tan, black and cream, rarely liver, tan and liver, tan and white. **Type:** Double, soft undercoat, long, harsh outercoat.	**Head:** Mesocephalic, scissor bite. **Ears:** Folded. **Tail:** Saber.

Genetic Diseases and Conditions:

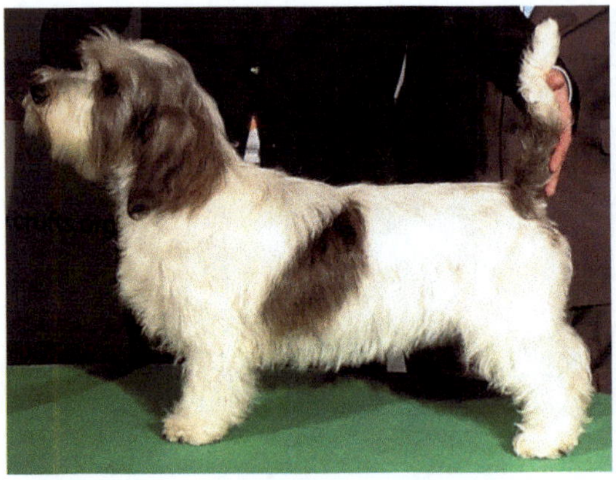

FIGURE 3.34 Petit Basset Griffon Vendéen

Petit Basset GriffonVendéen

Origins: This French hound was developed in the Vendéen region of France. They were bred down from the larger Vendéen hound to trail hare and rabbit. They are a recent addition to the AKC, being accepted in 1990.

Characteristics: The Petit Basset Griffon Vendéen (PBGV) are the party dogs of the hound world. They are not a hound that can be ignored. They will sing along with the crowd or do a solo just for the sheer joy of it. They tend to be mouthy and loud when at play, when on the trail, or just plain bored. Their deep chest allows the PBGV to put a lot of oomph behind its bark. This fun-loving character wins hearts and friends wherever it goes. They are excellent around people, making great therapy dogs. Small children can be injured unintentionally by a PBGV's boisterous play. The PBGV should be introduced to the family cat, and it must be made clear that the cat is part of the pack. Another thing to remember is that while on walks with a PBGV, its welcome bark and erect tail carriage may elicit an aggressive response from the passersby. This may be misinterpreted as an aggressive gesture, and if the passerby is aggressive, it may cause a fight to break out. Weekly brushings will remove dead hair, and with little shedding they are potentially good candidates for people with mild allergies to dog dander.

QUICK FACTS 3.18 Petit Basset Griffon Vendien

Quick Facts	Height	Weight	Coat	Features
Life Span: 12 years	12–15 inches 30–38 centimeters	31–40 pounds 14–18 kilograms	**Colors:** White with spots of orange, lemon, black, grizzle, or sable. **Type:** Harsh double coat, long, rough hair on body, softer fur on face and legs. A beard and mustache completes the look.	**Head:** Mesocephalic. **Ears:** Hanging. **Tail:** Saber, held erect.

Genetic Diseases and Conditions: Persistent pupillary membranes, retinal folds, glaucoma, aseptic meningitis, epilepsy, hypothyroidism, and food and skin allergies.

Non-Sporting Group

This group of dogs has something for everyone. From the rambunctious Dalmatian to the stoic Bulldog; from long, thick hair to short, fine hair; and tall or short, this group has substantial variety. Each breed in this group has its own personality, quirks, and some faces only a mother could love. On the whole, these dogs are the quintessential companions; they may have been bred for a different purpose, but when that purpose was eliminated, they were then refined to be man's best friend, and they take that job very seriously!

American Eskimo Dog

Origins: This "spitz type" dog is a mixture of German, Italian, and Japanese spitz, the white Keeshond, and white Pomeranian. They were very popular in the early 20th century as circus performers and were called American Spitz; it is unclear why their name was changed to the American Eskimo Dog. Perhaps it is because of their snowy white coat. The AKC recognized them in 1994.

FIGURE 4.00 American Eskimo Dog

Characteristics: Bright, alert, inquisitive, and daring are adjectives that describe the American Eskimo or "Eskies." Being a member of the Spitz family that were developed as watchdogs, they are very vocal dogs and need to be trained to stop barking once the family is alerted. They will investigate noises and suspicious goings on and will often insist their owners participate. They have been known to alert parents when

FIGURE 4.01 American Eskimo, Puppy

children are crying or in a predicament and are often credited with being good with children. They will learn tricks, good and/or bad, quickly and need socialization and a strong, consistent, confident owner. If this is lacking, the Eskie will quickly take over, which can lead to negative behaviors such as anxiety, obsessive barking, aggression, and hyperactivity such as spinning. Daily exercise and obedience training is a must with these high-energy dogs. One of their appealing characteristics is their thick, snowy white coat which does require regular brushing and bathing to reduce shedding and matting. They also come in three sizes, toy, miniature, and standard, to fit any household.

QUICK FACTS 4.00 American Eskimo Dog

Quick Facts	Height	Weight	Coat	Features
Life Span: 12–16 years	**Toy:** 9–12 inches 23–30 centimeters **Miniature:** 12–15 inches 30–38 centimeters **Standard:** 15–20 inches 38–51 centimeters	**Toy:** 6–10 pounds 2.7–4.5 kilograms **Miniature:** 10–17 pounds 4.5–7.7 kilograms **Standard:** 17–25 pounds 7.7–8.3 kilograms	**Colors:** Brilliant white or biscuit cream. **Type:** Double coat, undercoat is dense and thick, outer coat has long guard hairs that stand out from the body.	**Head:** Mesocephalic, lips and nose are black to dark brown as are the rims around the eyes, bite is scissor or pincher. **Ears:** Erect, set high with a blunt tip. **Tail:** Squirrel, covered with hair.

FIGURE 4.02 Bichon Frise

Genetic Disease/Conditions: Ophthalmic disorders, cataracts and progressive retinal atrophy, patellar luxation, obesity, hip dysplasia, tear staining, allergies, and dental issues.

Bichon Frise

Origins: First appearing in 14th century Mediterranean documents, they are a cross between the Barbet Water Spaniel and the Poodle, and they were initially traded by Spanish and Italian sailors. The Bichon gained and lost popularity over the decades, becoming a favorite of the 16th century French royal courts and Spanish painters. During one of their waning periods, they were "street" dogs that performed tricks with buskers. In 1933, the French adopted the official standard; it wasn't until 1971 when the Bichon Frise became eligible to be shown in the AKC's Miscellaneous Class, and finally in 1973 it was admitted to the Non-Sporting group.

FIGURE 4.03 Bichon Frise, Puppy

Characteristics: This white ball of poof is a sturdy, merry, playful, affectionate, independent, social, and energetic dog. These descriptions fit the Bichon perfectly, but the one to concentrate on is independent as they are full of it. Because of this independent streak, the Bichon can be difficult to house train and can take over a household if not firmly and consistently handled. They get along well with children if raised with them, and the children need to be taught not to get too rough with their play or to tease the Bichon. The Bichon won't hesitate to warn them off with a growl or a nip which can escalate into them becoming the boss over the child and even the household. Unfortunately, the Bichon became very popular in the 1980s, and as a result many were carelessly over-bred with some lines becoming very nippy. Their exercise requirement is a daily walk along with a romp in the park. They are often advertised as a non-allergenic breed, and to a point this is true. The undercoat is dense and thick with the curly outer hairs coming through which catch in the undercoat. However, they still have dander, they still shed, and they will lick themselves, so there is saliva as well, all of which are the usual allergen suspects. Regular home and professional grooming will be required to reduce these allergens.

QUICK FACTS 4.01 Bichon Frise

Quick Facts	Height	Weight	Coat	Features
Life Span: 14 years	**Males:** 9–12 inches 23–30 centimeters **Females:** 9–11 inches 23–28 centimeters	7–12 pounds 3–5 kilograms	**Colors:** Color is white, biscuit, and cream. **Type:** Soft, thick, dense undercoat; outer coat coarser and curlier; hair grows in length continuously.	**Head:** Mesocephalic, rim of eyes and lips are black, and nose is black, scissor bite. **Ears**: V-Shaped. **Tail**: Squirrel, covered with hair.

Genetic Diseases/Conditions: Some are prone to watery eyes, cataracts, glaucoma and dry eye, cancer, congenital cardiac disease, Cushing's disease, Legg-Calve-Perthes disease, allergies, deafness, epilepsy, and patellar luxation. They can be very sensitive to flea bites.

Boston Terrier

FIGURE 4.04 Boston Terrier, Brindle

Origins: Boston, Massachusetts is where the "Boston" was developed by crossing an English Bulldog with an English Terrier. Known as "roundheads," the French Bulldog was added to reduce their size. Initially, they were around 32 pounds, but now they should not exceed 25 pounds. They were admitted to the AKC in 1893.

Characteristics: This dog is not a fighter; although able to take care of himself, he is a gentleman or gentlewoman which has won them a new nickname as the "American gentleman" among dogs. As companions and house pets, they live up to that moniker by being gentle, affectionate, and well-mannered. Most of the terrier attributes were bred out of them long ago, but on occasion you may see a "Boston" become feisty and willful if handled harshly and inconsistently. They learn quickly and are willing to perform tricks. They don't tend to be barkers, but they can be quite vocal in other ways to get attention. Daily walks and a good romp with the children or other dogs in a fenced-in yard will be just what they need to keep in shape. They are good with people of all ages and do very well with other dogs and animals. They are good therapy dogs and will be kind with strangers. Bostons, have a short-hair coat thus require very little regular care.

FIGURE 4.05 Boston Terrier, Black and White

QUICK FACTS 4.02 Boston Terrier

Quick Facts	Height	Weight	Coat	Features
Life Span: 11–15 years	10–12 inches 25–30 centimeters	10–25 pounds 4.5–11 kilograms	**Colors:** Black, brindle, or seal with white markings on breast, fore and hind legs, across muzzle, blaze between eyes. **Type:** Short, smooth, fine textured.	**Head:** Brachycephalic, square, stop well defined. Short, square muzzle, level bite or undershot. **Ears**: Cropped or Bat. **Tail**: Docked - if undocked can be a gay, screw or crank tail.

Genetic Diseases or Conditions: Inhalant allergies, deafness, hemivertebra, patellar luxation, cataracts, glaucoma, dystocia, heart defects, esophageal achalasia, hydrocephalus, cleft palate, and demodectic mange. These dogs suffer heat stroke easily and may be prone to skin cancer.

Bulldog

Origins: The Bulldogs of Great Britain evolved to withstand the rigors of bull baiting, a sport in which a tethered bull or bear was set upon by Bulldogs. Their job was to grab the animal's nose and hang on until it was pinned to the ground. This of course required the Bulldog to have a powerful build, be aggressive towards animals, including other dogs, and have jaws powerful enough to hang on to the animal's nose. This "sport" was outlawed in 1835, and the breed was almost lost. However, a few fans of the dog decided to save the breed and make some fundamental changes to "tame the beast" and make it more of a companion than a fighter. Pugs were used to shorten the legs, widen the body, and mold the head into a brachycephalic form. The AKC recognized this breed in 1886.

FIGURE 4.06 Bulldog

FIGURE 4.07 Bulldog, Head

Characteristics: These dignified dogs should have an easy-going and kind disposition, be resolute and courageous but never vicious or aggressive. They will protect their own, and although they may be willing to chase off intruders, they don't have the stamina necessary for long hauls because of their facial features. The brachycephalic head is very large and round, with a much-shortened upper jaw that is almost nonexistent. This causes a malformation of the hard and soft pallets, which in turn interferes with normal breathing. If stressed by vigorous exercise or high heat and humidity, they can actually suffocate because they cannot get enough air into their lungs. Another issue caused by this enlarged, round head is they often cannot deliver their puppies naturally, so cesarean sections are often prearranged in order to save puppies and mothers. These dogs often have a face "only a mother could love" but are endearing to those who are fans. Grooming is easy as they have a short-hair coat; however, the creases in the face need to be wiped free of moisture to prevent bacterial infections.

QUICK FACTS 4.03 Bulldog

Quick Facts	Height	Weight	Coat	Features
Life Span: 8–10 years	12–16 inches 30–40 centimeters	40–50 pounds kilograms Females: pounds kilograms	**Colors:** Brindle, all white, red, fawn or fallow, piebald. **Type:** Short, straight and fine texture.	**Head:** Brachycephalic, very large in circumference, measuring as wide as the dog is tall. Undershot mandible with thick flews that completely overhang lower jaw. **Ears**: Rose ear, small, and thin. **Tail**: Straight or corkscrew.

Genetic Diseases or Conditions: Cleft palate, cryptorchidism, vaginal hyperplasia, walrus syndrome (subcutaneous edema—usually around head and thorax), wry jaw, dermatitis (especially around face), inhalant allergies, hemivertebrae, demodectic mange, lymphoma, mast cell tumors, elongated soft palate, cardiac disease, hip and elbow dysplasia, hypothyroidism, patellar luxation, entropion, cherry eye, ectropion, and distichiasis. These dogs often require artificial insemination and caesareans.

Chinese Shar-Pei

FIGURE 4.08 Chinese Shar-Pei

Origins: This ancient breed was developed in China with evidence of them going back as far as 200 BC. They were a multipurpose farm dog with many of the breed's characteristics related to that work: perimeter guard, vermin catcher, hunter, fighter, and herder. During the Communist Revolution in China many of them were killed. A few survived in Hong Kong and Taiwan with some of these exported to the United States in 1966. They were accepted into the AKC's Non-Sporting group in 1992.

Characteristics: The most distinguishing characteristic of the Shar-Pei is their skin. It is loose, abundant, and wrinkly. Puppies are said to grow into their skin as they tend to be very wrinkled, becoming more even as they reach adulthood. Their hair coat is also unique in that it is bristly. It, along with the loose skin, aids the Shar-Pei in fights. If a boar or another animal grabs the dog, the bristles make holding on uncomfortable, and it is said the Shar-Pei could turn and bite back because of the looseness of its skin. Shar-Peis are considered a fighting breed and so require obedience training and socialization as puppies. They are intelligent but are often stubborn and aloof towards strangers, but extremely loving and devoted to family. The don't tend to bark often, so when they do, it is a good idea to pay attention as they are most likely trying to alert you to intruders or some other impending disaster. Unfortunately, because they were a trendy dog in the 1970s, indiscriminant breeding occurred that produced dogs with more wrinkles, which in turn created skin-related issues and behavior problems. The breeder clubs have worked hard to return this breed to its original health and mental soundness. They have two types of hair coats: The bush coat sheds lightly year round, and the horse coat molts once or twice a year. During this molting, regular brushing will help remove the dead hair and allow the new hair to show through.

FIGURE 4.09 Chinese Shar-Pei, Puppy

QUICK FACTS 4.04 Chinese Shar-Pei

Quick Facts	Height	Weight	Coat	Features
Life Span: 8–10 years	18–22 inches 46–56 centimeters	**Males:** 55–60 pounds 25–27 kilograms **Females:** 40–55 pounds 18–25 kilograms	**Colors:** "Shar-Pei" means sandy coat, any other solid color. **Type:** Three coat types: horse-coat (short and harsh), brush-coat (a bit longer and smoother), and bear-coat (one-inch hair lengths).	**Head:** Brachycephalic, unique "hippopotamus" shape, broad muzzle, bluish-black tongue, scissors bite. **Ears**: Extremely small, erect ear, thick, equilateral triangles. **Tail**: Ring.

Genetic Disease or Conditions: Familial Shar-Pei fever, amyloidosis, entropion, cataracts, glaucoma, cherry eye, lens luxation, ear infections, hypothyroidism, patellar luxation, bloat, cancer, hip and elbow dysplasia, skin allergies, parasites, and conditions.

Chow Chow

Origins: Considered one of the oldest dogs known to man, genetic testing marks it as a direct descendent from the gray wolf. It is thought to have come from the northern parts of China, and with its densely thick coat it is not hard to believe. The Chow was used for all jobs, most particularly as a hunting dog but also for guarding, drafting, and herding. The name Chow Chow is said to have come about by Dutch sea captains using it to cover a variety of items in their holds. Things such as porcelain and ivory figurines and other items, including the dogs, were all classified as "knick knacks," and the Chinese translation for "knick knacks" is "chow chows." The AKC recognized this breed in 1903.

FIGURE 4.10 Chow Chow

Characteristics: Chows have very unique tongue, lips, and gum tissue; it is bluish/black in color. They have small, forward-facing eyes that results in poor peripheral vision. Because of this and a thick ruff of hair around the head, they should be approached from the front and addressed in a soft, reassuring voice. Chows are aloof and often do not appreciate being handled by strangers, so the owner should be present when handling a Chow. They are very strong, powerful dogs, with a lordly attitude. Chows bond with their owners and often prefer one member of the family over others. They require early and constant socialization to keep them from becoming aggressive with strange people and other dogs. They can be a bit stubborn and will not accept corrections if given too mildly or too harshly. A strong, consistent, confident owner is required to raise a Chow to become a good citizen. Chows tend to be lazy and must be taken on walks in order to maintain sound mind and body. They require regular grooming especially during seasonal molting to keep the coat from becoming compacted.

FIGURE 4.11 Chow Chow, Face

QUICK FACTS 4.05 Chow Chow

Quick Facts	Height	Weight	Coat	Features
Life Span: 9–15 years	17–20 inches 43–51 centimeters	**Males:** 55–70 pounds 25–32 kilograms **Females:** 45–60 pounds 20–27 kilograms	**Colors:** Red, black, blue, fawn, cream, white, solid throughout. **Two types:** Rough: double, soft, woolly undercoat, dense, straight outer coat. Smooth: the outer coat is shorter.	**Head:** Mesocephalic large, broad, flat skull, scissor bite, blue-black tongue, mouth, and lips. **Ears**: Prick. **Tail**: Ring tail.

Genetic Diseases or Conditions: Heat prostration, eye problems (entropion causing the most problems, plus glaucoma and cataracts), hip and elbow dysplasia, skin problems, lymphoma, diabetes mellitus, canine pemphigus, gastric cancer, melanoma, bloat, hypothyroidism, patellar luxation, and epilepsy. Intolerance of morphine, meperidine, and barbiturates is common.

Dalmatian

Origins: Shrouded in antiquity, as seen in engravings, paintings and writings in Europe, Asia, and Africa, these dogs may be descendants of dogs from Croatia; however, it is still up to debate.

FIGURE 4.12 Dalmatian, Black Spotted

FIGURE 4.13 Dalmatian, Liver Spotted

The name comes from Dalmatia, a province of Austria on the eastern shore of the coast of Venice. They have been classified as hounds, sporting dogs, performers, and mascots for firemen! The AKC recognized them in 1888.

Characteristics: The Dalmatian was developed to run along with carriages to act as a guardian of the horses and carriage. They were also used to run in front of horse-drawn fire wagons down the center of narrow, very crowded streets. They would bark, growl, plow into, or even bite people and other animals to get them to move so the fire wagons could pass. Their bright white coats and black or liver spots also helped them to stand out in a crowd. Today they are still extremely hardy and energetic. They require a firm and commanding owner who can establish the rules and is strong enough to enforce them. This includes establishing who the boss is in the family hierarchy and that everyone in the family is higher than the dog. If owners cannot give them enough exercise and attention, as they really love their families, then this is not the dog for them. Many young Dalmatians are surrendered because of behavior issues because they haven't gotten the exercise and proper leadership from their owners. Although the Dalmatian is friendly, he is not everyone's dog, and it is important to approach this breed with caution as they may not tolerate being handled by strangers. They have a short-hair coat that sheds year round and so it requires brushing to stay on top of the shedding hair. They tend to be neat and clean dogs, so bathing them is usually not necessary. Puppies are born all white with the spots developing later.

QUICK FACTS 4.06 Dalmatian

Quick Facts	Height	Weight	Coat	Features
Life Span: 12–14 years	19–23 inches 48–58 centimeters	50–55 pounds 23–25 kilograms	**Colors:** Black or liver brown spots on a body color of pure white. **Type:** Short, hard, dense but fine.	**Head:** Mesocephalic, long muzzle, scissor bite. **Ears**: V-shaped, with rounded point. **Tail:** Saber.

Genetic Diseases or Conditions: Deafness, uroliths, bacteriuria, dermatitis, allergies, and demodex.

French Bulldog

Origins: It is thought that the "Frenchie" was selectively bred from the Bulldog using the "toy" or smaller variation by English Lace Makers. During the Industrial Revolution, the displaced lace makers took them to France, where they became quite popular as a companion dog. American vacationers saw these toy bulldogs and imported them to America in the 1800s. Here they enjoyed the lap of luxury from being owned by America's wealthiest citizens. In 1905 they were recognized by the AKC as Bouledogue Français which was changed to its current name in 1912.

FIGURE 4.14 French Bulldog, White and Black

Characteristics: The French Bulldog's miniature size combined with a muscular upper body, a uniquely rounded head, and huge "bat" ears are sure to bring a smile when met by strangers. They proceed to then steal hearts with their welcome wiggle as it is their entire back end since their tails are cropped short. Because of these traits, they have often been described as "a clown in the cloak of a philosopher." They are well behaved with at least a bit of obedience training, adaptable, affectionate, and even tempered, getting along with almost everyone and everything. They will be playful, always alert, and will make themselves comfortable on their favorite person's lap. "Frenchies" are indoor dogs and require air conditioning in warm weather and heating in cold weather. They can suffer heat exhaustion in hot, humid weather and frostbite in cold. Exercise should be moderate; a sedate daily walk is really all they need to keep their figures in shape. It is important to slow them down if getting too excited as their "brachiocephalic" faces can result in breathing difficulties if they become over exerted. Watch them carefully around swimming pools as their muscular chests act very similar to an anchor.

FIGURE 4.15 French Bulldog, Black, Fawn, and White

QUICK FACTS 4.07 French Bulldog

Quick Facts	Height	Weight	Coat	Features
Life Span: 8–12 years	12 inches 30 centimeters	22–28 pounds 10–13 kilograms	**Colors:** Brindle, fawn, white, brindle and white. **Type:** Short, smooth, and fine.	**Head:** Brachycephalic, large with a slightly rounded forehead, flews hang over an undershot jaw. **Ears**: Bat. **Tail:** Straight or corkscrew.

Genetic Diseases and Conditions: Breathing difficulties related to their brachycephalic head shape, hypothyroidism, spinal issues such as hemivertebra, cherry eye, entropion, cataracts, patellar luxations, and skin disorders especially in the facial skin folds. This breed often requires artificial insemination and caesareans.

FIGURE 4.16 Keeshond

FIGURE 4.17 Keeshond Puppy

Keeshond

Origins: Keeshond, pronounced "kayz-hawnd," is an old breed that is thought to have come from the same lines as Huskies, Norwegian Elkhounds, Samoyeds, and Pomeranians. The Dutch are credited with their development as family dogs that were asked to act as a watchdog on farms, boats, and barges. The AKC recognized this breed in 1930.

Characteristics: Keeshonds are gentle with children, whom they adore. They interact with other people and pets well, and they are very loving, active, and alert. Because of their watchdog status, they can be barkers and so will have to be trained to stop when told. This is not a huge undertaking since they are quite smart and train easily. Keeshonds require daily exercise and will gladly romp in the yard with the children as well. If owners see their "Kees" spinning, they need to increase the amount of daily exercise. They have a special look called "spectacles" in that their eyes are rimmed in black, and the black continues from the eyes towards the ears, looking very much as if they are wearing glasses! Their thick, double coat has very long guard hairs that stand almost straight out, giving them a very fluffy appearance. They do require frequent brushing to keep the coat from matting. Expect heavy shedding at least twice a year, usually spring and fall.

QUICK FACTS 4.08 Keeshond

Quick Facts	Height	Weight	Coat	Features
Life Span: 12–15 years	16–19 inches 40–48 centimeters	33–44 pounds 15–20 kilograms	**Colors:** Mixture of gray, black, and cream that fades from light to dark. **Type:** Double with a thick ruff and "trousers."	**Head:** Mesocephalic, seems small for its body because of the profuse coat around the neck, scissor bite. **Ears**: Erect. **Tail**: Squirrel, light-colored with a black tips.

Genetic Diseases and Conditions: Primary hyperparathyroidism, patellar luxations, hip dysplasia, allergies, epilepsy, and skin disorders.

Lhasa Apso

Origins: The Lhasa Apso comes from Tibet, a country of extremes in weather, topography, and people. This little dog was bred to be inside and on alert while the mastiff was outside on patrol. The people of Tibet believe that the Snow Lion is a protector of their country, and when the lion comes down from the mountains, it turns into a dog, specifically the Lhasa Apso. That is why the Lhasa Apso is known as the Abso Seng Kye, or the "Bark Lion Sentinel Dog." They were recognized by the AKC in 1935.

FIGURE 4.18 Lhasa Apso, Side View

Characteristics: Over the years the Lhasa has not lost its watchdog proclivities and will be a loud, constant barker. Luckily, they are also very smart and with food incentives can be trained to stop on command. They are loving, gay, and assertive with their families but wary of strangers

FIGURE 4.19 Lhasa Apso, Adult

and lively children. Too much activity and loud noises will distress a Lhasa. They can become quite difficult to handle if left to believe that they are the top dog in the family. It is important that owners train their dogs to be submissive to all humans and socialize them so they do not become timid or aggressive towards other dogs. When well-adjusted they can be wonderful pets that travel well and do well in apartments, requiring only a daily walk for exercise or a rousing game of fetch indoors. Because of the harshness of Tibet they are a very hardy breed, staying quite healthy their entire lives. Their long coats require daily brushing; otherwise, they will mat up quickly. Some owners will have a professional groomer trim the coat short, which can help with the matting.

QUICK FACTS 4.09 Lhasa Apso

Quick Facts	Height	Weight	Coat	Features
Life Span: 15–18 years, some to 29 years!	10–11 inches 25–28 centimeters	18–22 pounds 8–10 kilograms	**Colors:** Shades of tan, dark grizzle, slate, smoke, parti-color, black, white or brown, with or without dark tips on the ears. **Type:** Thick, dense, straight and hard, hair grows in length continuously.	**Head:** Brachycephalic, narrow, not quite flat, fair length, nose about 1 1/2" long, level or slight undershot jaw. **Ears:** Pendant, heavily feathered. **Tail:** Ring tail, feathered, with or without a crochet.

Genetic Diseases or Conditions: Intervertebral disc disease, umbilical and inguinal hernias, dystocia, hot spots, eye problems (including cherry eye and eye dryness), uroliths, inhalant allergies, and kidney and skin diseases.

Poodle

Origins: In Germany, the ancestors of the modern standard Poodle were known as the Pudel or Canis Familiaris Aquatius and were used as water retrievers. France adopted the dog as their own and developed the breed as we know it today. There are three varieties. The standard poodle is the oldest of the three varieties and still retains its hunting instincts. The miniature poodle didn't care for the water and was used as a truffle hunting dog. The toy poodle was a "sleeve pet" and was bred to be a companion. The AKC recognized all three varieties in 1887.

FIGURE 4.20 Poodle, Miniature, Silver

Characteristics: Poodles of all sizes are intelligent, calm, and active. Because of indiscriminate breeding, you may see shy, aggressive, neurotic behaviors and fear-elicited aggression. It is unfortunate that this occurs as a well-bred poodle is a delight to its owner. Standards make excellent family companions. The more active the family, the better, and the more children, the better. Miniatures and Toys do prefer a much quieter atmosphere and do better in single households or ones where the children are older and more sedate. All three varieties are easily trained and excel in obedience and agility trials. Show a poodle what you want once, and it will be able to perform the task. It is one of the traits that made them circus performers and endeared them to the public. The poodle's hair coat does require professional grooming to keep it looking neat and mat free. Regular brushing will also help, but the coat will grow to long lengths that become more difficult to manage if not trimmed.

FIGURE 4.21 Poodle, Standard, Black

QUICK FACTS 4.10 Poodle

Quick Facts	Height	Weight	Coat	Features
Life Span: **Toys:** Up to 20 years **Miniatures:** 14–15 years **Standards:** 11–12.5 years	**Toys:** Up to 10 inches 25 centimeters **Miniatures:** 10–15 inches 25–38 centimeters **Standards:** Over 15 inches Over 38 centimeters (most 22–27 inches)	**Toys:** 6–9 pounds 2.7–4 kilograms **Miniatures:** 15–17 pounds 6.8–7.7 kilograms **Standards:** 45–70 pounds 20–31.8 kilograms	**Colors:** Solid color in shades of blue, gray, silver, brown, cafe-au-laits, apricot, and creams. **Type:** Curly, profuse, and dense, harsh texture. Can become corded. Hair grows in length continuously.	**Head:** Mesocephalic, moderately rounded, long, straight muzzle, scissor bite. Overshot seen in standard puppies but usually corrects itself as it matures. **Ears:** Hanging and thickly feathered. **Tail:** Straight or docked.

FIGURE 4.22 Poodle, Toy, Cafe-Au-Lait

Genetic Diseases or Conditions: All poodles: ear infections, Addison's and Cushing's disease, epilepsy, eye disorders, Von Willebrand disease's, and hypothyroidism. Minis and Toys: allergies, skin problems, and uroliths. Standards: hip dysplasia, eye disorders, bloat, gastric dilation volvulus, tracheal collapse, juvenile renal disease, malignant neoplasms, and sebaceous adenitis.

Schipperke

Origins: This lively little dog was developed in Belgium from the Leauvenaar which was a black sheepdog. The Schipperke (pronounced skip-er-key, meaning "little captain") was utilized on canal boats and by shoemakers along the canals as watchdogs and vermin catchers. They were brought to the U.S. in the late 1800s. It took the AKC until 1904 to recognize them.

FIGURE 4.23 Schipperke

Characteristics: Living up to their original purpose as watchdogs, Schipperkes will bark and bark and bark until told to stop. They are extremely curious and love to be in on the action and will put up a fuss if left out. Schipperkes require socialization and obedience training from an early age. It is recommended that they not be a first dog for a new owner as they will challenge you with their independence and willfulness. Owners must be aware that Schipperkes will bolt through open doors if not trained to stay back as they are opened. With that being said, once they are trained, you will have a lively and loving companion. They will do nicely in apartments, once trained to not bark at everything. They are agile and enjoy agility, and flyball games. They do well with children if socialized to them at an early age. Their coats require weekly brushing, and they will "blow" or shed their coats 2–3 times a year.

FIGURE 4.24 Schipperke, Head

QUICK FACTS 4.11 Schipperkes

Quick Facts	Height	Weight	Coat	Features
Life Span: 15 or more years	10–13 inches 25–33 centimeters	10–18 pounds 4.5–8 kilograms	**Color:** Black. **Type:** Double, distinctive ruff around neck, cape across shoulders, jabot over chest and between front legs, culottes on the back of rear legs.	**Head:** Mesocephalic, medium-width, slightly round head, narrowing to muzzle, scissor or level bite. **Ears**: Erect. **Tail**: Docked.

Genetic Diseases and Conditions: MPS 111B disease, Legg-Calve-Perthes disease, hypothyroidism, cataracts and entropion, patellar luxation, epilepsy, and skin disorders.

Shiba Inu

FIGURE 4.25 Shiba Inu, Trio

Origins: Descended from the primitive dogs of the ancient people of Japan, the Shiba was bred to hunt small wild game. The name Shiba is Japanese for ***brushwood***, after the breed's hunting terrain or the color of brushwood leaves in the fall, and Inu means ***dog***. They are the most popular companion dog in Japan, have national status, and are still used as hunting, tracking, and watchdogs. Their first appearance in the U.S. was in 1954, and they were first shown in the Non-Sporting Group in 1993.

FIGURE 4.26 Shiba Inu, White

Characteristics: This fastidious dog will clean itself much like a cat. Regular brushing is a must to help them stay clean which is good for their mental health. They are easy to house train and if socialized early, they will be good around children. Otherwise, they have a strong prey drive and don't do well with other small animals and birds, as well as other dogs showing any dog aggression tendencies. They are loving and loyal to those that have earned their respect, which can be quite an undertaking as they are independent and naturally reserved toward strangers. The Shiba adapts well to different living situations but must be exercised regularly on leash or in a secure area. A fun quirk Shiba Inu's have is a joyous shriek that they let loose when very happy such as when their owner comes home from a long absence. They also emit a similar sound that is described as a "Shiba Scream" when very, very unhappy.

QUICK FACTS 4.12 Shiba Inu

Quick Facts	Height	Weight	Coat	Features
Life Span: 12–13 years	**Males:** 14.5–16.5 inches 37–42 centimeters **Females:** 13–15.5 inches 34–39 centimeters	**Males:** 23 pounds 10 kilograms **Females:** 17 pounds 7.7 kilograms	**Colors:** Red, black/tan, and sesame. In addition the color must have Urajiro, cream to white, on the ventral surfaces. **Type:** Double coated with the outer coat being stiff and straight.	**Head:** Mesocephalic. **Ears**: Prick triangular shape and small. **Tail**: Squirrel or ring.

Genetic Diseases and Conditions: Allergies, chylothorax, pseudohyperkalemia, obsessive disorders such as spinning, glaucoma, cataracts, entropion, luxating patella, hip dysplasia, hypothyroidism, epilepsy, and cancers.

Tibetan Spaniel

FIGURE 4.27 Tibetan Spaniel

Origins: An ancient dog from Tibet, the Tibetan Spaniel was bred both in monasteries and towns and was prized as a representative of the "guardian lions of China." The breed shares the same ancestors as the Japanese Chin, Pekingese, Lhasa Apso, and Shih Tzu. The spaniel designation came from the French calling them "epagnuel," which meant companion and comforter, which they did by acting as companions and bed warmers to the monks. The first Tibetan Spaniel arrived in the U.S. in 1965, and the dog was shown for the first time in the Non-Sporting Group in 1984.

FIGURE 4.28 Tibetan Spaniel, Head

Characteristics: The "little lion" dog loves to be on "top of the world, looking down" which was one of their jobs in Tibet! They would perch on top of city and monastery walls to keep a lookout for approaching danger, and they would alert everyone with loud barks. They will still do this today perching on the backs of chairs and couches to see out the window. "Tibbies" are gregarious by nature and will often ignore calls to come in order to explore their surroundings. They love to be in the mix with their people and are quite loyal and loving. They will need basic socialization and obedience training (which may take a while as they can be stubborn), weekly brushing, and moderate exercise.

QUICK FACTS 4.13 Tibetan Spaniel

Quick Facts	Height	Weight	Coat	Features
Life Span: 12–15 years	10 inches 25 centimeters	9–15 pounds 4–6.8 kilograms	**Colors:** Any color. **Type:** Double coat, silky texture, long outer coat on body that lays flat.	**Head:** Brachycephalic, small in proportion to body, slightly domed, slight undershot or level bite. **Ears**: Pendant and feathered. **Tail**: Ring.

Genetic Diseases and Conditions: Portosystemic shunts, hernias, progressive retinal atrophy, allergies, cherry eye, patellar luxation, and hip dysplasia.

Tibetan Terrier

FIGURE 4.29 Tibetan Terrier

Origins: Another ancient breed of dog from Tibet that is known as the "Holy Dog of Tibet" and referred to as "the little people" dog. They were never sold or mistreated as it was thought to bring bad luck upon the family or village. They were good luck charms that also helped out on farms guarding sheep, acting as watch dogs, and being companions. They were first seen in the U.S. in 1956 and were recognized by the AKC in 1973.

Characteristics: Because they are "terrier sized," they were given the name of terrier by some of the first explorers to reach the "Lost Valley" in Tibet. Don't let their size fool you; they are big dogs in small bodies. They are agile, active, intelligent, laid back, sensitive, and empathic to owners. They respond very well to positive-reinforcement training and are smart enough to be taught virtually anything. They do well in apartments as long as they get regular exercise. For their size they have a commanding bark which they use very judiciously. Agility trials, obedience rallies, showmanship, flyball, and tracking are some of the activities in which they excel. Their moderate-length coats do require maintenance as it will grow quite long and so will require professional grooming.

FIGURE 4.30 Tibetan Terrier

QUICK FACTS 4.14 Tibetan Terrier

Quick Facts	Height	Weight	Coat	Features
Life Span: 12–15 years	14–17 inches 35–43 centimeters	18–30 pounds 8–13 kilograms	**Colors:** Any color and color with white. **Type:** Double coat, under coat is soft and wolly, outer coat is fine and profuse; hair grows in length continuously.	**Head:** Mesocephalic, slightly rounded, scissor or level bite. **Ears**: V-shaped and heavily fathered. **Tail**: Curled and heavily feathered.

Genetic Diseases and Conditions: Lens luxation, cataracts, progressive retinal atrophy, canine neuronal ceroid lipofuscinosis, hip dysplasia, patellar luxation, allergies to grains and dairy, deafness, hypothyroidism, and heart murmurs.

Sporting Group

There are four types of dogs in this group: pointers, retrievers, setters, and spaniels. They all were developed to hunt upland game which includes pheasants, grouse, prairie chickens, and/or water fowl, including geese and ducks. Their instinct for pointing and retrieving is such in some of them that puppies as young as eight weeks of age go into point. It may be at a leaf or a bug, but it is a point, and they are showing you there is something interesting to look at. All of these dogs will retrieve game that has been shot. The retrievers rarely need to be trained or shown how to retrieve; it is in their nature. The spaniels are a combination of all these traits in a compact body.

Overall, these dogs are easy to train as their whole reason for being is to please man. In addition to a natural hunting ability, they excel at obedience, agility, and flyball, and a number of them are used as guide or service dogs. Because of their love of man, they make great family pets. For the most part they are sweet, loving dogs that enjoy everyone in the family. They get along with other dogs, and strangers are just friends they haven't met yet. There is often a joke told about the sporting group: They will sound the alarm when a stranger approaches, but after realizing that it is a person, the dogs will show them where the good silverware is kept. Being a guard dog is not their specialty!

The sporting group comes in a variety of sizes, hair-coat lengths, and colors. If you want a dog that is totally devoted to you, is fun loving, is excellent with children, and is energetic, this is the group to pick from.

POINTERS

A pointer is usually a bigger dog that as it works through a field looking for birds will stop and point at the bird. This is usually a stance that includes the muzzle pointing at the bird with an intent stare, tail straight out, and sometimes a foreleg lifted. They will hold that stance until the bird flushes from nervousness or the hunter gives the signal for the dog to pounce to flush the bird.

FIGURE 5.00 English Pointer

English Pointer

Origins: Around the middle of the 1600s, wing-shooting birds came into vogue, and the pointer was used to find, flush, and then retrieve the bird as well as rabbits. It is thought that the English Pointer was developed using Greyhounds, Bloodhounds, and Foxhounds, plus some type of spaniel and setter to develop this gun dog. The AKC recognized the breed in 1884.

Characteristics: Pointers are high-energy dogs that will fit into a family that doesn't hunt as long as long walks or, even better, jogs are provided on a daily basis. If not exercised they can become destructive from sheer boredom. They are usually not dog aggressive but should be socialized at a young age to prevent timidness. Pointers are very loyal to their family and will bark to alert the family but will usually not go any further as far as guarding. They are great with children and have the stamina to romp with them for hours. They will take to water, although they were not developed for that particular task. They have a short-hair coat that is easy to maintain.

FIGURE 5.01 English Pointer, Head

QUICK FACTS 5.00 English Pointer

Quick Facts	Height	Weight	Coat	Features
Life Span: 13–14 years	**Males:** 25–28 inches 63–71 centimeters **Females:** 23–26 inches 58–66 centimeters	**Males:** 55–75 pounds 25–34 kilograms **Females:** 44–65 pounds 20–29 kilograms	**Colors:** White body with liver, lemon, black and/or orange markings that are either patched or speckled, solid and tricolored are also common. **Type:** Short, dense, smooth to the touch.	**Head:** Square shaped, mesocephalic, scissor or even bite. **Ears:** Large V-shaped. **Tail:** Saber that is held straight out when pointing.

Genetic Diseases and Conditions: They are really very genetically sound dogs and suffer from very few issues. Thyroid diseases, dwarfism, hip dysplasia, epilepsy, allergies, and cherry eye are the only issues seen.

German Shorthaired Pointer

Origins: It is unclear how the German Shorthaired Pointer came about other than a number of German and Spanish field dogs were used to start their development. English pointers were added to set the type and to add endurance and elegance to this all-purpose breed. The AKC admitted them to their stud book in 1930.

FIGURE 5.02 German Shorthaired Pointer

FIGURE 5.03 German Shorthaired Pointer

Characteristics: Versatility is the key word when talking about the German Shorthair Pointer. They have keen noses, making them excellent trailers, working equally well on pheasant, quail, grouse, partridge, duck, rabbit, coon, opossum, and deer. They are natural retrievers, and with a water-repellent coat and webbed feet they do quite well in water. "Shorthairs" are wonderful family dogs and are not usually aggressive towards other dogs. They are high-energy dogs and absolutely need an outlet for that energy, or they will entertain themselves. This can include digging, trying to escape, or becoming anxious, destructive, and/or nervous. They train easily but without proper leadership can become overbearing and not listen. Training should start young with enrollment in puppy kindergarten then advancing through obedience training. Their short hair is easy to maintain, and they are average shedders.

QUICK FACTS 5.01 German Shorthaired Pointer

Quick Facts	Height	Weight	Coat	Features
Life Span: 13–14 years	**Males:** 24–26 inches Centimeters **Females:** 21–23 inches	**Males:** 55–70 pounds **Females:** 45–60 pounds	**Colors:** Ticked; liver and white, liver patched and white ticked, or liver roan, can also be solid liver, liver and white spotted. **Type:** Short, hard, thick.	**Head:** Mesocephalic, scissor bite. **Ears:** Hanging. **Tail:** Docked.

FIGURE 5.04 German Wirehaired Pointer

Genetic Diseases or Conditions: False pregnancy, epilepsy, hip dysplasia, entropion, lymphedema (pitting edema), hermaphroditism, gastric torsion, and mast cell tumors.

German Wirehaired Pointer

Origins: This breed was created by the Germans as a multi-purpose hunting dog, by using other pointers, Foxhounds and Poodles to develop this outstanding dog. The breed arrived in the U.S. in 1920 and was finally admitted to the AKC in 1959.

FIGURE 5.05 German Wirehaired Pointer, Head

Characteristics: Loyal and affectionate, the German Wirehaired Pointer craves human companionship and bonds closely with its people. They do not do well kenneled. A high-energy and high-drive breed, the Wirehair must receive regular vigorous exercise and have a job to do; otherwise, their creativity and independence may get them into trouble. They can become willful and destructive if not allowed to expend their energy regularly. Wirehairs have a sound, reliable temperament that at times can be aloof but not unfriendly toward strangers. They are loyal and affectionate companions, eager to please and enthusiastic to learn. Wirehairs require early socialization and training by a confident and consistent owner. If these dogs feel superior to the owner, they will be difficult to deal with. Their coats need to be brushed weekly, and about twice a year the dead hair needs to be plucked.

QUICK FACTS 5.02 German Wirehaired Pointer

Quick Facts	Height	Weight	Coat	Features
Life Span:	**Males:** 24–26 inches centimeters **Females:** not > 22 inches	60–70 pounds 27–32 kilograms	**Colors:** Liver and white, liver and white spotted, liver roan, liver and white spotted with ticking and roaning or solid liver. **Type:** Double coated, wiry, weather resistant and water-repellent, dense undercoat, outer coat is straight, harsh, wiry, and flat.	**Head:** Long, squared muzzle, scissor bite. **Ears:** Hanging. **Tail:** Docked.

Genetic Diseases or Conditions: Sebaceous cysts, ear infections, eye diseases, cancers, Von Willebrand's disease, hip dysplasia, heart disease, and malocclusions.

Weimaraner

Origins: Germany developed this dog in the early 19th century; Bloodhounds, German Shorthaired Pointers, and other German pointers were believed to have played a role in its development. The Weimaraner was originally built to hunt a variety of game, but when it was imported to the U.S. by Howard Knight, it shifted to the upland game and water fowl dog we know today. The AKC recognized it in 1943.

FIGURE 5.06 Weimaraner, Mouse Gray

Characteristics: This uniquely colored dog is often referred to as the "Gray Ghost." This nickname comes from their eye color matching their hair coat color, resulting in a rather uncanny stare. As hunting dogs, their instinct for the job is quite good, and so the breed is a good choice for the novice hunter but not necessarily for the novice dog owner. The Weimaraner is a powerfully built dog, and without proper obedience training, it drags people on walks and knocks children and elderly people over with enthusiasm. Training must be consistent and firm but kind, as a "Weim" will shut down if treated harshly. Their hunting instinct is such that they will go after small game, which can

FIGURE 5.07 Weimaraner, Silver Gray

be the household cat or other small pet, even if introduced at an early age. They can be aggressive towards other dogs, especially the males. They are people dogs and do best when close to their people. Daily vigorous exercise is a must with this high-energy dog. Jogging, running, and a long game of fetch may just be enough. Without exercise, Weimaraners have been known to develop separation anxiety and destroy property or harm themselves in trying to escape. They have a short-hair coat that is easy to keep, and they are average shedders.

QUICK FACTS 5.03 Weimaraner

Quick Facts	Height	Weight	Coat	Features
Life Span: 13 years	**Males:** 25–27 inches 63–68 centimeters **Females:** 23–25 inches 58–63 centimeters	**Males:** 70–80 pounds 32–36 kilograms **Females:** 55–70 pounds 25–32 kilograms	**Colors:** Shades of mouse-gray to silver-gray. **Type:** Sleek, and short.	**Head:** Mesocephalic, scissor bite. Eyes match coat color. **Ears:** Hanging. **Tail:** Docked.

FIGURE 5.08 Wirehaired Pointing Griffon

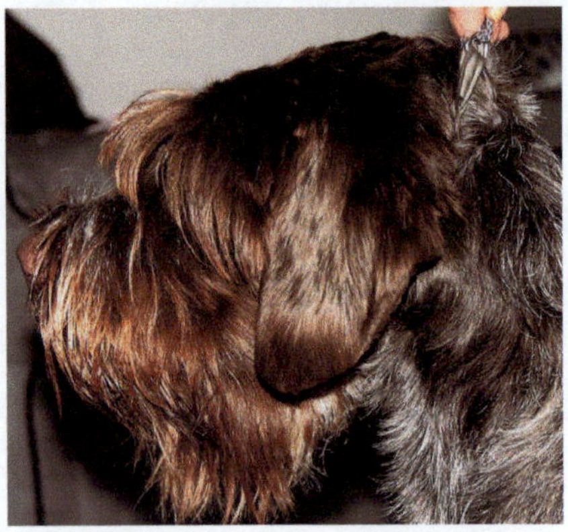

FIGURE 5.09 Wirehaired Pointing Griffon, Head

Genetic Diseases or Conditions: Bloat, entropion, mast cell tumors, cryptorchidism, hyperuricosuria, undershot jaw, elbow dysplasia, hypertropic osteodystrophy, hypothyroidism, distichiasis, progressive retinal atrophy, and Von Willebrand's disease.

Wirehaired Pointing Griffon

Origins: Dutchman Eduard Korthals is credited with being the founding father of the "WPGs." He wanted an all-purpose gun dog that would track, point, and retrieve on land and in water. Though not as popular in the U.S. as other gun dogs, the AKC recognized them in 1887.

Characteristics: The Wirehaired Pointing Griffon is an excellent sporting dog that retains its puppy-like energy well into adulthood. They are unafraid to plow through bushes, brambles, into cold water, or muck when on the trail or retrieving game. They are very social and get along with people and other animals quite well. They do require early socialization to avoid a nervous or timid dog, and, of course, no amount of exercise is too much. They need to be brushed twice weekly and the dead hair pulled from their coats at least twice a year.

QUICK FACTS 5.04 Wirehaired Pointing Griffon

Quick Facts	Height	Weight	Coat	Features
Life Span:	**Males:** 22–24 inches 56–61 centimeters **Females:** 20–22 inches 51–56 centimeters	50–60 pounds 22–27 kilograms	**Colors:** Steel gray with brown, roan, white and brown or white and orange. **Type:** Double, undercoat is thick, fine down; outer coat is harsh, wiry of medium length.	**Head:** Mesocephalic, scissor bite. **Ears**: Hanging, covered in fur. **Tail**: Docked.

Genetic Diseases and Conditions: Because of careful breeding they are a very hardy and healthy breed.

Vizsla

Origins: Developed in Hungary and cherished by the aristocracy, this breed became all but extinct after World War II. The country was scoured for breeding pairs which were imported into other countries. They came to the U.S. in the 1950s and were recognized by the AKC in 1960.

FIGURE 5.10 Vizsla

FIGURE 5.11 Vizsla

Characteristics: Vizslas are lively, gentle-mannered, natural hunters that are endowed with a good nose and above-average intelligence. Training must be consistent and gentle as overzealous correction can make this dog distrustful. They need to be socialized and exposed to a wide range of situations to prevent them from becoming shy. When well-adjusted they are a multi-purpose dog suitable for working upland game, rabbits, and waterfowl, retrieving even though they are essentially a pointer-type. They are sensitive though fearless when hunting and have a well-developed protective instinct that extends to their family. They are a rather lightly built dog that still has the energy and stamina of its heavier cousins. They are tractable and affectionate companions, so much so that Vizsla owners describe them as "Velcro dogs." Exercise is a must to not only keep them fit physically but mentally as well. Exercise has to be extensive to avoid nervous energy from building up which may come out as destructive behavior. Their short-hair coat is easy to maintain, and they are average shedders.

QUICK FACTS 5.05 Vizsla

Quick Facts	Height	Weight	Coat	Features
Life Span: 10 to 14 years	**Males:** 22 to 24 inches 55–63 centimeters **Females:** 21 to 24 inches 53–61 centimeters	**Males:** 45–66 pounds 20–30 kilograms **Females:** 40–55 pounds 18–25 kilograms	**Colors:** Varying shades of solid golden rust. **Type:** Short, smooth, dense.	**Head:** Mesocephalic, scissor bite. **Ears:** Hanging, thin and silky. **Tail:** Docked, longer than other sporting dogs.

Genetic Diseases or Conditions: Hemophilia A, hip dysplasia, craniomadibular osteopathy, facial nerve paralysis, and sebaceous adenitis.

RETRIEVERS

Chesapeake Bay Retriever

FIGURE 5.12 Chesapeake Bay Retriever

Origins: This breed was developed in the U.S. in the Chesapeake Bay area using Newfoundlands and Otterhounds with Flat and Curly-Coated Retrievers used to finish the breed. They were accepted by the AKC in 1878.

Characteristics: "Chessies" love the water, no matter how cold it is. If there is even a small pool, they will quite happily dive in and muck about. Their hair coats were developed to repel water, so with a few shakes they can remove most of the moisture; the mud, however, will stick. They were bred to retrieve ducks and geese from the rough, icy waters of Chesapeake Bay. Market hunters would often shoot and ask their Chessie to retrieve hundreds of ducks in a day. They are an all-purpose game dog and will gladly help out on upland game as well. Chessies are reputed to be hard headed and about as sensitive as a Sherman tank. However, it is usually a misunderstanding between the dog and the owner. They require socialization and obedience training at an early age. Once the leader is established, the Chessie will follow and obey quite willingly. This is not the dog for a novice owner as they need consistent, firm but fair rules, and if the owner waivers, the Chessie will gladly take over and be the boss. Chesapeakes can become aggressive with other dogs and so should not only be socialized at an early age but that socialization should continue its entire life. Again, exercise is a must for this breed as they will become bored and do naughty things to relieve that boredom. Chessies have an oily coat and develop a doggy odor. They should not be bathed too often because it can strip the hair coat of the oils that protect it from icy cold water.

FIGURE 5.13 Chesapeake Bay Retriever

QUICK FACTS 5.06 Chesapeake Bay Retriever

Quick Facts	Height	Weight	Coat	Features
Life Span: 10 - 13 years	**Males:** 23–26 inches 58–66 centimeters **Females:** 21–24 inches 53–61 centimeters	**Males:** 65–75 pounds 29–34 kilograms **Females:** 55–65 pounds 25–29 kilograms	**Colors:** Any shade of dead grass, from dark brown to faded tan or dull straw. **Type:** Double, thick, dense, short, woolly undercoat with wavy, medium-length outer coat. Hair on face and legs is very short and straight.	**Head:** Broad, round, medium-short muzzle, pointed not sharply. **Ears:** Small, hangs loosely. **Tail:** Otter tail.

Genetic Diseases or Conditions: These dogs are very hardy and healthy with just progressive retinal atrophy, entropion, alopecia, and hip dysplasia as reoccurring problems, although recently congenital ectodermal dysplasia or skin fragility syndrome has been appearing.

Flat-Coated Retrievers

Origins: Again, we look to England as the birthplace of this retriever. It is thought that Newfoundland, Irish Setter, Collie, and another type of water dog were all used to develop this breed. They were quite popular in the late 1800s, and when the AKC recognized them in 1915, they were at the height of their

FIGURE 5.14 Flat-Coated Retriever

FIGURE 5.15 Flat Coated Retriever, Head

popularity. Their numbers declined after WWII, but since 1960 they are finding their way back into the mainstream.

Characteristics: This excellent family dog is very intelligent and easy to train if the training schedule has a variety, as they easily get bored doing the same tasks over and over. They are great in obedience, agility, flyball, and, of course, as an all-around hunting companion. They will alert you to strangers but are more than ready to make friends than to make sure no one is walking off with the property. They are an easygoing dog that will love to play and hunt as much as a human could ever want. Their hair coat requires weekly brushing, and they are average shedders.

QUICK FACTS 5.07 Flat-Coated Retriever

Quick Facts	Height	Weight	Coat	Features
Life Span: 10 years average	**Males:** 23–24.5 inches 58–62 centimeters **Females:** 22–23.5 inches 55–59 centimeters	60–70 pounds 27–31 kilograms	**Colors:** Black or liver. **Type:** Flat, medium length, feathering on legs and tail.	**Head:** Mesocephalic, scissor bite. **Ears:** V-shaped covered with medium-length hair. **Tail:** Plume.

Genetic Diseases and Conditions: Cancer, epilepsy, diabetes, progressive retinal atrophy, glaucoma, and hip dysplasia.

FIGURE 5.16 Golden Retriever

Golden Retriever

Origins: Scotsman Dudley Marjoribanks, first Lord of Tweedmouth, is credited with developing the Golden Retriever. He crossed yellow Wavy-Coated Retrievers with a now extinct Tweed Water Spaniel and then threw in some Bloodhound and Irish Setter. They were originally called Yellow Flat-Coated Retrievers, but the name was changed to Golden Retriever later in its development. The AKC recognized the breed in 1925.

Characteristics: This is a dog that excels in everything: obedience, agility, hunting on land and in the water, tracking, serving as a guide dog for the blind, acting as a service and therapy dog, narcotic detection, and as just a plain, old awesome family dog. Like the Vizsla they tend to be "Velcro" dogs and will follow their owners everywhere, sometimes leaning so hard against them that it may feel like they want to crawl inside your skin because they just can't get close enough. As hunters they have a very "soft mouth," which means that they don't damage the prey as it is retrieved. They are usually happy dogs that get along well with other house pets and other dogs. These dogs are very eager to please their human owners, so training them is fairly easy. Kind, consistent, and fair treatment during training will result in a well-balanced, mindful dog that will be loyal until death parts it from your side. Exercise is still a must for the "Golden" as they tend towards obesity if fed too much and exercised too little. Weekly brushing of the hair coat is necessary, and one can expect an average amount of shedding.

FIGURE 5.17 Golden Retriever

QUICK FACTS 5.08 Golden Retrievers

Quick Facts	Height	Weight	Coat	Features
Life Span: 10–13 years	**Males:** 23–24 inches 58–61 centimeters **Females:** 21.5–22.5 inches 54–57 centimeters	**Males:** 65–75 pounds 29–34 kilograms **Females:** 60–70 pounds 27–31 kilograms	**Colors:** Pale to burnished gold. **Type:** Double, undercoat is dense, water-repellent, outer coat straight or lightly wavy, feathering on back of legs, chest, and underside of tail.	**Head:** Mesocephalic, scissor bite. **Ears:** Short, pendent with rounded tips. **Tail:** Plume carried merrily.

Genetic Diseases or Conditions: Cataracts, cancers, Von Willebrand's disease, hip and elbow dysplasia, progressive retinal atrophy, entropion, cardiac diseases, allergies, hemophilia, hotspots, inhalant allergies, and hypothyroidism.

Labrador Retriever

Origins: The Labrador Retriever was developed alongside of the Newfoundland. The dogs helped dock workers pull in nets, retrieve parts and fish, and actually "tolled" or paddled a foot to attract fish. Originally called St. John's Water Dogs, the name was changed to Labrador to reflect the point of origin in the country. They were recognized by the AKC in 1917.

FIGURE 5.18 Labrador Retriever, Black

Characteristics: The Labrador is the number-one dog in the U.S., and it isn't because they are used only for hunting. This highly intelligent dog is utilized for tracking, as a guide dog for the blind, for disabled-assistance, for therapy, for carting, for police work, as narcotic and bomb detectors, for water rescue, and, of course, for upland game and water fowl hunting. They are full of energy as a puppy, and that trait continues well into their third year. Because of that high energy level, they have to have obedience training to keep them from inadvertently injuring their owners, children, or elderly people with their antics. Because of their great stamina, they can keep up a game of ball forever, some becoming obsessed. They love the water and will swim in anything from a small mud pond to the sea. They have an amazing sense of smell and are very soft mouthed.

FIGURE 5.19 Labrador Retriever, Yellow

"Labs" are loyal and loving family pets that will alert you if a strange noise is heard but will rarely keep barking. If they do, it is important to investigate, as many a Lab has saved its owner's life by barking until someone comes to see what all the fuss is about. They want to please their humans, so training is relatively easy; however, if they are distracted by a scent or noise, you will want to gently get their attention back to you before expecting them to obey. Labs are easy keepers, so measured food amounts will keep the dog at a healthy weight. Their short-hair coats require an occasional brushing, and you can expect average shedding.

QUICK FACTS 5.09 Labrador Retriever

Quick Facts	Height	Weight	Coat	Features
Life Span: 10–12 years	**Males:** 22.5–24.5 inches 57–62 centimeters **Females:** 21.5–23.5 inches 54–59 centimeters	**Males:** 65–80 pounds 29–36 kilograms **Females:** 55–70 pounds 25–31 kilograms	**Colors:** Black, yellow, chocolate. **Type:** Short, dense, waterproof.	**Head:** Mesocephalic, scissor bite. **Ears:** V-shaped. **Tail:** Otter tail.

Genetic Diseases or Conditions: Progressive retinal atrophy; cataracts; corneal dystrophy; hereditary myopathy; copper toxicosis; shoulder, elbow, and hip dysplasia; osteochondrosis; luxating patella; autoimmune diseases; exercise-induced collapse; hypoglycemia; hypothyroidism; diabetes; deafness; and melanoma.

FIGURE 5.20 Labrador Retriever, Chocolate

Nova Scotia Duck Tolling Retrievers

Origins: This dog coming out of Nova Scotia has a checkered past. It is thought that a mixture of dogs that includes various gun dogs such as spaniels and setters, some retrievers and a farm dog or two went into the brew to make the "Tollers." Known as Nova Scotia's provincial dog, it didn't make an impact in the U.S. until the late 1970s. It was accepted into the AKC in 2003.

FIGURE 5.21 Nova Scotia Duck Tolling Retriever

Characteristics: The Tollers were named such as that is how they were used to hunt ducks and geese. They would toll or play in the water, attracting the attention of the water fowl to come over and investigate. Once the birds were close enough for a shot, the dog would be called into the duck blind, the hunter would stand up and scare the birds into flight and take the shot. The Toller would then retrieve the birds and start the process over again. Tollers love obedience trials, water sports, and agility trials, and because of their excellent sense of smell, they make great

FIGURE 5.22 Nova Scotia Duck Tolling Retriever, Head

search-and-rescue dogs. They are sociable with people and other dogs. They have a shrill bark that is a combination screech and howl that is used when they are super excited. They are always game for retrieving anything and will play with children for hours. Their smaller size makes them a good choice for people with smaller homes, but they still have all the great qualities of the retriever group. Weekly brushing of the double-hair coat will keep the oils spread evenly, which will keep the coat water repellent. They are average shedders.

QUICK FACTS 5.10 Nova Scotia Duck Tolling Retriever

Quick Facts	Height	Weight	Coat	Features
Life Span: 12–14 years	**Males:** 18–21 inches 45–53 centimeters **Females:** 17–20 inches 43–50 centimeters	30–51 pounds 13–23 kilograms	**Colors:** Shades ranging from golden to coopery red. **Type:** Double coat, water-repellent soft, dense undercoat, slightly wavy, medium-length outer coat.	**Head:** Mesocephalic, scissor bite. **Ears**: V-shaped with rounded tips. **Tail**: Plume tail.

Genetic Diseases and Conditions: This is a very hardy and disease free dog at current times; however, as the demand for these great dogs increases, indiscriminant breeding may start introducing the usual genetic conditions such as progressive retinal atrophy, autoimmune diseases, and thyroid conditions.

FIGURE 5.23 English Setter

SETTERS

English Setter

Origins: In the early 1800s Sir Edward Laverack brought two speckled setters from France and began an inbreeding program to develop the breed as we know it today. The term setter was how the dogs would set or position themselves when they found a bird. Before guns they would go into a praying-like pose so a net could be thrown over them and onto the bird. With guns, they go into more of a crouch and wait for the hunter to come within range before flushing the bird.

English Setters were accepted into the AKC in 1884.

FIGURE 5.24 English Setter, Head

Characteristics: The English Setter is mild mannered with a sweet disposition. They are empathic to their owner's moods and will always be up for a snuggle. They are a beautifully marked dog with a white body and orange, blue, or a combination of both spots that are clearly defined on the face and then seem to be smudged on the longer fur of the body. They are highly intelligent, being able to make decisions while on the hunt as to how to hold the different birds in place. Training must be consistent and kind as they will wilt under harsh treatment. Patience is required as they will avoid doing anything which might be wrong and could displease owners. It requires considerable exercise to keep it sane. They do well in obedience, agility, and flyball exercises and, of course, in hunting. There are two strains of English Setter. Those that are of the field variety usually have thinner hair coats and are lighter-framed dogs than those that are shown. The show variety will have a much thicker and longer hair coat, and their frame is a bit heavier.

QUICK FACTS 5.11 English Setter

Quick Facts	Height	Weight	Coat	Features
Life Span: 12–15 years	**Males:** 24–27 inches 61–69 centimeters **Females:** 23–26 inches 58–66 centimeters	**Males:** 55–80 pounds 25–36 kilograms **Females:** 45–70 pounds 20–32 kilograms	**Colors:** Belton; blue, orange, tri-color, lemon, liver. **Type:** Flat, straight, medium length, moderate feathering on legs, ears, and tail.	**Head:** Mesocephalic, scissor or even bite. **Ears**: V-shaped, long, slightly rounded. **Tail**: Plume.

Genetic Diseases or Conditions: Deafness, eclampsia, hypothyroidism, progressive retinal atrophy, tonsillitis, vaginitis, dermatitis, allergies, cancer, anesthetic reactions, and anaphylactic reactions to routine vaccinations.

FIGURE 5.25 Gordon Setter

FIGURE 5.26 Gordon Setter, Head

Gordon Setter

Origins: The development of the Gordon Setter is credited to Alexander, the 4th Duke of Gordon, Scotland, 1743–1827. However, historical documents also show that a Mr. Coke of Longford, Scotland also had "black and tan" setters, and the two men often traded dogs to infuse fresh blood into their bloodline. The AKC recognized them in 1884.

Characteristics: The Gordon Setter of today is a heavier dog than its cousins and has a distinctive coat of wavy black fur with tan points over the eyes, around the muzzle, and on the lower legs. They are known to work close to the hunter and are methodical when looking for birds. They have great energy and endurance with an admirable nose for tracking. Gordons are affectionate and devoted family dogs. They are quiet and rather sedate for a sporting dog, but that doesn't mean that they don't need obedience training and early socialization. They are sensitive dogs, so training should be consistent, kind, and patient. Without good training, they can occasionally be aggressive with other dogs. Gordons are great in agility and obedience trials and as show dogs. Many Gordons will develop a repertoire of hums, growls, whines, and barks to indicate that they need something.

QUICK FACTS 5.12 Gordon Setter

Quick Facts	Height	Weight	Coat	Features
Life Span: 10–12 years	**Males:** 24–27 inches 61–68 centimeters **Females:** 23–26 inches 58–66 centimeters	**Males:** 55–80 pounds 25–36 kilograms **Females:** 45–70 pounds 20–31 kilograms	**Colors:** Black coat with "kiss marks" and tan points. **Type:** Straight or slightly wavy, long hair on ears, under stomach, chest, fore and hind legs, and tail.	**Head:** Mesocephalic, scissor or level bite. **Ears:** Folded. **Tail:** Plume tail, carried gaily.

Genetic Diseases or Conditions: Hip dysplasia, thyroid imbalance, cataracts, progressive retinal atrophy, epilepsy, and bloat.

Irish Setter

FIGURE 5.27 Irish Setter, Field

FIGURE 5.28 Irish Setter, Show

Origins: Irish Setters were developed in Ireland by crossing English Setter, Irish Terrier, Irish Water Spaniel, Pointers, and a dash of Gordon Setter. Originally, they were red and white or all red with perhaps a white spot on the chest. The latter is acceptable to the AKC standards although you'll see both in Ireland today. The AKC accepted the Irish Setter in 1878.

Characteristics: This striking dog can be a stereotypical Irishman, with a wild, devil-may-care attitude. It makes them tremendously likable but also somewhat hard to control. This can be avoided by consistent, firm-but-kind training starting immediately after being weaned. They frequently require more training time than other setters, but once trained, they are trained for life. There are currently two strains of Irish Setters. The show strain tends to be bigger boned, taller and with a more dome-shaped head. The field strain is a smaller dog with a more square head and shorter, thinner hair coat. Both will have boundless energy which adds to his value as a bird dog in rough country and briars or for putting on a good show in the show ring. However, this excess energy often tags him as a goof-ball, which is usually a result of not enough exercise. They need extensive time exercising and mental stimulation to stay balanced. The Irish Setter is gregarious, gentle, lovable, loyal, good, and tough.

QUICK FACTS 5.13 Irish Setter

Quick Facts	Height	Weight	Coat	Features
Life Span: 11–15 years	**Males:** 26–28 inches 66–71 centimeters **Females:** 24–26 inches 61–66 centimeters	**Males:** 65–75 pounds 29–34 kilo-grams **Females:** 55–65 pounds 25–29 kilo-grams	**Colors:** Mahogany or chestnut red, white on chest, tips of toes or thin streak on skull. **Type:** Flat, moderate length on body, short, fine hair on head, forelegs, and ears, feathering on ears, forelegs, thighs, belly, brisket, and between toes.	**Head:** Mesoce-phalic, scissor bite, over and undershot jaws common. **Ears:** Hanging. **Tail:** Plume tail, carriage straight or curving slightly upward.

Genetic Diseases or Conditions: Dystocia, eclampsia, gastric torsion, eye problems, allergies, penicillin reactions, otitis, Von Willebrand's disease, epilepsy, celiac disease, Leukocyte adhesion deficiency, hypothyroidism, autoimmune disease, cancer, mammary tumors, entropion, pyometra, and melanomas.

SPANIELS

FIGURE 5.29 American Cocker Spaniel, Parti Colored

American Cocker Spaniel

Origins: Developed in the U.S. from the English Cocker Spaniel, they are the smallest member of the sporting group. Their name was derived from their proficiency in hunting the woodcock. They were officially recognized by the AKC in 1873.

Characteristics: This lively little dog is perfect for the family that has a hunter and children as they can fill both roles quite adequately with judicious training. The Cocker has to be shown who the leader of the family is, and when that is thoroughly understood, he is respectful of his master's authority. This will take starting obedience training with puppy classes that should continue into adulthood, making a great family dog that is trustworthy and adaptable. Because of their popularity, indiscriminant breeding has produced some dogs that are timid and fearful, which can evolve into possessive and/or fear-elicited aggression. These dogs can be good citizens if socialized early and have a strong, consistent owner. Though the Cockers have been pets more than hunting dogs in recent years, the dogs haven't lost their instinct for hunting. They are starting to make a comeback in field trials and are quite capable of flushing and retrieving game on land and in water. Cockers are not as high energy as the larger dogs, so care must be taken not to exhaust these little hunters. They can do well in obedience trials, agility, and flyball if looking for something fun to do. Cockers have thick, long hair that requires professional grooming with consistent brushing in between to keep their fur from matting. Their super long, heavy hair-covered ears are also prone to infections. To prevent the ears from becoming infected,

FIGURE 5.30 American Cocker Spaniel, ASCOB (any solid color other than black)

care must be taken to keep the ears dry whenever the dogs are bathed or dry the ears thoroughly after the dogs have been in water. At the first "smell" of yeasty odor coming from the dog's head, a trip to the veterinarian is in order as soon as possible. Cockers are also prone to obesity, so measured food and sufficient exercise is required to keep them trim.

QUICK FACTS 5.14 American Cocker Spaniel

Quick Facts	Height	Weight	Coat	Features
Life Span: 12–15 years	14–15 inches 36–38 centimeters	20–35 pounds 9–16 kilograms	**Colors:** Black, ASCOB, black and tans; parti-color (two or more colors) roans, tri-colors. **Type:** Short and fine on head; silky, flat or slightly wavy, medium-length hair on body, ears, chest, and abdomen, legs are well feathered.	**Head:** Mesocephalic, scissor bite, frequent malocclusions. **Ears**: Folded, long. **Tail**: Docked.

Genetic Diseases or Conditions: Otitis, obesity, cataracts, glaucoma, entropion, ectropion, cherry eye, elbow and hip dysplasia, skin problems, hypothyroidism, liver disease, cardiomyopathy, immune mediated hemolytic anemia, uroliths, and melanomas.

American Water Spaniels

Origins: The state dog of Wisconsin is a relatively rare breed with English Water Spaniel, Irish Water Spaniel, Curly Coated Retriever, Poodle, field spaniels and Native American dogs in their heritage. It was recognized by the AKC in 1940.

FIGURE 5.31 American Water Spaniel

Characteristics: This compact spaniel is known well in Wisconsin, Michigan, and Minnesota as an all-purpose dog for hunters of upland game and water fowl. They are high energy dogs that require exercise to maintain their sanity. They are a sensitive dog that responds well to firm, quiet, and persuasive training

FIGURE 5.32 American Water Spaniel

techniques. They may become defensive if treated harshly, and if the dog believes it is king of the castle, it will become snappish. Without a job or consistent exercise, they can turn to barking or hyperactivity. They have a medium-long hair coat that requires weekly brushing.

QUICK FACTS 5.15 American Water Spaniels

Quick Facts	Height	Weight	Coat	Features
Life Span: 10–12 years	15–18 inches 38–46 centimeters	25–40 pounds 11–18 kilograms	**Colors:** Shades of brown. **Type:** Double coat, with a curly or wavy outer layer.	**Head:** Mesocephalic, scissor or level bite. **Ears**: Pendant, covered with fur. **Tail**: Saber, thin, moderate feathering.

Genetic Diseases and Conditions: Progressive retinal atrophy, cataracts, epilepsy, allergies, diabetes, and hypothyroidism.

Brittany

FIGURE 5.33 Brittany

Origins: This breed is named for the region of France from whence it originated. It is thought by some that they are actually small Setters, but records from the mid-1800s talk about an orange and white bobtail spaniel that was used for upland game. They were accepted by AKC in 1907.

Characteristics: The Brittany is a moderately sized spaniel that is easy to care for, as their coats are not as thick or long as the Cocker Spaniel or English Springer Spaniel. It is a slightly more reserved dog than the other sporting dogs, but it is always eager to please and is a very active and enthusiastic hunter. It is easily trained and responds well to calm, confident, and consistent training. They can become timid if handled harshly. They can adapt to apartment living with plenty of daily exercise. They are sweet and playful, making them great family dogs if socialized at an early age. If not used for hunting, they are great at agility, obedience trials, and flyball.

FIGURE 5.34 Brittany, Head

QUICK FACTS 5.16 Brittany

Quick Facts	Height	Weight	Coat	Features
Life Span: 12–13 years	**Males:** 17–21 inches 43–53 centimeters **Females:** 18–20 inches 46–51 centimeters	**Males:** 35–40 pounds 16–18 kilograms **Females:** 30–40 pounds 14–18 kilograms	**Colors:** Orange and white or liver and white, roan. **Type:** Dense, flat, or wavy hair.	**Head:** Mesocephalic, scissor bite. **Ears**: V-shaped rounded very slightly. **Tail**: Docked or natural bobtail.

Genetic Diseases or Conditions: Hip dysplasia, cryptorchidism, canine discoid lupus erythematosus, luxating patellas, cancer, and epilepsy.

English Cocker Spaniel

Origins: One of the oldest types of land spaniel known, in England all dogs of this type were known as spaniels, whether they were large or small, long-bodied or short, fast or slow on their feet. Gradually the difference in size resulted in the larger dogs becoming the English Springer Spaniel and the smaller ones becoming the English Cocker Spaniel. Both breeds could be found in the same litters depending on the size of the pup. In 1874, the first stud books were published, and any dog under 25 pounds was placed in the Cocker breeding pool. In the U.S. the breed standard was established in 1935, and American and English Cocker crossbreeding was discouraged. The AKC recognized the English Cocker Spaniel in 1946.

FIGURE 5.35 English Cocker Spaniel, Show Strain

FIGURE 5.36 English Cocker Spaniel, Field Strain

Characteristics: There are two types of English Cockers, a field line and a show line. Both are sturdy, robust dogs that are faithful to master and household. They have a happy personality, which makes them good companions with a natural instinct for hunting. English Cockers are sensitive, and obedience training should be calm, firm, and consistent with a gentle hand. They should be socialized at an early age, exposing them to many situations. They do well with children if raised with children and not allowed to be dominant over humans. They can become barkers if not taught to be quiet. As with all sporting breeds they need plenty of exercise to maintain their mental stability.

QUICK FACTS 5.17 English Cocker Spaniel

Quick Facts	Height	Weight	Coat	Features
Life Span: 12–15 years	**Males:** 15–17 inches 38–43 centimeters **Females:** 14–16 inches 36–41 centimeters	**Males:** 28–34 pounds 13–16 kilograms **Females:** 26–32 pounds 12–15 kilograms	**Colors:** Solid colors, parti, roans, black and tans, tricolors. **Type:** Medium length, flat, or slightly wavy and silky, feathering on legs and abdomen. Show strains have longer hair.	**Head:** Mesocephalic, scissor bite. **Ears**: Hanging. **Tail**: Docked or left natural.

Genetic Diseases or Conditions: Retinal degeneration, cataracts (usually occurs at eight years old and will be totally blind by ten), glaucoma, ear infections, skin allergies, deafness, congestive heart failure, heart murmurs, and obesity.

FIGURE 5.37 English Springer Spaniel, Field Line

English Springer Spaniel

Origins: The largest of the spaniels got the name by "springing" at upland birds to get them to fly up for the hunter to shoot. In 1902, the Kennel Club of England recognized the Springer as a distinct breed with the AKC following suit in 1910.

FIGURE 5.38 English Springer Spaniel, Show line

Characteristics: The English Springer Spaniel makes a flashy show dog with its pure white body and liver or black saddle across its back and the same color over both ears. There are two types of the breed, a show line and a field line. The show line will have longer hair over its entire body and longer feathering on the ears and the back of the legs. The field line has more white on its body and less feathering. "Springers" are friendly, especially if socialized well when young, which should be with people and other dogs. They are eager to please, quick to learn, and very willing to obey. Training a Springer is relatively easy, and it needs to be consistent and fair. Harsh words will often shut a Springer down completely. They are good family dogs as they will romp and play fetch with children. They are great competitors at field trials, obedience, agility, flyball, and, of course, are great hunting dogs. A relatively new job Springers are being asked to fill is that of "sniffer dogs." They are being used to search for explosives, illegal mobile phones in prisons, bumblebee nests, narcotics, and illegal immigrants. They are also being used as search-and-rescue dogs with mountain rescue teams. They are also great as assistance, therapy, and visitation dogs. This is a lot of dog in a fairly small package. Remember, as with any of the sporting group exercise is one of the main ingredients for a well-balanced dog; not enough and you may get a dog with nervous energy that can build into separation anxiety or destruction of property.

QUICK FACTS 5.18 English Springer Spaniel

Quick Facts	Height	Weight	Coat	Features
Life Span: 12–15 years	**Males:** 19–21 inches 48–56 centimeters **Females:** 18–20 inches 46–51 centimeters	**Males:** 45–55 pounds 20–25 kilograms **Females:** 40–50 pounds 18–23 kilograms	**Colors:** Liver or black with white markings, tri-color, blue or liver roan; or white with tan, black or liver markings. **Type:** Double, flat or wavy, medium-length outer coat; short, soft, and dense under coat. Together they are water-, weather-, and thorn-proof.	**Head:** Mesocephalic, scissor, or even bite, undershot jaw not uncommon. **Ears**: Long, hanging, fairly wide. **Tail**: Docked and fringed with wavy feathers.

Genetic Diseases or Conditions: Otitis, retinal dysplasia, entropion, glaucoma, progressive retinal atrophy, hip dysplasia, epilepsy, phosphofructokinase deficiency, allergies, autoimmune diseases, and obesity.

Working Group

The working group can be broken into three categories based on the jobs they were developed to do and what they are used for today. The Sled Dog group includes the dogs that are good at working as a team and have the hair coats to survive very harsh winter conditions. With that being said, there are many dogs that have been used as sled dogs that don't have these characteristics. Guard Dogs are those that were developed to stand guard over home and flocks and also work alongside military personnel and police officers. Again, many dogs can be used for this purpose, such as the German Shepherd in the herding group. However, the dogs in the working group tend to be quite large with deep voices. Luckily, they are usually pretty docile and inflict damage only when provoked or commanded. The last few breeds are considered rescue dogs. Rescue Dogs were developed to look for people stranded in the mountains or to dive into the water and rescue people from drowning. They are gentle giants and can do double duty as guard dogs because an intruder would have to be insane to enter a home or yard with these fellows on the property. Again, there are plenty of dogs that will search or look for people, such as Bloodhounds or Labrador Retrievers.

FIGURE 6.00 Alaskan Malamute

SLED DOGS

Alaskan Malamute

Origins: This old breed of spitz-type dog can be traced back 2,000–3,000 years, named after the Inuit Malamutes (now known as the Kuuvangmiut people) from the upper-western part of Alaska. There is no record of how they were developed, but by their features it seems that the wolf was certainly a part of their development. They not only hauled possessions but helped to hunt bears and seals. They were recognized by the AKC in 1935.

FIGURE 6.01 Alaskan Malamute

Characteristics: Malamutes are powerfully built dogs with compact, muscular bodies and affectionate dispositions. They are loyal and devoted companions. They are exceptionally friendly to strangers and are playful when asked. They tend to be very vocal dogs. If bored, they will express their displeasure by howling to the moon and digging to the South Pole. They are very dignified at maturity, making a great family dog. They require extensive amounts of exercise and are ideal for the person that loves to do activities outdoors. They are still used as sled dogs but also are finding a use in outdoor events called skijoring or bikejoring. This is where one or two dogs are harnessed and pull people on skis or bikes. They are also a great dog to pick if interested in "canicross" which is cross country running with your Malamute. As the gold rush started, these dogs were crossed with other breeds to increase their speed for racing, an event in which they still excel today. Even with this cross breeding, they have retained much of their original instincts to hunt, so care must be taken with them being around house cats. Their coats are extremely thick with guard hairs rising through the undercoat. This keeps the snow from getting down into their very wooly undercoat, allowing them to live outside during even extreme winter weather. They can suffer greatly in extreme heat and must be given plenty of fresh water and shade, and, if possible, they should be brought inside during the hottest periods of the day.

QUICK FACTS 6.00 Alaskan Malamute

Quick Facts	Height	Weight	Coat	Features
Life Span: 12–15 years	23–28 inches 58–71 centimeters	75–90 pounds 34–41 kilograms	**Colors:** Light gray to black guard hairs, white on underbelly, legs/feet, and mask. **Type:** Double coat, undercoat is 1–2" dense, wooly, oily, outer coat of thick, coarse guard hairs.	**Head:** Mesocephalic, scissor bite. **Ears:** Prick. **Tail:** Squirrel.

Genetic Diseases or Conditions: Cancer, hip and elbow dysplasia, chondrodysplasia (dwarfism), cataracts, progressive retinal atrophy, diabetes, epilepsy, bloat, hypothyroidism, kidney and heart disease, and skin disorders.

Samoyed

FIGURE 6.02 Samoyed

Origins: DNA analysis shows that the Samoyed is one of the oldest breeds of dogs dating back over 3,000 years. They are descended from a spitz-type dog from Siberia/Northern Asia, of nomadic tribes that used them for sledding, herding, guarding, and companionship. They were recognized by the AKC in 1906.

FIGURE 6.03 Samoyed, Head

Characteristics: Pronounced sam-uh-yehd, "Sammys" are gentle companions that have an unusual intelligence which is attributed to being so entrenched with people during their development. The tribes valued them as herding dogs as the reindeer herds were virtually their only means of protein. With very little game to hunt, the Samoyed has virtually no hunting instinct. They excel as sled dogs and have been used for both Arctic and Antarctic expeditions. They require obedience training if you don't wish to be "hauled" down the sidewalk as they are very powerful and will do so from pure instinct. They make excellent watchdogs as they are loyal to their people and like to bark, but for the most part they are "people dogs" and do not do well as guard dogs. They will also bark and dig if left alone for long periods of time and become bored. Brushing is a must to keep their beautiful white coats free from mats. As they shed for the summer heat, many people will keep the wooly hair from the undercoat as it can be woven into beautiful clothes. They suffer in hot weather and should be allowed inside when temperatures and humidity are high. These dogs will do well in agility, obedience trials, and, of course, any of the pulling events including skijoring and bikejoring.

QUICK FACTS 6.01 Samoyed

Quick Facts	Height	Weight	Coat	Features
Life Span: 12–15 years	**Males:** 21–23.5 inches 53–60 centimeters **Females:** 19–21 inches 48–53 centimeters	**Males:** 45–65 pounds 20–30 kilograms **Females:** 35–50 pounds 16–20 centimeters	**Colors:** White, cream, biscuit, or white and biscuit. **Type:** Double coated, undercoat soft, short, wooly thick, dense, outer coat long and harsh guard hairs growing through it, thick ruff around neck.	**Head:** Mesocephalic, scissor bite. **Ears**: Erect, small, round at tips. **Tail**: Squirrel.

Genetic Diseases or Conditions: Renal disease called "Samoyed Hereditary Glomerulopathy," diabetes, progressive retinal atrophy, bloat, hip dysplasia, pulmonary stenosis, sebaceous adenitis, urinary infections, tonsillitis, hypothyroidism, impacted anal glands, and dermatitis.

FIGURE 6.04 Siberian Husky

Siberian Husky

Origins: Developed by Chukchi people of Northeast Asia to pull sleds, herd reindeer, and act as a watchdog, Siberian Huskies are part of the group of ancient spitz-type of dogs. Known for their great endurance and speed, they are one of the principle dogs used in the Iditarod Trail Sled Dog Race to commemorate the 1925 diphtheria serum run from Nenana to Nome, Alaska. The Chukchi people maintained pure breeding lines up to the 19th century. The AKC recognized them in 1930.

Characteristics: Today these hardy dogs are still used to pull sleds and as companions. They have an even temper, being friendly and outgoing toward people. They are not possessive, suspicious, or aggressive towards other dogs and strangers. They will bark but will quiet down very quickly. They do not have a doggy odor as they are fastidiously clean, cleaning themselves like a cat. They can adapt well to all kinds of living conditions but must have a lot of exercise and/or a job to do such as obedience trials, agility, or even skijoring or bikejoring as they have great endurance carrying or hauling light loads. They do great with other dogs and house pets as well as children. They do shed and so need to be brushed often to keep the hair from overwhelming the house. These dogs are fifth least affected in all breeds from hip dysplasia. Whelping takes them longer with frequent rests of five to six hours halfway through parturition. A litter of six may take the Siberian 10–12 hours to deliver. It is also common for the bitch to have diarrhea during her entire lactation period.

FIGURE 6.05 Siberian Husky, White and Red

QUICK FACTS 6.02 Siberian Husky

Quick Facts	Height	Weight	Coat	Features
Life Span: years	**Males:** 21–23.5 inches 53–60 centimeters **Females:** 20–22 inches 51–56 centimeters	**Males:** 45–60 pounds 20–27 kilograms **Females:** 35–50 pounds 16–22.5 kilograms	**Colors:** All black to all white, reds. **Type:** Double coat with medium-long guard hairs growing through the dense, wooly undercoat. They have hair that grows between their toes.	**Head:** Mesocephalic, scissor bite, blue or brown eyes, sometimes one of each. **Ears**: Prick. **Tail**: Sickle.

Genetic Diseases or Conditions: Glaucoma, cataracts, progressive retinal atrophy, corneal dystrophy, gastric disease and ulcerations, bronchitis, and hypothyroidism.

GUARD DOGS

Akita

FIGURE 6.06 Akita

Origins: The Akita came from the northern mountains of Japan and was bred as a hunting and fighting dog. They are considered a national monument and carry spiritual significance as a protector of the home and are given as gifts to provide good health. U.S. servicemen brought them to the U.S. after World War II. The AKC recognized them in 1973.

FIGURE 6.07 Akita

Characteristics: Akitas are not a dog for the novice dog owner. They require a firm, consistent, and confident owner that can train and socialize them while maintaining their standing in the family hierarchy. An Akita that believes it is in charge will become very aggressive toward people and other dogs and will be food possessive as well. When they know their place, Akitas are affectionate with family members and will be protective of the children. However, they may not be good with children that do not understand how to behave around dogs as Akitas do not tolerate being teased. They can be aggressive toward other dogs and require lots of socialization as puppies. Even then, it is important to know that these dogs are not good candidates for dog parks. Akitas are used as seeing-eye dogs, police dogs, beach patrol, and draft/sled dogs.

QUICK FACTS 6.03 Akita

Quick Facts	Height	Weight	Coat	Features
Life Span: 10–12 years	**Males:** 26–28 inches 66–71 centimeters **Females:** 24–26 inches 61–66 centimeters	**Males:** 75–120 pounds 34–54 kilograms **Females:** 75–110 pounds 34–50 kilograms	**Colors:** Any color, white, brindle, or pinto. **Type:** Double, thick, soft dense undercoat, straight, harsh hairs of outer coat about 2" in length.	**Head:** Mesocephalic, scissor or level bite. **Ears**: Erect. **Tail**: Curled full, ¾, or double curl.

Genetic Diseases or Conditions: Hip and elbow dysplasia, entropion, progressive retinal atrophy, multiple congenital ocular defects, deafness, intussusceptions, autoimmune diseases, bloat, cruciate-ligament injuries, patellar luxation, pemphigus, sebaceous adenitis, unique juvenile-onset polyarthritis syndrome, uveodermatological syndrome, and Von Willebrand's disease. Some are very sensitive to anesthetics, hypothyroid, and hyperthyroid.

FIGURE 6.08 Bernese Mountain Dog

Bernese Mountain Dog

Origins: One of four varieties of dogs from the Swiss Mountain region. They were bred as watchdogs and herders and were used as draft dogs, pulling milk and basket carts for weavers. The AKC recognized the breed in 1937.

FIGURE 6.09 Bernese Mountain Dog, Head

Characteristics: A "Berner" is friendly, self-confident, alert and good natured, making them great with families. They are never sharp or shy and are easily trained because they have remarkable memories and superior intelligence. Training must be done with a gentle touch as they can be sensitive. Berners need to be with their people; they become despondent and unstable if they are left in the backyard alone. Their devotion to their families makes them great watchdogs but not so great as guard dogs because they like everyone. They can be taught to pull a cart or small wagon, but they don't have the stamina found in sled dogs. They have a thick coat of fur that sheds all year round, requiring twice-weekly brushing. However, it is worthwhile to keep these beautiful dogs in good shape. The thick coat does make them uncomfortable in hot weather, and care must be taken to keep them from getting heat stroke.

QUICK FACTS 6.04 Bernese Mountain Dog

Quick Facts	Height	Weight	Coat	Features
Life Span: 6–8 years	**Males:** 24–28 inches 61–71 centimeters **Females:** 23–27 inches 58–69 centimeters	**Males:** 85–110 pounds 38–50 kilograms **Females:** 80–105 pounds 36–48 kilograms	**Colors:** Tri-colored; black over most of the dog, rich rust "kiss marks" on legs and chest, white blaze and muzzle band, white on underbelly and tip of toes. **Type:** Moderately thick, long and slightly wavy or straight.	**Head:** Mesocephalic, scissor bite. **Ears:** Pendent. **Tail:** Plume.

Genetic Diseases or Conditions: Cancers (even in young dogs), hip and elbow dysplasia, hypothyroidism, cataracts, progressive retinal atrophy, entropion and ectropion, umbilical hernias, cerebellar degeneration, osteochondrosis dissecans, Von Willebrand's disease, cardiac disease, bloat, and fragmented coronoid process.

Boxer

Origins: Boxer-like dogs have been known all throughout Europe, but its greatest development was in Germany. Two, now-extinct, mastiff-like dogs were crossed with the bulldog to develop the Boxer. The AKC recognized the Boxer in 1904.

FIGURE 6.10 Boxer, Uncropped Ears and Natural Tail

FIGURE 6.11 Boxer, Brindle, Cropped, and Uncropped Ears

Characteristics: The original Boxers were all-purpose dogs used for carting, herding, and hunting, as well as in bull baiting and dog fighting. They were the first dogs selected by the Germans to be police dogs because of their intelligence, courage, and tenacity if threatened.

Today's Boxer is spirited and energetic, making it a great active family dog. They love their families and will play with the children endlessly. They do need to be trained to not jump on people as they have a habit of using their front feet much like a human boxer. This can cause some injuries, especially to small children if they catch a paw in the face. Boxers are responsive to positive reinforcement and consistent training, actually thriving on this guidance. They will need to be socialized to other pets and critters but even then should not be left alone with them. Their dignified and self-assured manner can thwart stranger's intent on harm although they respond gracefully to honest and friendly overtures. Most Boxers will whelp without problems; however, some bitches will need help severing the umbilical cord. They have a short coat that is easy to care for, and they tend to self-clean. Boxers require exercise or something to do to prevent boredom which is expressed by digging, chewing or licking themselves. They have been used in search and rescue, obedience trials, agility, tricks and a good, old-fashioned game of fetch. Boxers snore and drool. Some have an issue with flatulence especially if fed something other than a high-quality dog food.

QUICK FACTS 6.05 Boxer

Quick Facts	Height	Weight	Coat	Features
Life Span: 11–14 years	**Males:** 22–25 inches 56–63 centimeters **Females:** 21–24 inches 53–61 centimeters	**Males:** 60–70 pounds 27–32 kilograms **Females:** 53–65 pounds 24–29 kilograms	**Colors:** Fawn, brindle, mahogany, black, white, black mask, white on feet and chest, blaze on face. **Type:** Smooth, short, shiny.	**Head:** Brachycephalic, underbite. **Ears**: Cropped or pendent. **Tail**: Docked, short.

Genetic Diseases or Conditions: Tumors (both benign and cancerous), epilepsy, arthritis, gastric torsion, hip dysplasia, acquired and genetic cardiac issues, hypothyroidism, degenerative myelopathy, unilateral cryptorchidism, and ulcerative keratitis.

Bullmastiff

Origins: The Bullmastiff was developed by crossing mastiffs and bulldogs in England in the 1700s. They were used by game keepers to keep poachers out of game preserves by keeping the trespassers pinned until the keeper arrived. The Bullmastiff was recognized by AKC in 1933.

FIGURE 6.12 Bullmastiff

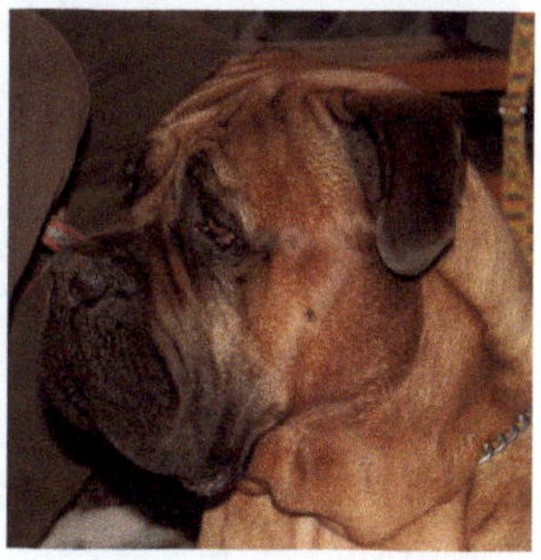

FIGURE 6.13 Bullmastiff, Head

Characteristics: The Bullmastiff is a loyal dog that needs no training to be on guard. These dogs will protect their families and property with devotion and courage. They do not fight unless provoked, and once the intruder is pinned to the ground they will not bite. Their training must be done by an owner that can be authoritative but not harsh, and they must have variation in their training as they do get bored and stop paying attention which is often mistaken for stubbornness or a lack of intelligence. Because of their great size they need to be taught not to pull on the leash as they can pull most people off their feet. For their size they are gentle and patient with children. The Bullmastiff can be a bit aggressive towards other dogs and must be socialized early if they are to be around others. These dogs have a short-hair coat, and they drool, slobber, and snore. They need to be where their people are and will not be healthy minded if left in a kennel. They require careful feeding as adults to prevent bloat. Three small meals per day in elevated dishes is recommended, and as puppies they require a quality feed that is formulated for giant breeds to prevent them from growing too quickly. It takes the Bullmastiff almost 3–3.5 years to reach its full maturity.

QUICK FACTS 6.06 Bullmastiff

Quick Facts	Height	Weight	Coat	Features
Life Span: 7–9 years	**Males:** 25–27 inches 63–69 centimeters **Females:** 24–26 inches 61–66 centimeters	**Males:** 110–133 pounds 50–60 kilograms **Females:** 100–120 pounds 45–54 kilograms	**Colors:** Fawn, red or brindle with a black mask on the face. **Type:** Short, dense, slightly rough.	**Head:** Mesocephalic, level or undershot bite. **Ears**: V-Shaped. **Tail**: Saber, long.

Genetic Diseases or Conditions: Bloat, tumors, cancer, progressive retinal atrophy, hip and elbow dysplasia, allergies, cardiac diseases, hypothyroidism, arthritis, osteochondritis dissecans, hypertrophic osteodystrophy, panosteitis, and kidney disease.

FIGURE 6.14 Doberman Pinscher, Black

Doberman Pinscher

Origins: Developed by Karl Friedrich Louis Doberman, a tax collector and dog pound warden, who needed a dog that would be loyal, intelligent, and fearless to accompany him on his tax-collection rounds. It is thought that several breeds were used to create the Doberman including the German Pinscher, Beauceron, Rottweiler, Greyhound, Great Dane, and Manchester Terrier among others. After his death in 1894 the Germans named the breed in his honor. Others took up the development of the breed, and the AKC recognized it in 1908.

Characteristics: Originally these dogs were used exclusively as guards and watchdogs. They were trained to attack on command and are so reliable in obeying commands that the breed was pressed into police and military service. It has a good nose and has been used as a search-and-rescue, patrol, or policeman's dog to follow criminals and as a hunting dog. The modern American versions have been tamed down, and while not losing their intelligence they have gained in affection and are devoted to home and family. A properly bred, socialized, and trained "Dobie" will have a sane mind and the heart and spirit of a gentleman. Their loyalty is legendary and is commemorated with a life-sized statue outside of the war dog cemetery at the U.S. naval base in Guam entitled, "Always Faithful." The Doberman is good in agility and obedience trials, as guide dogs, and as therapy and visitation dogs. Giving these dogs a purpose is important to keep them balanced, which can also be accomplished with daily exercise and training.

FIGURE 6.15 Doberman Pinscher, Black and Red

QUICK FACTS 6.07 Doberman Pinscher

Quick Facts	Height	Weight	Coat	Features
Life Span: years	**Males:** 26–28 inches 66–72 centimeters **Females:** 24–26 inches 61–68 centimeters	**Males:** 75–100 pounds 34–45 kilograms **Females:** 59–90 pounds 27–41 kilograms	**Colors:** Black, red, blue, and fawn with kiss marks and rust markings on throat, chest, legs and feet, and below tail. **Type:** Smooth, short, hard and thick.	**Head:** Mesocephalic, scissor bite. **Ears**: Cropped, pendant if uncropped. **Tail:** Docked, whip if natural.

Genetic Diseases or Conditions: Gastric torsion, Wobbler's syndrome (rear leg incoordination to complete paralysis), cervical vertebral instability, cardiac disorders, hip dysplasia, hypothyroidism, progressive retinal atrophy, Von Willebrand disease, alopecia or skin disorders (especially the fawn and blue colors), and obesity.

FIGURE 6.16 Doberman Pinscher, Fawn

Dogue de Bordeaux

Origins: An example of the molossoid heritage, these large built, solid dogs can be found in history as far back as Julius Caesar's Rome. The French are credited with keeping the dog in its current form. The AKC recognized this massive dog in 2008.

Characteristics: A person would think that with an ancestral history as guardians for businesses, homes, and livestock, as fearless hunters of bears and boars, and as fighters of bulls,

FIGURE 6.17 Dogue de Bordeaux

FIGURE 6.18 Dogue de Bordeaux, Head

bears, jaguars, and men, this breed would be a very aggressive and difficult dog to have around. However, the Dogue de Bordeaux is not aggressive! They make excellent companions with a calm, easy going temperament. They are patient with children and become very attached to their families. Because of their phenomenal strength, they are not for the novice dog owner. Training must be done with a calm, assertive, consistent approach. Their socialization is also of the utmost importance, and it must be made clear that all humans are higher up in the hierarchy than the dog. The Dogue de Bordeaux is a super "drooler," and because of the deep wrinkles in their faces, they often have skin infections that need to be addressed immediately to keep them under control. They have a short-hair coat that is easy to care for, and they require only moderate daily exercise.

QUICK FACTS 6.08 Dogue de Bordeaux

Quick Facts	Height	Weight	Coat	Features
Life Span: 5–10 years	23–30 inches 58–75 centimeters	120–145 pounds 54–65 kilograms	**Colors:** Shades from fawn to mahogany. **Type:** Short, soft.	**Head:** Brachycephalic, undershot bite. **Ears**: V-shaped with round tips. **Tail**: Saber.

Genetic Diseases or Conditions: Hip and elbow dysplasia, osteochondrosis, patellar luxation, cardiac disease, gastroenteritis, flatulence, hypothyroidism, Von Willebrand's disease, and hyperkeratosis. These dogs often require cesareans and have a high rate of stillborn or neonatal deaths.

Giant Schnauzer

Origins: German Pinschers found in southern Germany would whelp both smooth and coarse-haired pups. The Germans decided to split the two and established the Standard Schnauzer. These were crossed with black Great Danes and Bouvier des Flandres to develop the "***Reisenschnauzer***" or Giant Schnauzer. The AKC recognized the Giant Schnauzer in 1930.

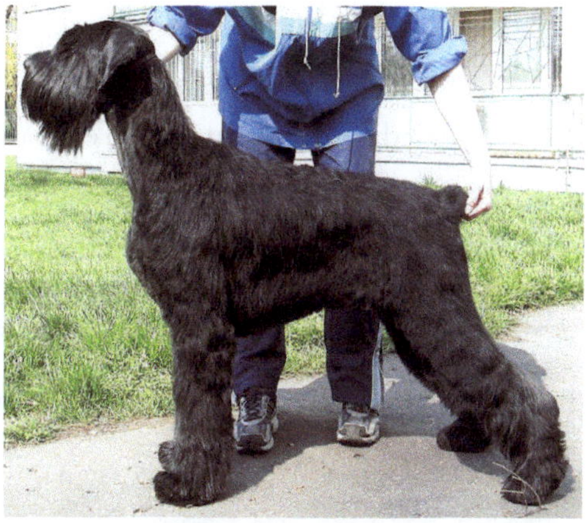

FIGURE 6.19 Giant Schnauzer, Black

FIGURE 6.20 Giant Schnauzer, Salt and Pepper

Characteristics: Originally used as a farm dog that would help drive and protect herds as well as pull a cart, they were quickly adopted by the police because of their extreme intelligence. The modern Giant Schnauzers still possess high intelligence and loyalty, making them easy to train. They can be very protective of hearth and family, so early socialization is important for them to discern between friend and foe. They are high-energy dogs and require a lot of daily exercise. They do very well with children and will romp and play willingly for hours in addition to their daily walks. Giant Schnauzers do well in agility, obedience trials, Shutzhund, carting and tracking. They do not do well kenneled as they need to be near their people. It takes this dog about two years to reach its full growth, and as with most large, deep-chested dogs they should be fed a good quality food formulated for their size. Their coats do require professional grooming with weekly brushing in between. Their beards collect food and moisture and need to be cleaned often.

QUICK FACTS 6.09 Giant Schnauzer

Quick Facts	Height	Weight	Coat	Features
Life Span: 10–12 years	**Males:** 26–28 inches 66–71 centimeters **Females:** 23–26 inches 58–66 centimeters	**Males:** 60–105 pounds 27–48 kilograms **Females:** 55–75 pounds 25–34 kilograms	**Colors:** Solid black or pepper and salt. **Type:** Double coat; outer coat is hard, wiry, very dense; undercoat is soft and mats easily. Coarse hair on top knot, beard, and eyebrows.	**Head:** Mesocephalic, scissor bite. **Ears**: cropped, uncropped medium-length V-shaped. **Tail**: Docked.

Genetic Diseases or Conditions: Cancer, cardiac diseases, epilepsy, hip and elbow dysplasia, osteochondrosis dissecans, Addison's and Cushing's disease, hypothyroidism, hyperkeratosis, skin disorders, cataracts, entropion, ectropion, progressive retinal atrophy, retinal dysplasia, urinary tract infection, and gastroenteritis. Births often require cesareans.

FIGURE 6.21 Great Dane, Fawn

Great Dane

Origins: Great Dane-like ancestors appear in reliefs, paintings, and coins in Egyptian monuments dating back to 3000 BC. Modern-day Great Danes were developed in Germany using English Mastiffs, Irish Wolfhounds, and Greyhounds. They were used as bear, wolf, and boar hunting dogs, as bodyguards, and as fighting dogs. The AKC recognized them in 1887.

Characteristics: This "Gentle Giant" thinks it is a lapdog. Danes have stable minds and playful attitudes towards people and other dogs. They can adjust to apartment living if given long, twice-daily walks or a long walk with a romp in the dog park. Exercise and diet are important all through their lives but especially as they grow. The wrong diet and too much exercise can cause bone-growth issues. As adults their meals should be divided into three meals a day and be elevated off the floor to prevent bloat. They have a short hair coat which is easy to keep up, although this is a LOT of dog, so it will still take time. Danes need obedience training and socialization from an early age. They should be trained not to jump on people as their great weight can cause inadvertent injuries. Calm and consistent handling by all members of the family will ensure that you have a "great" Great Dane for its entire life span, which is unfortunately short with the average being around ten years.

FIGURE 6.22 Great Dane, Harlequin

QUICK FACTS 6.10 Great Dane

Quick Facts	Height	Weight	Coat	Features
Life Span: 6–10 years	**Males:** 30–34 inches 76–86 centimeters **Females:** 28–32 inches 71–81 centimeters	**Males:** 120–200 pounds 54–90 kilograms **Females:** 100–130 pounds 45–59 kilograms	**Colors:** Brindle, fawn, blue, black, harlequin. **Type:** Short, thick, smooth, and glossy.	**Head:** Mesocephalic, scissor bite. **Ears:** Folded, or cropped. **Tail:** Saber.

Genetic Diseases or Conditions: Gastric torsion, hip and elbow dysplasia, mast cell tumors, cancers, cardiac disease, hypothyroidism, cataracts, entropion, Wobbler syndrome, hypertrophic osteodystrophy, and panosteitis (bone development problems: usually caused by poor nutrition or over-dosing vitamins). Low tolerance for tranquilization.

Mastiff (English Mastiff)

Origins: Early signs of the Mastiff type have been found on Egyptian monuments dating back to 3000 BC. During the Roman Empire they impressed Caesar with their courage and strength. He brought the breed to Rome and added them to the games at the Coliseum where they fought bulls, lions, and gladiators. They became very popular with English peasants and were used to protect homes and offer companionship. The AKC recognized the English Mastiff in 1885.

FIGURE 6.23 Mastiff, Fawn

Characteristics: An instinctive guardian, the Mastiff will quietly watch over hearth and home. He is a massive dog with a gentle spirit unless his people are being attacked. If this should occur, the Mastiff will knock the perpetrator down and hold him pinned to the ground until told to release. This is often done without a sound or even a severe bite! Mastiffs excel in police and military work as well as search and rescue and in weight-pulling trials. They require socialization and training to keep them from becoming dominant over the household. A

FIGURE 6.24 Mastiff, Brindle

firm, consistent, authoritative owner is required for these gentle giants. They are known to be pillars of patience with children and will allow all sorts of transgressions from them. Owners need to be willing to teach their children not to abuse this great dog. The Mastiff can be a couch potato and will be healthier if made to go out for daily exercise. They tend to suffer from obesity and bloat, so measure food and feed two or three meals per day to help them stay healthy. Mastiffs drool, snore, and wheeze, so they are not for the meticulous owner. They have a short-hair coat that is easy to brush; however, they are not so easy to bathe as they are big.

QUICK FACTS 6.11 Mastiff

Quick Facts	Height	Weight	Coat	Features
Life Span: 7–10 years	**Males:** 30 inches 75 centimeters **Females:** 27.5 inches 69 centimeters	**Males:** 150–250 pounds 68–113 kilo-grams **Females:** 120–200 pounds 54–91 kilograms	**Colors:** Fawn, apricot fawn, silver fawn, or brindle. Muzzle, ears, and nose dark. **Type:** Double coat; outer coat straight, coarse, short; undercoat dense, short, and close lying.	**Head:** Mesoce-phalic, scissor bite. **Ears**: V-shaped, rounded at tips. **Tail**: Saber.

Diseases/Genetic Conditions: Obesity, hip and elbow dysplasia, cruciate-ligament rupture, degenerative myelopathy, hypertrophic osteodystrophy, panosteitis, spondylosis, Wobbler's Syndrome, bloat, ectropion, entropion, progressive retinal atrophy, persistent pupillary membranes, vaginal hyperplasia, pyometria, arthritis, calluses on elbows, nephritis, hypothyroidism, cancer, cystinuria, Von Willebrand's disease, allergies, and heart problems.

Rottweiler

Origins: The "Rottie" is definitely a descendent of a Mastiff, and Germany is credited for developing the breed as we see it today. They were originally herding dogs that also doubled as guard dogs since robbers were the rule rather than the exception when driving herds to market. They not only protected the herds but also the cattlemen and their purses. After selling their herd the cattlemen would tie their purses around the dog's neck as few highwaymen cared to tangle with this powerfully built dog. On the farm they were also put to work as draft animals. The AKC recognized them in 1931.

FIGURE 6.25 Rottweiler

Characteristics: The Rottie is a great family dog that is calm, protective, trainable, and devoted to family. Their owners must be confident, calm, firm, and consistent in their training and socialization. They have to be shown who the leader of the household is and where they stand in the hierarchy. This must be maintained in order to keep everyone safe which includes family friends and other pets. Today they are still great guard dogs as well as fine companions that can be used in tracking, search and rescue, as guide dogs, in police work, for carting, and in obedience and agility trials. They are high-energy dogs that require a large amount of exercise to maintain their health and state of mind. They will swim, bike, or play fetch or whatever their humans wish to do. Their short-hair coats are easy to care for, and they are average shedders.

FIGURE 6.26 Rottweiler

QUICK FACTS 6.12 Rottweiler

Quick Facts	Height	Weight	Coat	Features
Life Span: 8–10 years	**Male:** 24–27 inches 61–69 centimeters **Females:** 22–25 inches 56–63 centimeters	**Males:** 110–130 pounds 50–60 kilograms **Females:** 77–106 pounds 35–48 kilograms	**Colors:** Black, with kiss marks, and rust on cheeks, muzzle, chest, and legs. **Type:** Double coated; short, coarse, and flat.	**Head:** Mesocephalic, scissor bite. **Ears**: V-shaped. **Tail**: Docked.

Genetic Diseases or Conditions: Hip dysplasia, shoulder injuries, ACL injuries, predilection to Parvo Virus, entropion, ectropion, cancer, and obesity.

FIGURE 6.27 Great Pyrenees

FIGURE 6.28 Great Pyrenees, Head

RESCUE DOGS

Great Pyrenees (Pyrenees Mountain Dog)

Origins: Although it is believed they came from Central Asia or Siberia, fossil remains found on the slopes of the Pyrenees Mountains between Spain and France date the Great Pyrenees to the Bronze age around 1800–1000 BC. The French nobility appreciated their devotion to home and great courage to fend off bears and wolves. The AKC recognized them in 1933.

Characteristics: The brave and loyal Pyrenees became an indispensable part of home life in the mountains. They wore iron, spiked collars that when combined with their naturally long heavy coats made them invincible. They are still used for draft work and as guard dogs for flocks. It is a practice to place a pup with the flock early in life, so it develops a bond with the flock as it grows. Besides these jobs, the Pyrenees have a precocious sense of smell and keen eyesight making them useful as avalanche-rescue dogs, and they are also used as pack dogs on ski trips. They are great companions that are intelligent, gentle, faithful, obedient, and very loyal to family. Their training needs to be firm, consistent, and fair, and their socialization should be done early as they tend to be wary of strangers because of their strong guarding instinct. This breed NEEDS exercise and loves to romp with the family. They can slobber and will need weekly brushing to keep their coats in good shape. They have a slow metabolism which makes them require less food than dogs of the same weight. This is also important to remember when they are going under anesthesia, which if not reduced could cause them to have longer recovery periods than other dogs.

QUICK FACTS 6.13 Great Pyrenees

Quick Facts	Height	Weight	Coat	Features
Life Span: 10–12 years	**Males:** 27–32 inches 69–81 centimeters **Females:** 25–29 inches 63–74 centimeters	**Males:** 100 pounds 45 kilograms **Females:** 85 pounds 38 kilogram	**Colors:** White or white with patches of reddish-brown, gray, or pale yellow spots. **Type:** Double coated; undercoat is heavy, fine, white; overcoat is long, coarse, flat, and thick.	**Head:** Mesocephalic, scissor, or level bite. **Ears**: V-shaped, rounded at tips. **Tail**: Saber, well feathered. **Dewclaws:** Single on fronts and doubles on back - should NOT be removed.

Genetic Diseases or Conditions: Entropion, hip and elbow dysplasia, patella luxation, achondroplasia, degenerative myelopathy, osteochondritis dissecans, sebaceous adenitis, Addison's disease, cancers, congenital cardiac disease, Glanzman's thromboasthenia, swimmer puppies, and dermatitis (especially in hot weather). Some may develop cataracts by age 2, but it usually doesn't progress to blindness.

Newfoundland

FIGURE 6.29 Newfoundland, Black

Origins: Newfoundland, Canada, is where the original Newfoundland was developed. They were used to help fishermen with lines and nets, rescue items and people that fell overboard, and to guard boats at dock. The Newfoundland and the Labrador Retriever have common ancestors and were often bred to each other in olden days. The AKC recognized them in 1886.

Characteristics: This massive dog is as at home in the water as it is on land. They love the water and with their webbed toes, utilizing all four feet to swim, and an otter-like tail that acts as a rudder, they are powerful swimmers even in rough seas. "Newfs" are great family dogs that are totally devoted and tolerant of children of all sizes, undertaking the duties of nursemaid without training. Their strength and courage, intelligence, and sweet dispositions have won them many fans, even though they tend to drool and slobber. The "Newfie"

FIGURE 6.30 Newfoundland, Copper

has a rolling gait that brings a chuckle as you see him romp with the children. Training these dogs for obedience is relatively easy with calm, consistent, and firm instructions. They have a thick, water-resistant double coat that requires weekly brushing to avoid dermatitis. They too have the slow metabolism that other giant breeds possess, so veterinary staff must be careful with anesthetic dosages. They have a tendency towards bloat, so two or three small meals in elevated dishes will help to prevent this from happening. The Newfoundland often takes a long time to whelp, and the last puppy may be stillborn or weak. They also have a tendency to retain the placenta.

QUICK FACTS 6.14 Newfoundland

Quick Facts	Height	Weight	Coat	Features
Life Span: 9–15 years	**Males:** 27–29 inches 69–74 centimeters **Females:** 25–27 inches 63–69 centimeters	**Males:** 130–150 pounds 59–68 kilograms **Females:** 100–120 pounds 45–54 kilograms	**Colors:** Black, bronze and black, and white (Landseers). **Type:** Oily, water-resistant double coat; outer coat is moderately long with slight wave, undercoat is soft and dense.	**Head:** Mesocephalic, scissor, or level bite. **Ears:** Small V-shaped with rounded tip. **Tail:** Otter tail. **Toes:** Webbed.

Genetic Diseases or Conditions: Hip and elbow dysplasia, gastric dilation/torsion, heat stroke, hot spots, skin allergies, heat stroke, ectopic ureters, uroliths, osteosarcoma, degenerative myelopathy, hypothyroidism, geriatric onset laryngeal paralysis polyneuropathy, and subvalvular aortic stenosis.

Portuguese Water Dog

FIGURE 6.31 Portuguese Water Dog, Black

Origins: These dogs are from Portugal and rode along on the fishing boats to retrieve lost tackle and broken nets, men gone overboard, as a courier between ships or ship to shore, and for herding fish into the nets. They are thought to have the same ancestors as the Poodle and were used to establish the Kerry Blue and Irish Water Spaniel. The AKC recognized this breed in 1983.

FIGURE 6.32 Portuguese Water Dog, White

Characteristics: A spirited yet obedient dog with a robust, medium build suited for a full day's work in and out of the water, Portuguese Water Dogs (PWD) are high-energy dogs with webbed feet and a muscular build, making them amazing swimmers and divers. They retain their working spirit and will transfer that energy into being a great active family pet. The PWD are athletic and will require vigorous daily exercise. They are easily trained, and owners have to be aware that they have to be one step ahead of the PWD, or the dogs may outsmart them! They will do great in agility and obedience trials. They love to swim and will retrieve a stick from the water for hours on end. If left to their own devices and become bored, they can be destructive, so make sure they have plenty of chew toys. Their hair coat requires regular maintenance and professional grooming in what is called a lion clip where the face and back quarter is clipped short with the tip of the tail left long. A second clip is the retriever clip where the coat is left approximately one inch in length.

QUICK FACTS 6.15 Portuguese Water Dog

Quick Facts	Height	Weight	Coat	Features
Life Span: 10–14 years	**Males:** 20–22 inches 50–57 centimeters **Females:** 17–20 inches 43–52 centimeters	**Males:** 42–55 pounds 19–25 kilograms **Females:** 35–49 pounds 16–22 kilograms	**Colors:** Black, white, shades of brown, silver, gray, and parti colors **Type:** Curly or wavy single layer, profuse, waterproof; will continue to grow in length, requires professional grooming.	**Head:** Mesocephalic, scissor, or level bite. **Ears:** Hanging. **Tail:** Otter tail.

Genetic Diseases or Conditions: Gangliosidosis, progressive retinal atrophy, cataracts, microphthalmia, distichiasis, hip dysplasia, GM-1 Storage Disease, congenital kidney disease, allergies, juvenile dilated cardiomyopathy, and Addison's disease.

Saint Bernard

FIGURE 6.33 Saint Bernard, Smooth Coat

Origins: The monks of the St. Bernard de Menthon developed the Saint Bernard by crossing Tibetan Mastiffs, Great Danes, Greater Swiss Mountain dogs, and Great Pyrenees. The dogs were developed to rescue stranded, lost, or injured travelers in the mountains. They would be sent out in groups of threes, and when a victim(s) was found, two would lie down next to the individuals to revive them by licking their faces and warm them with their bodies while the third dog would head back to the monastery for help. The AKC recognized the breed in 1885.

FIGURE 6.34 Saint Bernard, Rough Coat

Characteristics: This breed is an ***acromegaly*** breed, which means it has enlarged body structures and all of the inherent problems that go along with that condition. As the Saint Bernard puppy grows, it is important to feed it a good quality food designed for large-breed dogs and provide only mild exercise so that it doesn't develop growth issues. Training is relatively easy as they do like to please their owners, but it must be done early when they are easier to handle. They should also be socialized at an early age to prevent any possible aggressiveness. Once trained and socialized, the "Saint" lives up to its name as the breed is loving, loyal, and protective of its family. They do eat a lot and can be expensive to feed, plus they slobber and drool. There are two types of hair coats available: a short coat and a rough coat. The rough-coated Saint needs to be brushed weekly.

TABLE 6.16 Quick Facts – Saint Bernard

Quick Facts	Height	Weight	Coat	Features
Life Span: 8–11 years	25.5–27.5 inches 61–70 centimeters	110–200 pounds 50–91 kilograms	**Colors:** White with red, brown, mahogany, brindle, and black in many combinations, with a white blaze. **Type:** Smooth and rough coated.	**Head:** Mesocephalic, scissor, or level bite. **Ears:** V-shaped. **Tail:** Otter tail.

Genetic Diseases or Conditions: Bloat, hip and elbow dysplasia, osteosarcoma, lip fold pyoderma, eczema, entropion, ectropion, dilated cardiomyopathy, diabetes mellitus, epilepsy, heat stroke, and Wobbler Syndrome.

Terriers

Terriers were developed to catch vermin such as rats, mice, rabbits, snakes, and other critters. They were expected to work independently and to be unrelenting until the prey was vanquished or escaped. They are feisty, independent, stubborn, and sometimes vicious—hopefully always towards prey and never towards people! These dogs must have firm, consistent, and confident training that includes the command "drop it" so they release anything they are holding. Socialization must be started early and continued throughout their lifetimes with everything and everyone in order to have a trustworthy companion. Terriers kill their prey by violently shaking their heads from side to side. The "terrier-shake" can be rather frightening to behold, but don't let that scare you off. With good obedience training and exercise, owners will have a great dog that is fun and energetic and will fill their days with laughter and smiles.

Airedale Terrier

Origins: Historians attribute the Airedale's development as taking place in the Aire River Valley in England. They resemble an extinct black-and-tan terrier that is linked with the development of the Irish, Fox, and Welsh terriers. It is thought that Otterhound was added to increase their swimming ability and improve their personality. The Airedale Terrier was accepted by the AKC in 1888.

FIGURE 7.00 Airedale Terrier

Characteristics: The Airedale was developed to hunt all kinds of game either with or without a hunter. They hunted river otters and rats, birds, rabbits, and were even used to hunt big game in Africa, India, and Canada. They were one of the first dogs to be used as police and war dogs in England during WWII. Modern Airedales have sweet dispositions; they are loyal, intelligent, courageous, and protective of their family and property. They can be indifferent towards strangers and can quickly take over a household if they think their humans are weak. Early socialization to children and other pets and exposure to various situations are important to help them be a well-balanced dog. In addition they need daily exercise and other mental stimulation. They do train easily but get bored or distracted quickly, so variation is important. Firm, consistent, and fair training techniques work well with the Airedale. They may not pick a fight with another dog, but if matched against a dog of equal size, they usually come out on top. They have a "broken" coat that is hard, stiff, and wiry. Some have undercoats that require stripping 2–3 times a year as they do not shed.

FIGURE 7.01
Airedale Terrier, Head

QUICK FACTS 7.00 Airedale Terrier

Quick Facts	Height	Weight	Coat	Features
Life Span: 11–14 years	**Males:** 22–24 inches 56–61 centimeters **Females:** 22–23 inches 56–58 centimeters	**Males:** 50–65 pounds 23–29 kilograms **Females:** 40–45 pounds 18–20 kilograms	**Colors:** Tan with a black or dark grizzle saddle over the back and sides. **Type:** Crinkled or slightly wavy hard, dense.	**Head:** Mesocephalic, scissor bite. **Ears:** Button ear. **Tail:** Docked, carried gaily.

Genetic Diseases or Conditions: Malocclusion, malignant neoplasms, lymphoma, cardiac, colonic and urinary disease, hip dysplasia, dermatitis, and hypothyroidism.

American Staffordshire Terriers

FIGURE 7.02 American Staffordshire Terrier

Origins: The Bulldog and either the English Terrier or Black-and-Tan Terrier was used to develop the Staffordshire Terrier in England. The Bulldog of that area looked very much like the Staffordshire we know today. These dogs were imported into the U.S. in 1870 and were known by many names such as Pit Bull Terrier, American Bull Terrier, Pit Dog, or Yankee Terrier. As they became better known and dog fighting was banned in the U.S., Americans started to breed a dog that was heavier and a bit taller. These breeders petitioned the AKC to call them American Staffordshire Terriers. In 1936 the breed was accepted as the American Staffordshire Terrier.

FIGURE 7.03 American Staffordshire Terrier, Head

Characteristics: Two strains of this dog have developed over the years, a show-quality strain called American Staffordshire Terriers and a non-show strain referred to as American Pit Bull Terriers. The AKC does not recognize the American Pit Bull Terrier. "Amstaffs" are intelligent, outgoing dogs that are stable and confident. They excel as guardians and can easily discriminate between strangers who mean well and those who do not. They must be socialized at an early age to other animals and people in order to become a friendly and trustworthy companion. They do need to have a strong owner that will establish the hierarchy of the household and help other family members stay above the dog in the "pack." They are easily trained, but in addition to the usual obedience commands they especially need to be taught to release objects on command as they have a tendency to hold on to something once they have bit down. If provoked, they will fight, and if cornered will fight to the death especially if their family is being attacked. Amstaff owners must have a stoutly, fenced-in yard, and they must be exercised on a double leash if out of the yard, not only for protection of the public but also for their protection as well. They are very powerful, and if they suddenly pull, a double leash will give the owner another chance to maintain control. Amstaff enjoy the challenge of obedience trials, agility, police work, guard duty, and pulling competitions. This breed is often included in "breed bans" by municipalities and may be on insurance company lists as bad risks, so homeowner's policies may be denied or revoked if this breed is brought into the home.

QUICK FACTS 7.01 American Staffordshire Terriers

Quick Facts	Height	Weight	Coat	Features
Life Span: 9–15 years	**Males:** 17–19 inches 43–48 centimeters **Females:** 16–18 inches 41–46 centimeters	57–67 pounds 25–30 kilograms	**Colors:** Any color, solid, parti, or patched. **Type:** Glossy, short, close, stiff to the touch.	**Head:** Mesocephalic, very pronounced cheek muscles, scissor bite. **Ears:** Cropped or uncropped rose. **Tail**: Short, whip.

Genetic Diseases or Conditions: Heart murmurs, congenital heart disease, autoimmune thyroiditis, cerebellar ataxia, hip and elbow dysplasia, umbilical and inguinal hernias, tumors, epilepsy, and allergies.

FIGURE 7.04 Bedlington Terrier

FIGURE 7.05 Bedlington Terriers, Blue and Sandy

Bedlington Terrier

Origins: In Northumberland, England, gypsies, nailers (blacksmiths who only made nails), and miners took a fancy to the breed as they were great vermin catchers with lots of "pluck." Bedlingtons can trace their lineage all the way back to 1782. It is thought that the Whippet was used to add speed and endurance to the breed, and it certainly added to its unique body shape. The AKC recognized this unique breed in 1886.

Characteristics: The Bedlington never shies away from any kind of vermin and can more than hold his own at drawing a badger out of its hole or at ratting. This dog was once rather feisty, but when nobility and women took a fancy to them, they were bred towards being calmer. Having said that, they are not trouble seekers but will fight to the death if attacked or cornered. Also, many today will still fill the role of hunting dog by tracking, pointing, retrieving, swimming, and "going to ground," which is where they will go into a den or burrow after vermin. They are amazing swimmers, some saying they rival even Newfoundlands. They have a great heart and are very loyal to their masters. Bedlingtons are suited to apartment living if given long daily walks to keep them fit both physically and mentally. Training a

Bedlington will take someone with patience and persistence as the dog can be quite stubborn. They should be socialized early and trained to "drop it" if they have something in their mouths. If they catch something, the "terrier shake" will most likely put an end to it, be it once alive or your favorite shoe.

QUICK FACTS 7.02 Bedlington Terrier

Quick Facts	Height	Weight	Coat	Features
Life Span: 12–15 years	**Males:** 16–17 inches 41–43 centimeters. **Females:** 15–16 inches 38–41 centimeters	18–23 pounds 8–10 kilo-grams	**Colors:** Blue, sandy, liver, with or without tan markings. **Type:** A mixture of hard and soft hair standing out from the skin, crisp but not wiry, hair continuously grows in length.	**Head:** Mesoce-phalic, level, or scissor bite. **Ears:** Filbert with rounded tips. **Tail:** Whip.

Genetic Diseases or Conditions: Copper toxicosis which affects their livers, retinal dysplasia, renal cortical hypoplasia, hepatitis, and thyroid disease.

Border Terriers

Origins: Bred in the Cheviot Hills between England and Scotland, the Border Terrier is one of the oldest terriers in Great Britain. They were recognized by the AKC in 1930.

FIGURE 7.06 Border Terrier

FIGURE 7.07 Border Terrier, Jumping Over a Rail

Characteristics: The Border Terrier's ancestors were farm dogs that helped out by keeping the vermin under control. They excelled at hunting rats and mice as well as fox, martens, badgers, and otters. They would go to ground to flush their prey from the den and then kill it. The Border Terrier of today isn't quite as driven and is an excellent companion dog. They excel at agility and obedience trials and still have some hunting and tracking abilities. They can be escape artists as they can climb over fences of any height; some have been known to use corners and wedge walk their way over high walls. They are also great diggers. Exercise and a firm but fair hand in obedience training will make the Border Terrier a delightfully active and well-mannered pet.

QUICK FACTS 7.03 Border Terrier

Quick Facts	Height	Weight	Coat	Features
Life Span: 13–14 years	**Males:** 13–16 inches 33–41 centimeters **Females:** 11–14 inches 28–36 centimeters	**Males:** 13–16 pounds 6–7 kilograms **Females:** 11–14 pounds 5–6 kilograms	**Colors:** Red, tan and grizzle, or tan and blue, wheaten. **Type:** Double coat, undercoat is short and dense; outercoat is very wiry and broken.	**Head:** Mesochepalic, scissor bite. **Ears**: V-shaped. **Tail**: Gay, short.

Genetic Diseases and Conditions: Epilepsy, Canine Epileptiod Cramping Syndrome, Perthes disease, heart defects, cataracts, and progressive retinal atrophy.

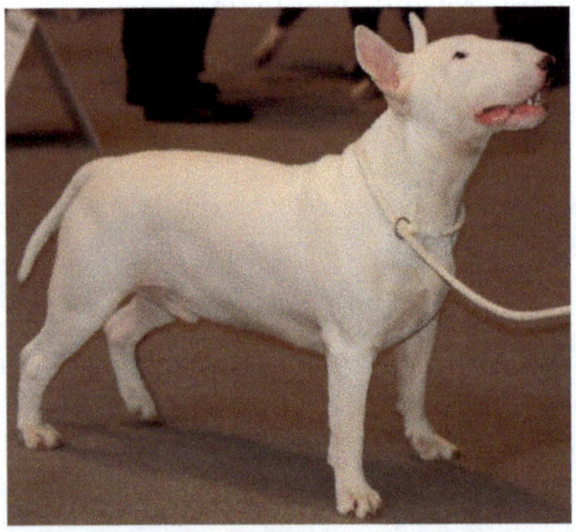

FIGURE 7.08 Bull Terrier, White

Bull Terrier

Two color varieties: White and Colored, two sizes: Standard and Miniature

Origins: James Hicks of England is credited with developing this breed by crossing Bulldogs with White English Terriers and later Dalmatians. These dogs were all white and were referred to as the White Cavaliers. They were later crossed with Staffordshire Bull Terriers to develop the colored variety. They were welcomed into the AKC in 1885. The ancestors of the Bull Terrier would often be fairly small in the 15–16 pound range with some even smaller. They weren't accepted into the AKC until 1991 as it was felt that they didn't breed true to size and form until then.

FIGURE 7.09 Bull Terrier, Colored

Characteristics: At first, they were developed to be rough and tough to fight bulls or each other in the pit. Modern Bull Terriers are more noted to be "cavaliers" or lovers rather than fighters. They will defend themselves but will not provoke fights with other dogs. This does come after intense socialization and training from a firm and consistent master. They are also known to be a bit stubborn sometimes, so patience is also a virtue as an owner. They are exceedingly friendly dogs and thrive on and will demand affection. They need that companionship and attention; otherwise, they have been known to become unstable or develop anxieties. Care must be taken to teach children not to tease the Bull Terrier as they don't tolerate that treatment. They are good watchdogs and will defend their territory and people.

QUICK FACTS 7.04 Bull Terrier

Quick Facts	Height	Weight	Coat	Features
Life Span: 10–12 years	**Standard:** 20–24 inches 51–61 centimeters **Miniature:** 10–14 inches 25–33 centimeters	**Standard:** 45–80 pounds 20–36 kilograms **Miniature:** 24–33 pounds 11–15 kilograms	**Colors: White** - colored markings on the head permissible. **Colored** - any color other than white, or any color with white markings, brindle preferred. **Type:** Short, flat, harsh.	**Head:** Mesocephalic, oval-shaped, somewhat Roman nose, level, or scissor bite. **Ears**: Candle flame. **Tail**: Saber, carried horizontally.

Genetic Diseases or Conditions: Eclampsia, umbilical and inguinal hernias, "stud tail" caused by excessive secretion of the sebaceous glands, epilepsy, compulsive behaviors, kidney disease, heart defects, zinc deficiency, and obesity. Severe skin disease in puppies is usually a lethal condition. Cesarean births are usually required.

FIGURE 7.10 Bull Terriers, Standard-White and Miniature-Colored

Cairn Terrier

Origins: In order to tell the story of the Cairn Terrier, we have to talk about the West Highland White and Scottish Terrier as well. They came to us from the Isle of Skye and the Highlands of Scotland. Scotland had grouped all the terriers together until 1873. At that time they were divided into the Dandi Dinmont and Skye Terrier groups. The Cairn, West Highland White, and Scottish Terriers were placed with the Skyes. All three terriers could be found in the same litter. White puppies were highly valued, but any other pup with white was considered flawed. The pups that were not white or long bodied were developed into the Cairn Terriers. England set the standard for them in 1923. The AKC recognized them in 1913.

FIGURE 7.11 Cairn Terriers, Sandy and Grey

FIGURE 7.12 Cairn Terrier, Head

Characteristics: Cairn Terriers got their name from flushing foxes, otters, and vermin from cairns which are rock memorials or markers. As working terriers, they have a ton of courage for scrambling into the nooks and crannies of the cairns to make their prey bolt then bark to alert the farmer. Modern-day Cairns are very adaptable to either country or city living and have maintained much of their cheeky, sometimes exasperating character of old. Cairns are very affectionate and totally devoted to owners. To help them develop this devotion, owners must provide training that is firm and consistent but without harshness. Plenty of exercise and inclusion in the family activities will keep this spunky little dog mentally healthy as well.

QUICK FACTS 7.05 Cairn Terrier

Quick Facts	Height	Weight	Coat	Features
Life Span: 12–15 years	**Males**: 10–13 inches 25–33 centimeters **Females:** 9–12 inches 23–30 centimeters	**Males**: 14–18 pounds 6–8 kilograms **Females:** 13–17 pounds 6–8 kilograms	**Colors:** Any color except white. **Type:** Double coated, profuse, harsh weather-resistant outercoat and short, soft, close furry undercoat.	**Head:** Mesocephalic, scissor bite. **Ears**: Erect, small, pointed. **Tail**: Gay, short.

Genetic Diseases or Conditions: Krabbe-type disease (a demyelinating disease of the nerves), inhalant and flea allergies, entropion, hyperthyroidism, luxating patella, portosystemic shunt, soft-tissue sarcoma, progressive retinal atrophy, cryptorchism, epilepsy, Legg-Calve-Perthes disease, cataracts, glaucoma, Von Willebrand's disease, craniomandibular osteopathy, and obesity.

FIGURE 7.13 Fox Terrier, Smooth

Fox Terrier Smooth

Origins: An old English breed, the Smooth Fox Terrier was created by crossing the smooth-coated black-and-tan terrier, Bull terrier, Greyhound, and Beagle. For many years the Smooth and Wire Fox Terriers were bred together to produce the breed as we know it today. However, that is no longer necessary and is frowned upon. The AKC admitted this dog in 1985. From the Smooth Fox Terrier, the Toy Fox Terrier was developed utilizing the Italian Greyhound, Manchester Terrier, Miniature Pinscher, and Chihuahua. The Toy Fox Terrier was admitted to the AKC in 2003.

FIGURE 7.14 Fox Terrier, Smooth, Head

Characteristics: The Smooth Fox Terrier has a keen nose and for a time was shown in the sporting group as they are great hunters. They protected farms from foxes and rats, actually following them down into holes or into drain pipes. They are most happy when their "people" are close, and they are included in the activities of the family. They require intense socialization to be trustworthy around other dogs and children; however, cats and other small pets are never safe around a fox terrier. Their training must be firm and consistent without waver as they will quickly take over the household as leader. Once that happens, they will become quite difficult as they will fight to keep that standing. The Smooth Fox Terrier requires daily exercise and is most happy off leash in a well-fenced yard. They will play ball and romp with the children for as long as you could possibly want. They are very devoted and loyal to their families and are quite affectionate. They have an easy-to-care-for short coat that can be cleaned with a wet towel or dry shampoo.

QUICK FACTS 7.06 – Fox Terriers

Quick Facts	Height	Weight	Coat	Features
Life Span 13–15 years	**Smooth Fox Terrier:** 12–15.5 inches 30–39 centimeters **Wire Fox Terrier:** 14–15.5 inches 35–39 centimeters **Toy Fox Terrier:** 9–11 inches 23–28 centimeters	16–18 pounds 7–8 kilograms **Toy Fox Terrier:** 3.5–7 pounds 1.5–3 kilograms	**Colors:** Predominately white with brindle, red, or liver markings. **Type:** Smooth-flat coat is hard, dense, and abundant. Wire coat is tight, soft undercoat.	**Head:** Mesocephalic, level bite. **Ears:** Button. **Tail:** Gay, docked.

Genetic Diseases or Conditions: Deafness, goiter, Legg-Clave-Perthes disease, lens luxation, distichiasis, cataracts, mast cell tumors, colonic disease, and epilepsy.

Fox Terrier Wirehaired

FIGURE 7.15 Fox Terrier, Wirehaired

Origins: The Wire Fox Terrier is the old rough-coated black-and-tan working terrier of Wales, Derbyshire, and Durham terriers. They were admitted to the AKC in 1985 along with the Smooth Fox Terrier.

Characteristics: The Wire Fox Terrier is similar in nature to the Smooth Fox Terrier in that it is loving, affectionate, and very devoted to family. These dogs need to be properly socialized and introduced to all types of dogs in order to avoid fights. Both Fox Terriers are very high-energy dogs and have a habit of springing up and down from pure joy and excitement. They are very smart, and unless they are stimulated both physically and mentally, they can become quite stressed. They can also become very dominant if not given firm and consistent training. Excessive barking, bolting, jealousy, separation anxiety, destructiveness, and aggression towards their humans and other pets can occur if they feel they are the leaders in the household. Both Fox Terriers do well in agility and hunting. They are excellent watchdogs and are easily taught tricks. By keeping the Fox Terrier busy and engaged, you will have a happy lovable dog for many years.

FIGURE 7.16 Fox Terrier, Wirehaired, Head

Genetic Diseases and Conditions: Epilepsy, deafness, patellar and lens luxation, distichiasis, cataracts, Legg-Calve-Perthes syndrome, and mast cell tumors.

Rat Terrier

FIGURE 7.17 Rat Terrier

Origins: An American-bred dog, the Rat Terrier was a rabbit- and rat-hunting dog used during the Dust Bowl era. It was originally developed from the Smooth Fox Terrier and Manchester Terrier in Great Britain. When it came to the U.S., breeders crossed them with Beagles or Whippets to increase their bulk and speed, as well as trailing and hunting prowess. Later on, more Smooth Fox Terrier and Chihuahua was added to develop the different sizes. A breed standard was written by the AKC and put into effect in 2010.

FIGURE 7.18 Rat Terrier

Characteristics: Of all the terriers, the Rat Terrier is most easily trained and ready to obey commands. They are fully devoted to their families, are full of energy, and are a lot of fun. Even though they are a short-haired breed, they do enjoy the outdoors, and if given proper protection, they can stay outside for much of the time. They do, however, love to dig and can often escape as they do dig quite quickly. They love to be with their families and are up for any adventure including swimming, playing ball, or just snuggling up on the couch. There are three sizes of rat terrier, and with their short-hair coat, they are easy to maintain. These dogs should not be confused with the Manchester Terrier as they have a similar appearance.

QUICK FACTS 7.07 Rat Terrier

Quick Facts	Height	Weight	Coat	Features
Life Span:	**Standard:** 14–23 inches 35.5–58.5 centimeters **Mid-Sized:** 8–14 inches 20–35.5 centimeters **Toy:** 8 inches 20 centimeters	**Standard:** 12–35 pounds 5.5–16 kilograms **Mid-Sized:** 6–8 pounds 3–3.5 kilograms **Toy:** 4–6 pounds 2–3 kilograms	**Colors:** Pied color pattern, with or without tan markings. **Type:** Short, close, shiny.	**Head:** Mesocephalic, scissor, or level bite. **Ears:** Erect. **Tail:** Docked or natural.

Genetic Diseases and Conditions: A very hardy the breed seeing some patellar luxation, cardiac abnormalities, hip dysplasia, and Legg-Calve-Perthes syndrome.

Russell Terrier – Parson, Jack, or Russell Terrier

FIGURE 7.19 Russell Terrier, "Jack," Smooth Coated

Origins: Named for the respected huntsman Reverend John Russell, who wanted a terrier that would go to ground to bolt fox without the drive to kill said fox. He used the various terriers of the day to develop the breed. After Reverend Russell's death, some of his bloodlines were crossed with the Welsh Corgi, Bull Terriers, and Chihuahuas, creating dogs ranging in size and working ability. What resulted was the Jack Russell Terrier, which is still considered a "working" terrier; the Parson Russell Terrier, which is considered a "show" terrier; and the Russell Terrier, the toy-sized terrier. They were essentially the same dog with differences in height and weight. The Jack Russell Breeders Association petitioned the AKC for recognition in 2001. AKC decided to recognize the breed as the Parson Russell Terrier, and the association changed its name to the Parson Russell Terrier Association of America.

FIGURE 7.20 Russell Terrier, "Parson," Rough coated

Characteristics: The "Russell" is a merry, fearless, willful breed of dog that requires a great deal of mental stimulation and outdoor exercise. This breed is not for a novice dog owner as the dogs require a long-term commitment to firm, consistent, and creative training and daily activity often extending well into their mid-teens. They will continue to test their limits throughout their life span, including who is in charge of the household. They are not good for apartments as they enjoy barking and if left for long periods of time will dig and/or destroy things. They must be socialized continuously from an early age for them to be with other dogs, and even then they shouldn't be left unsupervised with other dogs, especially those bigger than themselves. Many a Russell has been injured or killed by fighting with other dogs. It is commonly known that they will harass, injure, or kill other small pets, such as cats, birds, rabbits, mice, rats, etc., simply due to their strong natural hunting instinct. Raising a puppy with a cat does not guarantee the cat's life-long safety. They ***will not put up with even unintended abuse from a child.*** This must be carefully considered, particularly with children under the age of six.

QUICK FACTS 7.08 Russell Terriers

Quick Facts	Height	Weight	Coat	Features
Life Span: 15–17 years	**Jacks:** 10–15 inches 25–38 centimeters **Parsons:** 12–14 inches 30–36 centimeters	**Jacks:** 14–18 pounds 6–8 kilograms **Parsons:** 13–17 pounds 6–7.7 kilograms	**Colors:** Predominantly white, with or without black or tan markings, tri-color. **Type:** Broken or smooth.	**Head:** Mesocephalic, scissor bite. **Ears:** Button. **Tail:** Docked, gay.

Genetic Diseases or Condition: Ataxia, luxating patellas, arthritis, cataracts, glaucoma, ectopia lentis, Legg-Calve-Perthes syndrome, deafness, and Von Willebrand's disease.

Schnauzer – Miniature and Standard

FIGURE 7.21 Schnauzer, Miniature, Salt and Pepper

Origins: Germany brought us the Standard Schnauzer, which the Miniature and Giant Schnauzer were modeled after. The Standard Schnauzer was developed by crossing Poodles and Spitz-types with wirehaired Pinscher breeds. The Miniatures were developed by crossing Affenpinschers and Poodles with smaller Standard Schnauzers. The AKC recognized the Standard Schnauzer in 1904 and the Miniature in 1926.

FIGURE 7.22 Schnauzer, Standard, Black

Characteristics: By vocation the Schnauzer was a yard and stable dog acting as a guard dog and destroyer of vermin. The typical Schnauzer is a hardy, active, and intelligent dog that is fond of children. The Miniature is suitable for apartment living, and the Standard is a good house dog if it has a fenced-in yard. They get along well with children and strangers although their true love is the leader of the household. Schnauzers do need to be socialized, and, of course, obedience training is also a must to develop this breed's great personality.

They seldom roam too far from home; however, the Schnauzer is a noise maker. The bark is not a true bark as much as a drawn-out howl. Schnauzers of both sizes will still go after a rat, so keep them separate from pocket pets. They do well in competitive obedience trails and agility and are quick to pick up tricks. Their coats require professional grooming; otherwise, they look very un-kept.

QUICK FACTS 7.09 Schnauzer – Miniature and Standard

Quick Facts	Height	Weight	Coat	Features
Life Span: **Standard:** 12–13 years **Miniature:** 12–15 years	**Standard:** 18.5–19.5 inches 47–49 centimeters **Miniature:** 12–14 inches 30–35 centimeters	**Standard:** 32–34 pounds 14.5–15 kilograms **Miniature:** 11–18 pounds 5–8 kilograms	**Colors:** Salt and pepper, black and silver, and solid black. **Type:** Double coat, hard, wiry outer coat and soft undercoat.	**Head:** Mesocephalic, scissor bite. **Ears:** Cropped or small, V-shaped. **Tail:** Docked.

FIGURE 7.23 Schnauzer, Standard, Cropped and Uncropped ears, Salt and Pepper

FIGURE 7.24 Scottish Terrier, Black

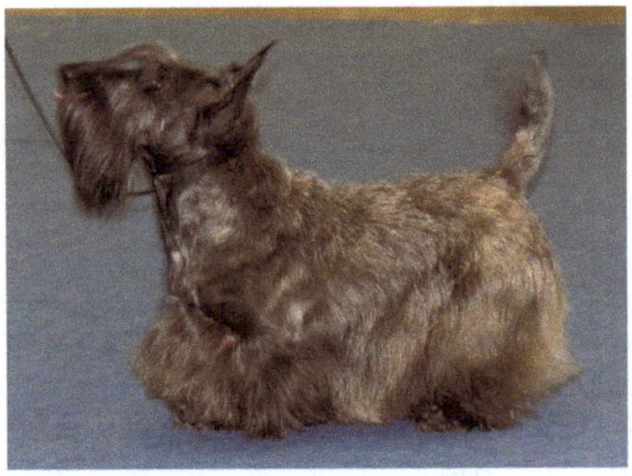

FIGURE 7.25 Scottish Terrier, Brindle

Genetic Diseases or Conditions: Hyperlipidemia, pancreatitis, diabetes, uroliths, eye problems, urinary tract problems, calcium oxalate stones, liver disease, diabetes, dermatitis, food and inhalant allergies epilepsy, Von Willebrand's disease, obesity, esophageal achalasia, and hip dysplasia.

Scottish Terrier

Origins: In order to tell the story of the Scottish Terrier, we have to talk about the Cairn and West Highland White Terrier as well. They came to us from the Isle of Skye and the Highlands of Scotland. Scotland had grouped all the terriers together until 1873. At that time they divided the terriers into the Dandi Dinmont and Skye Terrier groups. The Cairn, West Highland White, and Scottish Terriers were placed with the Skye Terrier group. All three terriers could be found in the same litter. Eventually the all-black, long-bodied pups were singled out and developed as a separate breed. The AKC recognized them in 1913.

Characteristics: The lovable Scottie is a great family dog as long as he knows his place in the family. Obedience training that the entire family participates in is the best way to develop that hierarchy. A Scottie will take over quickly if it feels the owner isn't the boss. This proud little dog likes to be in the thick of things and will be a very unhappy dog if left out. Their mode of showing that unhappiness is to dig. Their feelings do get hurt with harsh treatment or corrections which in turn can make them snappish, but consistent and fair discipline allows his character to shine and loyalty to blossom. Games like a tug-of-war or wrestling are also a no-no with the Scottie since he can get overwrought and bite. Scotties require professional grooming at least two to three times a year.

QUICK FACTS 7.10 Scottish Terrier

Quick Facts	Height	Weight	Coat	Features
Life Span: 12–15 years	10–11 inches 25–28 centimeters	19–23 pounds 8.5–10.5 kilograms	**Colors:** Black, wheaten, brindle. **Type:** Broken, outer coat hard and wiry, dense undercoat.	**Head:** Mesocephalic, square muzzle, scissor, or level bite. **Ears**: Prick, rather pointed. **Tail**: Gay, 7" long.

Genetic Diseases or Conditions: "Scottie Cramp," hematomas of the ears, uroliths, allergies, Von Willebrand's disease, dystocia, and mast cell tumors.

Soft Coated Wheaten Terrier

FIGURE 7.26 Soft Coated Wheaten Terrier

Origins: The Soft Coated Wheaten Terrier has been known in Ireland for over 200 years; however, its prior history is unknown. They were the peasant's all-purpose farm dog, acting as a watchdog, vermin and other small-game hunter, and sheep herder. It is recorded to have been used to develop the Kerry Blue Terrier. "Wheatens" were recognized by the AKC in 1973 and have been growing in numbers ever since.

Characteristics: Wheatens are a medium-sized, sensitive, responsive, playful dog. They require a positive training technique that is fair and consistent. Their good humor is often shown by jumping straight up to give their owner or friends a lick in the face. It is important to train them to not do this as it is a dominance behavior. They do bond to their owners, and with socialization to children, other dogs, and cats they will be good citizens. They do bark to alert but are not ones to go on after the humans have paid attention. Young Wheatens grow into their hair coats and start off fairly dark, fading as they mature which takes them about three years. They have a single coat of soft, silky, and slightly wavy hair covering the entire body. Brushing the hair will make them fuzzy, so it is recommended that a comb be used daily to keep them from matting.

FIGURE 7.27 Soft Coated Wheaten Terrier

QUICK FACTS 7.11 Soft Coated Wheaten Terrier

Quick Facts	Height	Weight	Coat	Features
Life Span: 12–15 years	**Males:** 18–20 inches 46–51 centimeters **Females:** 17–19 inches 43–48 centimeters	**Males:** 35–45 pounds 16–20 kilograms **Females:** 30–40 pounds 14–18 kilograms	**Colors:** Shades of wheaten. **Type:** Single-layered, abundant, silky, soft, and wavy; hair grows in length continuously.	**Head:** Mesocephalic, scissor, or level bite. **Ears:** Button. **Tail:** Docked, gay, or natural.

Genetic Diseases or Conditions: Protein-losing nephropathy, protein-losing enteropathy, dermatitis, heart disease, renal dysplasia, Addison's disease, hip dysplasia, cataracts in older dogs.

FIGURE 7.28 Staffordshire Bull Terrier

FIGURE 7.29 Staffordshire Bull Terrier

Staffordshire Bull Terrier

Origins: A Bulldog along with either the English Terrier or the black-and-tan terrier was used to develop the Staffordshire Terrier in England in the 19th century. They were admitted to the AKC in 1975.

Characteristics: The ancestor of this breed was used to bait bulls and bears and then later in the pits against other dogs. This was outlawed in the 1800s, and the breed dropped out of favor. However, a group of enthusiasts decided to save this breed and worked to make it into the companion dog we know today. The modern dog is still courageous and highly intelligent, as were its past ancestors, but now it is affectionate toward its family, in particular children, and friends. They are not wary of strangers, treating most as friends until proven otherwise. "Staffs" do require good socialization and obedience training to keep them stable. Owners need to protect their dogs from themselves. They are so curious that they may walk off the end of a dock, and they are not good swimmers because of their heavy front ends. They are chewers and need to have lots of appropriate chew toys while they are growing up, and many can be difficult to house train.

QUICK FACTS 7.12 Staffordshire Bull Terrier

Quick Facts	Height	Weight	Coat	Features
Life Span: 10–16 years	**Males:** 14–16 inches 36–41 centimeters **Females:** 13–15 inches 33–38 centimeters	**Males:** 25–38 pounds 11–17 kilograms **Females:** 23–35 pounds 10–16 kilograms	**Colors:** Red, fawn, white, black or blue with white, or brindle with or without white. **Type:** Smooth, short, and close.	**Head:** Mesocephalic, scissor bite. **Ears:** Rose or half-pricked, uncropped. **Tail:** Undocked, saber.

Genetic Diseases or Conditions: Bilateral cataract, distichiasis, persistent hyperplastic primary vitreous, mast cell tumors, gas issues, and cleft palate or lip. May require cesareans at birth.

Welsh Terriers

Origins: This breed was originally known as the Old English Terrier or Black-and-Tan Wirehaired Terrier. It hasn't changed much over the years and played a role in the development of a number of other terrier breeds. It was admitted to the AKC in 1888.

FIGURE 7.30 Welsh Terrier

FIGURE 7.31 Welsh Terrier

Characteristics: A smaller version of the Airedale Terrier with many of the same attributes as its larger cousin, the Welsh Terrier is loving, patient with children, smart, has boundless energy, and is always on guard. They train easily if the owner is calm and authoritative without being harsh. This terrier, like all the others, will take over a household if not shown his place in the family hierarchy. Because of their intelligence, training needs to be varied, but expectations need to be consistent. Socialization is always good, and if done early, the Welsh will be a well-balanced, healthy minded dog. They do well as a city or country dog as long as they have plenty of exercise and mental stimulation. Welsh terriers love to dig, swim, and play fetch. They can be barkers, so it is important to teach them to stop on command.

QUICK FACTS 7.13 Welsh Terrier

Quick Facts	Height	Weight	Coat	Features
Life Span: 10–12 years	14–15 inches 35–38 centimeters	20–22 pounds 9–10 kilograms	**Colors:** Tan with a black saddle. **Type:** Hard, wiry, dense.	**Head:** Mesocephalic, scissor, or level bite. **Ears:** Button. **Tail:** Docked, gay.

Genetic Diseases or Conditions: Glaucoma, allergies, epilepsy, and hypothyroidism.

West Highland White Terrier

Origins: In order to tell the story of the West Highland White Terrier, we have to talk about the Cairn and Scottish Terrier as well. They came to us from the Isle of Skye and the Highlands of Scotland. Scotland had grouped all the terriers together until 1873. At that time they were divided into the Dandi Dinmont and Skye Terrier groups. The Cairn, West Highland White, and Scottish Terriers were placed with the Skye Terrier group. All three terriers could be found in the same litter. White puppies were highly valued, but any other pup with white was considered flawed. Soon the white terriers were bred to produce all-white terriers becoming the West Highland White Terrier in 1909. The AKC recognized them in 1908.

FIGURE 7.32 West Highland White Terrier

Characteristics: The sociable, spunky, smart "Westy" is a great family dog for both city and country living. They are devoted to their families and will alert you when they perceive danger. They will need to be trained to stop barking when given the command. They are easy to train with a firm, fair, and consistent hand. Because they are so smart, it is important to change up the training routine so that they do not get bored and think of ways to get out of the routine. An occasional Westy can be intolerant of children and other dogs, so it is important to socialize them early and reinforce it throughout their lifetime. This is a tough little dog and often masks or hides pain from an owner. It is important to carefully watch them for any signs of pain or illness as when

FIGURE 7.33 West Highland White Terrier

they finally give signs, it may be too late to do anything for them. When you have a Westy, you have a friend for life.

QUICK FACTS 7.14 – West Highland White Terrier

Quick Facts	Height	Weight	Coat	Features
Life Span: 12–16 years	**Males**: 10–12 inches 25–30 centimeters **Females:** 9–11 inches 23–28 centimeters	**Males**: 15–22 pounds 7–10 kilograms **Females:** 13–16 pounds 6–7 kilograms	**Colors:** White, with black pigment on lips, eye rims, pads of feet, nails, and skin. **Type:** Double coated, straight, hard and stiff outer coat, soft, dense undercoat.	**Head:** Mesocephalic, scissor, or level bite. **Ears:** Erect, sharp point. **Tail:** Naturally short, gay.

Genetic Diseases or Conditions: Inguinal hernia, liver disease, copper toxicosis, Legg-Calve-Perthes syndrome, craniomandibular osteopathy, variations in blood serum proteins, deafness, cherry eye, atopic dermatitis, globoid cell leukodystrophy, idiopathic pulmonary fibrosis, Addison's disease, and White dog shaker syndrome.

Toy Group

The Toy Group is a diverse group in look and personalities but not in size. Potential owners have their pick of terriers, spaniels, hounds, and working breeds all shrunk down to apartment size. All of the following dogs were developed to be "lap" dogs filling the role of ultimate companion. They can be lovely, fun little companions or obnoxious, horrible devil dogs. What could cause this wide range of behaviors? Humans! Because they are so tiny, it is assumed that they should be treated differently than other dogs. They are not made to behave, they are coddled, and they are allowed to get away with behaviors that would not be tolerated in any other dog. A dog is a dog no matter how big or small, and dogs need to be shown their place in the hierarchy of the family. To achieve this they must have obedience training and socialization in order for them to be a well-balanced pet. Another issue with this group of dogs is they believe they are mastiffs and often pick fights with all dogs no matter what size. Unfortunately, they often come out the loser when this happens. There are two things owners must do to protect their dogs. The first is to socialize them to all situations and other dogs from an early age and throughout their lifetimes. The second is to use a harness verses a collar while on a walk; that way if an altercation looks like it's about to happen, a quick lift will bring the dog into the owner's arms, and the dog won't slip its collar or choke.

FIGURE 8.00 Brussels Griffon, Rough Coat

Brussels Griffon

Origins: In Belgium a small Pinscher-Schnauzer-like dog was commonly found keeping rodents down in the stables. It is unsure what the origins were, but it is sure that the Affenpinscher, Pug, King Charles, and Ruby Spaniels were crossed to create the Brussels Griffon. They were recognized by the AKC in 1910.

Characteristics: Brussels Griffons are charming little dogs that can steal your heart with a look. They have very human-looking facial expressions and are often referred to as a "little monkey" or "ape face." They have great personalities and will charm their owners into getting what they want. They can be finicky eaters, and if the owner should succumb to offering table scraps, they become even pickier. They are affectionate towards their owners, choosing one in particular, and will get along well with other dogs and cats. They do well around children as long as the children don't get too rough with them. Some are difficult to house train and will be barkers. They are sensitive dogs, so training must be consistent and fair. The rough-coated variety needs to be brushed twice weekly and professionally trimmed every three to four months.

FIGURE 8.01 Brussels Griffon, Smooth Coat (Petit Brabancon)

QUICK FACTS 8.00 Brussels Griffon

Quick Facts	Height	Weight	Coat	Features
Life Span: 12–15 years	7–8 inches 18–20 centimeters	6–12 pounds 2.5–5.5 kilograms	**Colors:** Red, belge (black and reddish brown mixture), black and tan, black. **Type:** Two types: rough coat is wiry and dense, smooth coat is short and flat.	**Head:** Brachycephalic, overshot bite. **Ears:** Cropped or button. **Tail:** Docked, gay.

Genetic Diseases or Conditions: Syringomyelia, Chiari-like malformations, hypothyroidism, cleft palate, cataracts, lens luxations, glaucoma, corneal lacerations, heat stroke, and luxated patella. Births often require cesarean sections.

Cavalier King Charles Spaniel

Origins: The Cavalier King Charles Spaniel of today is named after King Charles II because of his love for this little dog. They are descended from the English Toy Spaniels and were quite common as ladies' pets. However, paintings of King Charles II show the small spaniel with a flat head, high-set ears, almond eyes, and a rather pointed nose. An American by the name of Roswell Eldridge is credited with saving this breed. He traveled to England looking to purchase a breeding pair and found only the English Toy Spaniel type. He offered a reward for the "old-style" of toy spaniel. This started the interest in reviving the "Cavalier" breed. They were accepted into the AKC in 1996 and have grown in popularity ever since.

FIGURE 8.02 Cavalier King Charles Spaniel, Blenheim

FIGURE 8.03 Cavalier King Charles Spaniel, Color Trio

Characteristics: Cavaliers are affectionate dogs with people of all ages. They are content to snuggle on the couch or go for long walks. They are intelligent and train easily, even learning tricks quickly as they wish to please their owners. They are social dogs but will benefit from being exposed to all situations and other animals to cement that friendliness. They can adapt to apartment or country living, and if in the country, they will enjoy hunting as they retain their spaniel instincts. They have a fairly fine coat that does require brushing to keep mats from forming. The Cavalier comes in four coat colors; Blenheim, Ruby, Tri-colored, and Black and Tans. The term Blenheim is in honor of John Churchill, 1st Duke of Marlborough, and Blenheim Palace. These dogs do well in obedience and agility trials and as therapy dogs. Their hair coat is kept looking nice with biweekly brushing which will also keep the shedding under control.

QUICK FACTS 8.01 Cavalier King Charles Spaniel

Quick Facts	Height	Weight	Coat	Features
Life Span: 9–14 years	12–13 inches 30–33 centimeters	13–18 pounds 6–8 kilograms	**Colors: Blenheim:** pearly white body with chestnut markings. **Tricolor:** pearly white body with jet black and chestnut markings. **Ruby:** rich red body. **Black and Tan:** jet black body with tan markings. **Type:** Moderate length, silky, slightly wavy feathering on ears, chest, legs, and tail.	**Head:** Mesocephalic, scissor bite. **Ears:** Pendent, short. **Tail:** Plume, well feathers, docked is optional.

Genetic Diseases or Conditions: Cardiovascular problems, thrombocytopenia, macrothrombocytosis, episodic falling, cleft palette, luxated patella, cryptorchidism, hip dysplasia, keratoconjunctivitis sicca, corneal dystrophy, cherry eye, hereditary cataract, epilepsy, skin irritation, deafness, otitis, and syringomyelia.

FIGURE 8.04 Chihuahua, Short Hair

Chihuahua

Origins: Pottery and toys dating back to 100 AD found across Central America depict small dogs resembling the Chihuahua of today. Progenitors of these dogs were bred and raised for companions, religious rites, and food in the Chihuahua state of Mexico. The AKC recognized the Chihuahua in 1904.

Characteristics: Chihuahuas have a reputation for being shy, timid, overprotective of their owners, and bad with children because they tend to bite. This is true if the Chihuahua has been allowed to become that way without socialization and obedience training. In reality, they are sweet, charming, loyal dogs that are great fun and great friends if taught to be such. Socialization of the Chihuahua must begin at an early age. Exposure to everything and everybody will show them that the world isn't a big bad place with danger lurking in every corner. They are more comfortable around dogs their own size and often become snappish to warn overly inquisitive dogs off. This often happens even with well socialized dogs, so supervised meetings are important so this doesn't escalate into a full-on fight. Training will be successful with food treats, positive reinforcement, and consistency. Some can be difficult to house train; get them on a schedule and utilize the crate training process. "Chis" tend to select one member of the household as their favorite but are willing to snuggle up to others if that person isn't available. They love to burrow and create dens to rest in and can be found under the covers. Daily walks are important for this breed to maintain a healthy mind and body. In cold climates it is important to provide a doggy sweater and booties to protect them from frostbite during their walks. They are average shedders, and the long-coated variety will require weekly brushings.

FIGURE 8.05 Chihuahua, Long Hair

QUICK FACTS 8.02 Chihuahua

Quick Facts	Height	Weight	Coat	Features
Life Span: 12- 20 years	6–10 inches 15–25 centimeters	4–6 pounds 1.8–2.7 kilograms	**Colors:** Any color. **Type:** Smooth - glossy, soft, close. Long - soft texture, either flat or slightly curly, with undercoat preferred.	**Head:** Mesocephalic, "apple" dome skull, level, or scissor bite. **Ears:** Erect. **Tail:** Sickle carried up or out or in a loop over the back, longhair tail is plumed.

Genetic Diseases or Conditions: Puppies suffer from hypoglycemia when stressed, molleras, epilepsy, hydrocephalus, blood disorders, eye disease, skin infections, demodectic mange, heart problems, collapsed trachea, deafness, Legg-Calve-Perthes disease, obesity, periodontal disease, and cryptorchidism.

Chinese Crested

FIGURE 8.06 Chinese Crested, Hairless

Origins: Hairless mutations are shown to have originated in Africa, Mexico, and China. Hairless dogs captivated the attention and fancy of ancient dog lovers and soon could be found in most important ports. Chinese dogs accompanied their masters on trade vessels throughout the world, leaving mementos of their visits behind, and true to the dominant genetics, hairlessness would show up in subsequent litters. The famed

Gypsy Rose Lee became a fan of the breed which helped to promote them in the U.S. The AKC recognized the Chinese Crested in 1991.

FIGURE 8.07 Chinese Crested, Powderpuff

Characteristics: The Chinese Crested comes hairless or with a full coat of hair (referred to as Powderpuffs). Both varieties will be born in the same litter, and both carry one gene for hairlessness and one gene for a full coat. Puppies with two hairless genes don't survive. The hairless variety does have varying amounts of a single layer of long hair on its head, neck, from the ankles down, and the entire tail. The Powderpuff has a double-hair coat with a long, straight, silky outer layer and a soft, dense undercoat. Both varieties are loving, sweet, lively, playful, cuddly companions and are good with people and other pets in general. Chinese Cresteds do crave constant attention, with owners claiming they smile and give hugs. They become very attached to their owners and can have trouble adjusting to a new one, but with patience they will come around. They are affectionate with children that do not get too rough with them, not because they are fearful but because their skin can be injured easily if played with too roughly. "Cresteds" are entertaining and actually like to "sing" with their people, although they are not barkers. They can be shy around strangers, so it is important to socialize a puppy to loud noises, situations, adults, and children when they are young. They do not like to be left in kennels and can develop behaviors to escape from them such as climbing and digging holes. Overall, this dog is intelligent, learning tricks easily and excelling in agility and obedience trials, and even a rousing game of fetch. The hairless variety requires skin care, especially sunscreen and lotion for dry skin. A "Puff" will require daily brushing and bathing to prevent mats.

QUICK FACTS 8.03 Chinese Crested

Quick Facts	Height	Weight	Coat	Features
Life Span: 10–12 years	11–13 inches 28–33 centimeters	5–12 pounds 2–5 kilograms	**Colors:** Any color hair - skin pink to black, mahogany, blue, lavender, or copper, solid or spotted; powderpuff, any color combination. **Type:** Hairless – varying amounts of soft, silky hair, skin is smooth and soft. Powderpuff - double-coat, long, straight, moderately dense, silky outercoat, silky undercoat.	**Head:** Mesocephalic, scissor, or level bite. **Ears:** Erect, large. **Tail:** Scimitar, carried merrily.

Genetic Diseases or Conditions: Blackheads, sunburn, allergies to wool and lanolin, primitive mouth (hairless variety - teeth extend forward with missing teeth, and they tend to lose teeth as young adults and should not be given bones to chew), obesity, lens luxation, glaucoma, progressive retinal atrophy, luxating patella, Legg-Calve-Perthes disease, cerebellar ataxia, and autoimmune disease.

Havanese

Origins: Originating in Cuba as the "Blanquito del la Habana" (little white dogs of Havana), these dogs became popular in Europe after being brought in on trade ships. As the culture shifted in Cuba, they adapted as well, becoming great family dogs, watchdogs, and herders of the family's poultry. They were brought to the U.S. during the Cuban revolution. The AKC recognized them in 1996.

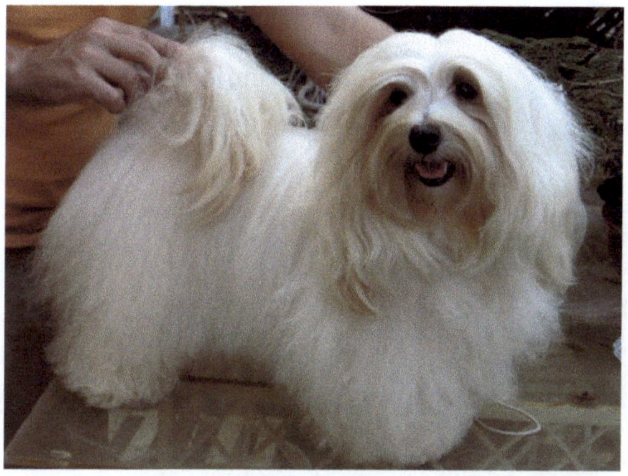

FIGURE 8.08 Havanese

Characteristics: This smart, easily-trained dog is good with any type of family from a single individual to a large family. They get very attached to their families and will often pick one person in particular but will be happy with others until reunited with their favorite. They have a great need to be close to their person and will snuggle up whenever possible as well as follow their person around the house. They can be trained to use a litterbox, so they are often ideal for apartment living. Because of their intelligence they will learn tricks and love to entertain their people.

FIGURE 8.09 Havanese

Obedience training does have to be gentle and consistent as they are sensitive dogs and will shut down if treated harshly. They are not barkers but will bark to alert you then quiet right down as they see their owner welcome the guest. The Havanese has long, silky hair that can be kept clipped short by a professional groomer. If the hair is allowed to grow, twice-weekly brushing is a must or it can be made to go into cords; however, the Havenese are not a naturally corded dog. The short-hair variety of Havanese, called a Shavanese, has a short coat that has rather natural shaggy look.

QUICK FACTS 8.04 Havanese

Quick Facts	Height	Weight	Coat	Features
Life Span: 14–15 years	8–11 inches 20–28 centimeters	7–13 pounds 3–6 kilograms	**Colors:** Any color and pattern. **Type:** Double-coated but being long, silky, soft, and abundant.	**Head:** Mesocephalic, scissor bite. **Ears:** Pendent, covered with hair. **Tail:** Ring.

Genetic Diseases or Conditions: Luxating patella, cataracts, retinal dysplasia, progressive retinal atrophy, juvenile heritable cataracts, chondrodyplasia, Legg-Calve-Perthes disease, unilateral and bilateral deafness, sebaceous adenitis, and epilepsy.

FIGURE 8.10 Italian Greyhound

Italian Greyhound

Origins: Miniature greyhounds appear in the ancient art of the Mediterranean countries dating back 2,000 years. During the sixteenth century this dog was so highly sought by Italians and Southern Europeans that it acquired the name Italian Greyhound. The breed made its way throughout Eurasia, steadily gaining in popularity, especially in England in the seventeenth century. The AKC registered the first one in 1886.

Characteristics: The Italian Greyhound is a true miniature greyhound that resulted from selective breeding. They are adaptable to city and country life. They are mild mannered, calm, and non-confrontational, but that can mean that they are timid and shy by nature. It is very important to socialize them to all situations, people, and other dogs to ensure that they are suitable for families of school-age or older children. They rank very low on the barking scale and so are good for apartment living. They do enjoy getting out and running free, although this should be in well-fenced-in yards or dog parks. They do quite well with agility trials as long as they are not distracted by a bunny. Their sight-hound heritage often takes over, and they will give chase. "IGs" suffer horribly in colder climates and will require coats or sweaters and boots if living in the colder areas of the country. They should never be left outdoors as they do not have the hair coat necessary to survive low temperatures. House training can be difficult, and patience and consistency is a key to success. They do not respond to harsh treatment nor will they listen to anyone they feel is weaker than themselves.

FIGURE 8.11 Italian Greyhound

QUICK FACTS 8.05 Italian Greyhound

Quick Facts	Height	Weight	Coat	Features
Life Span: 12–15 years	12–15 inches 30–38 centimeters	6–10 pounds 3–5 kilograms	**Colors:** Any color and markings. **Type:** Short, soft to the touch.	**Head:** Mesocephalic, scissor bite. **Ears:** Rose. **Tail:** Whip.

Genetic Diseases or Conditions: Monorchidism, cataracts, progressive retinal atrophy, Legg-Calve-Perthes disease, liver shunt, Von Willebrand's disease, autoimmune hemolytic anemia, Blue Doberman syndrome/color mutant alopecia, epilepsy, leg fractures, broken tails, luxating patellas, periodontal disease, and hypothyroidism.

Japanese Chin

Origins: Ancient dogs of China and Japan as depicted in their art and recorded histories, they were especially prized by the nobility of Japan and were a favorite gift to diplomats and foreign dignitaries. The Japanese Spaniel was accepted into the AKC in 1888, and its name was changed to the Japanese Chin in 1977.

FIGURE 8.12 Japanese Chin

Characteristics: Bred to be a good companion that is bright and alert, entertaining with tricks and singing, they are likened to cats in that they use their

FIGURE 8.13 Japanese Chin, Head

paws to wash their faces and love to perch up high. "Chins" are social by nature, and with socialization they will have steady behavior in different situations. This makes them great therapy and visitation dogs. They are sensitive dogs with definite likes and dislikes, but rarely, if ever, do they forget friend or foe. They are average shedders and require brushing daily to keep their fine hair mat free. Their eyes and ears should be cleaned as needed to help to keep them free of debris that can cause infections.

QUICK FACTS 8.06 Japanese Chin

Quick Facts	Height	Weight	Coat	Features
Life Span: 12–14 years	7–11 inches 18–28 centimeters	4–15 pounds 2–7 kilograms	**Colors:** Black and white or red and white. **Type:** Long, straight, silky, hair with long feathering on thighs and tail.	**Head:** Brachycephalic, undershot bite. **Ears**: V-shaped, covered with long hair. **Tail**: Ring, covered with long hair.

Genetic Diseases or Conditions: Monorchidism or cryptorchidism, achondroplasia (shortened limbs), depressed nasal bridge, and shortened maxilla.

Maltese

FIGURE 8.14 Maltese

Origins: An ancient dog from the island of Malta, records going back to 3500 BC show the Maltese much as it is today. About 40 AD they were described in poems and depicted in art and pottery. Ladies of the aristocracy cherished these little dogs. An actual spaniel and not a terrier, as they were called in prior years, they were admitted to the AKC in 1888.

Characteristics: Healthy, spirited, and very tiny describes the Maltese. As little as he is, the Maltese should be without fear, very trustworthy and affectionate, gentle mannered, yet lively and playful as well as vigorous, and great with children. Care must be taken when playing with children as an inadvertent belly flop can seriously injure a Maltese. Care must also be taken when they are puppies that they should not be allowed to play so hard that they develop hypoglycemia. If this should happen, a bit of honey or corn syrup under the tongue will put them to rights, and a trip to the veterinarian is in order to make sure there is nothing seriously wrong. The Maltese has a thick, silky, single-layered hair coat that can grow to lengths of 10–12 inches. Most families choose to keep their Maltese in a "puppy cut," which is rather short, so that a weekly brushing is all that is necessary. If showing the Maltese, its hair coat must be kept long and mat free. This will require many hours of grooming on the part of the owner. These great little dogs are a joy and will do well in obedience trials, agility, or as an awesome therapy and visitation dog.

FIGURE 8.15 Maltese

QUICK FACTS 8.07 Maltese

Quick Facts	Height	Weight	Coat	Features
Life Span: 12–14 years	8–10 inches 21–25 centimeters	6.5–9 pounds 3–4 kilograms	**Colors:** White, light tan, or lemon on ears. **Type:** Single layer of silky long hair, grows in length continuously.	**Head:** Mesocephalic. **Ears:** Hanging, covered with hair. **Tail:** Ring, covered with hair.

Genetic Diseases or Conditions: Hydrocephalus, meningoencephalitis, liver shunt, open fontanel, hypoglycemia, patellar luxation, reverse sneezing, cryptorchism, deafness, blindness, periodontal disease, and "white-shaker dog syndrome" (a generalized tremor which responds to corticosteroids). First-litter whelping is often difficult.

Miniature Pinscher

FIGURE 8.16 Miniature Pinscher, Black and Tan

Origins: It is thought that the Miniature Pinscher was developed by the crossing of Dachshund, Italian Greyhound, and German Pinschers.

They were used as ratters in stables and still have the attitude that no vermin is its friend. They were first recognized by the AKC in 1925 as a terrier but reclassified as a Toy in 1930 and underwent an official name change from Pinscher (Toy) to Miniature Pinscher in 1972.

FIGURE 8.17 Miniature Pinscher, Red and Stag Red

Characteristics: This dog is not for a first-time dog owner. Their self-confidence, intelligence, and assertiveness is a match for even an experienced dog trainer. They require loads of patience, firmness, consistency, and creativity to convince them to go along with the rules. They are high-energy dogs and escape artists that should not be allowed to run unsupervised unless the area has a cover. They can be difficult to housetrain, requiring patience and consistency to get it through to them where it is they should eliminate. Heaps of praise and food treats should accompany the celebration when they "get it." They are protective of their owners, so socialization to many people and situations will help them to not become over protective. Small children are not a good match with the "Min Pin" as they do not tolerate teasing or inadvertent tugs on ears, tails, or skin. They will learn tricks easily and are quite the clowns. They have a very short hair coat, so protection from cold weather is a must.

QUICK FACTS 8.08 Miniature Pinscher

Quick Facts	Height	Weight	Coat	Features
Life Span: 13–15 years	**Males:** 10–12.5 inches 25–31 centimeters **Females:** 10–11 inches 25–28 centimeters	**Males:** 8–10 pounds 4–5 kilograms **Females:** 8–9 pounds 3.6–4 kilograms	**Colors:** Red, "stag" red, black with "kiss marks." **Type:** Short, smooth, glossy, close.	**Head:** Mesocephalic, scissor bite. **Ears:** Cropped or natural erect. **Tail:** Docked or whip.

Genetic Diseases or Conditions: Relatively healthy, with some epilepsy, cervical disc disease, Legg-Calve-Perthes disease, and patellar luxation.

Papillon

FIGURE 8.18 Papillon, Erect-Eared

Origins: The French started this breed, but it was the Italians and Spanish that developed the breed extensively. It was used as trade goods being sent over the Alps on the backs of pack mules. This dwarf spaniel got its name from the French word for butterfly. Ladies of nobility in the 16th century were often painted with these elegant little dogs by their sides. The AKC recognized both the erect- and drop-eared varieties in 1915.

Characteristics: For a little dog they are hardy and can adapt to a country home or apartment living as long as regular exercise is in the mix. They are fairly easy to train, although some have difficulty with housetraining. They can become possessive and jealous of their special person, so socialization is a must. Without socialization they become fairly obnoxious around other people, dogs, and especially children. "Paps" like to be clean and will often groom themselves like cats. They are very athletic dogs, doing great in agility and obedience trials. If socialized well, they are great therapy and visitation dogs. The two varieties are often born in the same litter, and the difference is simply how their ears look. A drop-eared Pap will have a large ear that sets low like a Rose ear on their heads but stands straight out. The erect-eared dog has the same large ear that stands straight out from the head with a bit lower set than most erect ears on other breeds. Both will have a flow of long hair that takes about three years to grow in completely. The hair on the ears must be a color other than white and cover the entire ear that continues down around the eyes. This looks like a "butterfly" with extended wings and thus their name.

FIGURE 8.19 Papillon, Drop-Eared

QUICK FACTS 8.09 Papillon

Quick Facts	Height	Weight	Coat	Features
Life Span: 12–15 years	8–11 inches 20 -28 centimeters	**Males:** 8–10 pounds 3.6–4.5 kilograms **Females:** 7–9 pounds 3–4 kilograms	**Colors:** Parti-colored, colored hair has to cover both ears and eyes without interruption. **Type:** Single layers, long, abundant, straight, silky, profuse.	**Head:** Mesocephalic, scissor bite. **Ears**: Erect or drop, large and rounded tips, covered with feathers. **Tail:** Ring, covered with hair.

Genetic Diseases or Conditions: Eclampsia, deafness, seizures, periodontal disease, progressive retinal atrophy, allergies, luxating patellas fontanel, and intervertebral disc disease.

Pekingese

FIGURE 8.20 Pekingese

Origins: The Pekingese was an ancient dog of China, where it was held sacred and kept only by the imperial family. Peasants would have to bow to the dogs, and if they were caught stealing one, it was punishable by death. The "little lion dog" was given as gifts to visiting dignitaries, and it was once thought that there were more Imperial Palace Pekingese in the U.S. than anywhere else in the world. The AKC recognized these dogs in 1906.

Characteristics: Dubbed the "small lion dog" of China, the Pekingese's purpose in life is to provide affection and loyalty to his owners. They have many lion-like traits including boldness, high self-esteem, and unfortunately combativeness. They are not dainty or delicate in any way, showing lion-like independence and courage. They absolutely require discipline and socialization as youngsters, or they will take advantage of their owners and rule the household. They work very hard to keep this position and can become quite naughty. This can be reversed, but it is very difficult, and it takes a very patient owner to show the Pekingese how to be a dog in its rightful place. Never give "Pekes" table scraps as they have been known to refuse to go back to dog food. They tend to gain weight easily, so it is important to measure their food and be firm with not giving them "treats." Pekes can do well in obedience trials but because of their brachycephalic facial features do not have the respiratory capacity for agility or other strenuous activities. If socialized well, they are great as therapy or visitation dogs. Daily walks around the neighborhood will be just right for their needs. They have a very thick hair coat that requires daily brushing. Some owners will have their dogs clipped in a puppy cut to reduce the amount of daily care.

FIGURE 8.21 Pekingese

QUICK FACTS 8.10 Pekingese

Quick Facts	Height	Weight	Coat	Features
Life Span: 10–18 years	6–9 inches 30–45 centimeters	8–10 pounds 3.6–4.5 kilograms "Sleeve" Pekes are under six pounds.	**Colors:** Red, fawn, black, black and tan, brindle, white, and parti. Black masks and spectacles are desirable. **Type:** Double coated, thick, wooly undercoat, long guard hairs for outer coat.	**Head:** Brachycephalic, undershot bite. **Ears:** Filberts, carried low, with long feathering. **Tail:** Ring, covered with long hair.

Genetic Diseases or Conditions: Cleft palates and lips, obesity, elongated soft palate, bronchial dysplasia, heart failure, skin allergies, uroliths, heat stroke, intervertebral discs disease, eye problems including ophthalmorrhexis, progressive retinal atrophy, and cryptorchidism. They often require cesareans at birth.

FIGURE 8.22 Pomeranian

Pomeranian

Origins: It is thought that the Pomeranian is the descendant of the Spitz-like sled dogs of Iceland and Lapland. It is named after the area between Poland and Germany called Pomerania, not because it originated there, but that was where it was most likely bred down in size. The royalty of England developed a fondness for them and made further changes to the breed. The AKC recognized them in 1888.

FIGURE 8.23 Pomeranian

Characteristics: "Poms" are tiny dogs with big personalities. They are very loyal, eager to learn, and have a vivacious spirit that makes them amazing pets and companions. Obedience training is easy, especially with food treats or toys used as rewards. They are fairly sensitive dogs, so training techniques should be gentle yet firm and consistent. They are fairly easy to housebreak and will litterbox train for apartment living. They can become clingy to their owners and must be socialized early on in their lives. Exposure to children and other dogs is important. They do great in agility and obedience trials and are wonderful visitation and therapy dogs, again, if socialized well. They can be picky eaters and are barkers that have to be taught to stop. They have a very thick undercoat and can overheat easily. They do need to be brushed two or three times a week. The undercoat will shed at least twice a year.

QUICK FACTS 8.11 Pomeranian

Quick Facts	Height	Weight	Coat	Features
Life Span: 15–16 years	7–12 inches 18–30 centimeters	3–7 pounds 1–3 kilograms	**Colors:** Any solid color or parti-colored. **Type:** Double coated, short, soft, thick undercoat, longer, course guard hairs for the outer coat, a profuse frill around the neck and pantaloons on the back legs.	**Head:** Mesocephalic, scissor bite. **Ears**: Erect. **Tail**: Ring, covered in long fur.

FIGURE 8.24 Pug

FIGURE 8.25 Pug

Genetic Diseases or Conditions: Hydrocephalus, patellar luxations, cryptorchidism, hypoglycemia, tracheal collapse, patent ductus arteriosus, periodontal disease, skin diseases, Cushing's syndrome, and hypothyroidism. Long labor is normal; many have unnecessary cesareans.

Pug

Origins: It is commonly agreed that the Pug came from China and was imported by the Dutch in the 16th century. The European nobility adopted it quickly, and it soon became a favorite to all. The Pug made its way to the U.S. in the 19th century with the AKC recognizing it in 1885.

Characteristics: The Pug is many things in a little package. Willful but tractable, happy but docile, these dogs become bored easily but are highly trainable, spirited, and loyal. They are sensitive dogs that require fair, patient, calm, firm, and varied training. They need to be shown their place in the family, and when they are socialized well they do great around children. They can roughhouse with the children as they are sturdily built. Pugs make good watchdogs and will be easily trained to stop barking when told. With their brachycephalic head shape some issues develop, especially if the weather is hot and humid and if they are allowed to become overweight. Obesity is a huge issue with Pugs, so measuring their food and taking daily walks or long romps at the dog park are important to maintain the Pug's health. Their head shape also causes them to snore and wheeze, which can be annoying to the uninitiated. Pugs are great at obedience, agility (take care during warm weather), and as visitation and therapy dogs. They have a short hair coat, so maintenance is minimal.

QUICK FACTS 8.12 Pug

Quick Facts	Height	Weight	Coat	Features
Life Span: 12–15 years	**Males:** 12–14 inches 30–36 centimeters **Females:** 10–12 inches 25–30 centimeters	**Males:** 13–20 pounds 6–9 kilograms **Females:** 13–18 pounds 6–8 kilograms	**Colors:** Fawn, black, black muzzle. **Type:** Short, smooth, glossy, and soft.	**Head:** Brachycephalic, undershot bite. **Ears**: Rose and button. **Tail**: Curled.

Genetic Diseases or Conditions: Cleft lips and palates, elongated palates, slipped stifles, intertrigo, intolerance to heat, allergies, hemivertebra, mast cell tumors, encephalitis, eye injuries, entropion, keratitis, cherry eye, urolithiasis, demodectic mange, obesity leading to arthritis, hip dysplasia, patellar luxation. and heart disease. Birth by cesarean section is often necessary.

Shih Tzu (Sheet-zoo)

FIGURE 8.26 Shih Tzu

Origins: These dogs were brought from Tibet to the imperial court of China where they were the favored pets of royalty. They were almost destroyed completely during the revolution because of their association with royalty, but some were saved and sent to Europe where they were preserved. They were recognized by the AKC in 1969.

Characteristics: Their nickname is the "Chrysanthemum-faced dog" because of the way the hair on their faces grows in all different directions. Shih Tzus

FIGURE 8.27 Shih Tzu

are very active, lively, and alert little dogs with a big-dog attitude. They love to romp and play, although they will become a bit impatient with children if played with too roughly. Unlike the Lhasa Apso, they are not barkers and are content to be close to their human companions. Socialization and obedience training is important from a young age. They generally do well with other dogs and will play if taken to the dog park. They do snore and wheeze a bit as they breathe. The hair is very long requiring daily brushing or professional grooming to keep the coat short.

QUICK FACTS 8.13 Shih Tzu

Quick Facts	Height	Weight	Coat	Features
Life Span: 12–14 years	9–11 inches 22–28 centimeters	9–16 pounds 4–7 kilograms	**Colors:** Any color, nose and eye rims black, hair. **Type:** Double, undercoat is woolly, the long outer coat is dense, slightly wavy, grows in length continuously.	**Head:** Brachycephalic, undershot bite. **Ears**: Hanging, covered with hair. **Tail**: Ring, covered with hair.

FIGURE 8.28 Silky Terrier

Genetic Diseases or Conditions: Renal cortical hypoplasia leading to uremia and hyperparathyroidism, achondroplasia, liver shunt, inguinal hernia, periodontal disease, cataracts, cherry eye, cleft lip and palate, patellar luxation, inhalant allergies, disc disease, otitis, obesity, and hypertrophic pyloric gastrophy.

Silky Terrier

Origins: Developed in Australia during the 1800s, the Silky Terrier was created by crossing Yorkshire Terriers with

Australian Terriers. In the early times the Silky, Yorkshire, and Australian Terriers could be found in the same litter. The Silky types were bred to each other, and eventually the breed standard was fixed. They were recognized by the AKC in 1959.

Characteristics: Although a toy in size, the "Silky" has a true terrier personality, and they are of sufficient substance to hunt and kill vermin. They have a quick manner that is friendly and responsive to their owners. They will go on a walk, snuggle on the couch, or romp around the yard or dog park. They will alert you to strangers and need to be trained to stop barking when told. Owners need to be calm, consistent, and firm to keep these little guys in line. "Silkys" travel well, and their social nature makes them friends with whomever they meet. It never hurts to reinforce their social attitudes by exposing them to situations and other people at a young age. They have a long hair coat that requires daily brushing, and their topknot is parted down the middle.

FIGURE 8.29 Silky Terrier

QUICK FACTS 8.14 Silky Terrier

Quick Facts	Height	Weight	Coat	Features
Life Span: 12–15 years	9–10 inches 23–25 centimeters	8–11 pounds 4–5 kilograms	**Colors:** Blue (silver, pigeon, or slate) and tan, hair grows in length continuously. **Type:** Single layer, long, glossy and soft. Topknot on head.	**Head:** Mesocephalic, scissor bite. **Ears**: Erect. **Tail**: Docked, carried gay.

Genetic Diseases or Conditions: Yeast dermatitis, allergies, Cushing's disease, intervertebral disc disease, elbow dysplasia, tracheal collapse, cryptorchidism, hydrocephalus, epilepsy, patella luxation, Legg-Calve-Perthes disease, cataracts, and diabetes mellitus.

FIGURE 8.30 Yorkshire Terrier

Yorkshire Terrier

Origins: In the cloth mills and mines of Lancashire and Yorkshire counties, England, the Yorkshire Terrier was developed by possibly crossing the Scotch Terriers (Skye or Dandie Dinmont Terrier types) with the Maltese. They were recognized by the AKC in 1885.

FIGURE 8.31 Yorkshire Terrier

Characteristics: When not restrained this itty bitty dog will play hard, so hard, in fact, it can become hypoglycemic. This is usually more of an issue with puppies, but even some adults will be affected. A little corn syrup or honey under the tongue will help, and a trip to the veterinarian to make sure the dog is OK is in order. The "Yorkie" believes it is a mastiff and will often be the aggressor against large dogs. Many of these altercations end up in tragedy at the worst with them being killed or at best with just a bit of surgical repair. Owners need to take care of these little dogs when around other dogs so that these terrible accidents don't occur. Even though Yorkies are little, they are mighty fun, loving, and sensitive dogs that need gentle, firm, and consistent training. Daily walks are a must, even with these little dogs, as it keeps them mentally and physically fit. Their super-long coat requires daily combing and brushing and will tangle quite easily, so many adults are under strict surveillance to avoid tangles that can occur when at play. The average owner will have a professional groomer give the dog a puppy cut which is more manageable.

QUICK FACTS 8.15 – Yorkshire Terrier

Quick Facts	Height	Weight	Coat	Features
Life Span: 12–15 years	7–9 inches 17–23 centimeters	5–7 pounds 2–3 kilograms	**Colors:** Born black and tan, fade to steel-blue and tan. **Type:** Single layer, long, silky, fine, hair grows in length continuously.	**Head:** Mesocephalic, scissor, or level bite. **Ears:** Erect, small. **Tail:** Docked.

Genetic Diseases or Conditions: Eye infections, periodontal disease, fractures, cesarean sections, eviscerated puppies, inguinal hernias, hypoglycemia often misdiagnosed as epilepsy, patella luxation, lameness, diarrhea, vomiting, hydrocephalus, calcium oxalate stones, urolithiasis, fatty liver syndrome, liver shunt, and cryptorchidism.

CHAPTER 2

Felines

Breeds

The domestic cat can be traced back to the Neolithic period about 9,500 years ago. They share a common ancestry with wildcats that are classified as ***Felis silvestris*** and so are capable of crossbreeding. This crafty feline realized early on that humans were good for a meal and a warm place to live. However, they still can live quite well without us as evidenced by the millions of feral cats found around the world.

QUICK FACTS 9.00 Cat Nomenclature

Genus Species	Intact Male	Intact Female	Young	Group	Partuition
Felis catus	Tom	Queen	Kitten	Growl or stare	Queening

A purebred cat is one that has a known ancestry and contains only the same breed as sire (males) and dams (females). A cat of mixed or unknown origins is called a "domestic" cat usually with a description of hair length added for identification, for example, domestic shorthair, domestic medium hair, or domestic long hair. The length of hair, body type, and origins are often used to differentiate between the different breeds. There are four body types: cobby, standard, oriental or foreign, and giants. The cobby body is one that has a round head and a compact, deep-chested body with thick-boned short legs and a short tail. The oriental or foreign body is on the opposite side of the scale. The head is usually triangular, with a long slender body, legs, and tail. The standard body is in between the cobby and oriental, and the head can be either round or triangular. The giant body is usually one that is long with heavily boned legs and long, thick tails. The table below will be used throughout this chapter as a point of reference.

QUICK FACTS 9.01 Cat Body Types

Body Type	Breeds
Oriental: long thin body, legs, and tail. Triangular shaped head, with oval shaped eyes and large ears.  Figure 9.00: Oriental body	Balinese Burmese Rex – Cornish and Devon LaPerm Oriental Shorthair Oriental Longhair Siamese
Standard: slender to muscular body, legs, and tail of moderate length. Head can be a wedge or round shape, with almond, round, or oval-shaped eyes, ears in proportion to head.  Figure 9.01: Standard body	Abyssinian American and European Shorthairs Bengal Burmese Egyptian Mau Havana Brown Japanese Bobtail Korat Ocicat Russian Blue Sphynx Tonkinese Turkish Angora
Cobby: body length is short, rounder, and often rotund. Legs and tail tends to be shorter, and head and eyes are round.  Figure 9.02: Cobby body	American Bobtail British Shorthair Bombay Birman Exotic Shorthair Himalayan Manx Persian Scottish Fold Singapura

Giants: cats that are overall large with heavy, muscular bodies, head shape from wedge to rounded with ears in proportion to head, long legs, and tail, round to oval eyes.  Figure 9.03: Giant body	Maine Coon Cat Norwegian Forest Cat Ragdoll Turkish Van

Cats come in a variety of coat colors and types. The coats of cats are usually made up of long guard, spikey awn, and soft down hairs. Some cats will have long guard and down hairs like the Persian, while others will not have any guard hairs like the Rex. Cat colors can almost be limitless and are ruled by about a dozen different genes.

The natural color of cats is the brown tabby; they have a brown or gray body color with black or dark brown stripes. A blue tabby has a gray or buff body color with slate or lighter gray stripes. The red (ginger, orange, marmalade) tabbies have a cream-colored body with red (dark), orange (medium), or light (marmalade) stripes. A cream tabby has a paler cream body with sand or peach colored stripes. Silver tabbies have a white or cream body color (roots are white) and black stripes; they can be found in blue, cream, or red silver, depending on the color of the stripes. The best way to tell the color of a tabby is to look at the color of the stripes and the tip of the tail. The lighter-colored hair on all tabbies will have the agouti pattern making it hard to determine color.

Tabbies come in four different patterns. All will have "penciling" or thin lines of color on the face usually around the eyes, along the cheeks, and forehead. See Quick Facts Table 9.02, Figures 9.05–9.08.

- Tiger-striped or mackerel tabby has narrow stripes running parallel down its sides.
- Classic or blotched tabby has large blotches or swirling patterns on its sides like marble cake.
- Spotted tabby has spots, which are broken stripes long its sides. Some will have rosettes which are spots with black rings around them.
- Ticked or agouti tabby has 2–4 bands of color on each hair shaft, without stripes or spots on its body; however, they may have the tabby "M" or "scarab" on the forehead.

Shaded cats will have a chinchilla-colored body that is a light or white-colored coat. The guard hairs are tipped with a darker color. They can be various shades, and each has its own designation. If the color is 1/8 of the length of the hair shaft, they are chinchilla or shells; if the color is 1/4 the length, they are shaded; and if the color is 1/2 the length, they are smokes. Smoke-colored cats have a solid black or gray coat, except the roots are white. They are considered a solid version of the silver tabby. See Quick Facts 9.02 – Figures 9.08–9.09.

Solid or self-colored cats are of a single body and may or may not have a spot of white on the chest, called a locket, or white on the belly, called a button. This is usually not excessive; otherwise, the cat would be a bi-color. The colors seen are black, blue (slate, medium, or ash gray), and white. Solid-color cats have a recessive gene that suppresses the tabby pattern. Sometimes in certain lights a faint tabby pattern can be seen in the solid-colored cats. This gene is not effective on red or cream cats, so you will not see solid-colored reds or creams. Solid white cats carry a gene that suppresses all color, but sometimes this color isn't completely suppressed, and a gray smudge will be present on top of the head. See Quick Facts 9.02 – Figures 9.10 a and b.

Tortoiseshell, Patched Tabby, or Calico are randomly patched cats with different colors in stripes, patches, or both. They come with or without white markings which are usually on the underbelly, feet, and tip of tail. See Figure 9.11 a and b – 9.12.

Tortoiseshell or "Tortie" have random patches of red, black, and cream fur in either mixed or distinct areas of color. A blue and or dilute tortie has soft blue- and cream-colored fur. A patched tortie or "torbie" has brown and red patches of tabby stripes which can also be in the blue or dilute color with patches of blue and cream tabby. Tortoiseshell and white or blue-cream tortie and white cats will have small areas of white on the chest, abdomen (called buttons), or feet. A cat with more white on the body is called a Calico. Their patches tend to be larger and more distinct with both solid black and red tabby patches of various sizes scattered across the body. Calicos are also seen in the dilute blue-cream color as well. See Quick Facts 9.02 – Figure 9.13 a and b.

Color-Point cats have dark fur on the face, legs, and tail. This comes from a gene that affects the melatonin uptake, so the cooler parts of the cat have darker fur. This fur can be seal (dark brown), chocolate (light brown), blue (gray), lilac (pink gray), red (orange), lynx (tabby), and tortie color. The body color varies from light cream to fawn or beige to light brown or light gray depending on the point color. Kittens are all born white with the point color starting to come in at about four weeks of age. The body color will also darken as the cat ages. Some breeds of cats will have white marks on their heads, chest, feet (referred to as being "mitted"), and tail tip. See Quick Facts 9.02 – Figures 9.14.

Cats that are white with distinct patches of one color are designated by the color and white, for example, black and white or gray and white. Some folks will refer to a cat that is about half and half as a "bi-color." A cat that is mostly white with large patches of a single color is a "harlequin" color, and a "van" is a white bodied cat with color patches on the head and tail only. An informal term for a black and white cat that has black over the entire back and a white under belly (white on the face is often present as well) is a "tuxedo." See Quick Facts Table 9.02 – Figures 9.15–9.18.

QUICK FACTS Table 9.02 Coat Color

Figure 9.04: Tabby (Classic or Blotched): broad stripes on a lighter agouti ground. Colors can be brown, blue, red, cream, or silver. (Diluted colors chocolate and lilac are also seen.)

Figure 9.05: Tabby (Striped or Mackerel): narrower dark stripes on a lighter agouti ground. Same colors as Classic.

Figure 9.06: Tabby (Spotted): the lines are broken forming spots.

(a) Agouti or Ticking

(b) Abyssinian in the Agouti Pattern

Figure 9.07 (a): Tabby (Agouti): ruddy, reds (sorrel), blue, fawn (b): blue Abyssinian.

Figure 9.08: Smoke: black or gray fur with white roots, notice the faint tabby stripes.

Chinchilla

Figure 9.09: Shaded or colored silver, chinchilla, red or cameo, golden.

QUICK FACTS Table 9.02 Coat Color (Continued)

9.10: Solid in black (a), blue (b), chocolate brown, cinnamon, lavender, white
(a) Black

(b) Blue

(a) Tortoiseshell

(b) Tortie and white
9.11: Tortoiseshell (a) and Tortie and white (b): black, red, cream fur in mingled or distinct patches. Dilutes are blue and cream.

9.12: Torbie: brown and red patches in a tabby pattern.

(a) Calico

(b) Dilute Calico
9.13: Calico (a): white base coat with black and red. Dilute calico (b): the same colors only lighter.

(a) Seal point

(b) Lilac point
9.14: Color point: pale cream, fawn, or brown body with darker color on ears, nose, feet, and tail. Seal (a), blue (b), chocolate, lilac, red, tabby and tortie.

9.15: black and white or "tuxedo"

QUICK FACTS Table 9.02 Coat Color (Continued)

 9.16: orange and white tabby	 9.17: black and white or "bi-color"	 9.18: "van"

Cats that do not live with people are called feral cats and because of their high rate of reproduction contribute to the overpopulation of cats worldwide. Queens can have 2–3 litters per year producing up to 150 kittens in their life span of an average of ten years. This is one of the reasons there are so many cats in shelters, and millions are euthanized every year. Another common reason cats are surrendered or just abandoned is failure to use the litterbox. Cats have a natural instinct to cover their urine and feces, so using a litterbox is not a stretch for them. However, they do like to have a clean area in which to eliminate, and when humans do not take care of the litterbox promptly, some cats will stop using it. Another reason a cat may stop using the litterbox is if there are two cats, and one is dominant and guards the litterbox, not allowing the submissive one to use the box. That cat has no alternative but to eliminate elsewhere. For this reason in households with two cats it is recommended that there be a litterbox for each cat. If the household has more than two, it is recommended that there be one litterbox for every two cats. When a cat has stopped using the litterbox, it should be checked for a urinary infection or blockage. This is a common occurrence in cats, and it should be ruled out or confirmed and treated before doing any behavior modification to retrain the cat to using the litterbox.

Cats are ***seasonally polyestrous*** which means they come into ***heat*** many times throughout the year usually starting in early spring and ending late fall. Cats are induced ***ovulators*** which means they need to be bred before they can release their ovum or eggs. The gestation period for cats is between 64–67 days, and the size of the litter can be 5–7 on average. Kittens are generally weaned at 6–8 weeks and are usually ready for new homes by 12 weeks of age. Cats can be spayed or neutered (removal of reproduction organs) by six months of age which is usually right before the first estrous cycle. This is not only an effective means of controlling pet populations, but it changes many behavior issues as well.

Intact male cats tend to mark their territory by spraying urine, which is not appreciated inside the home. They also tend to become quite agitated when they smell a female in estrus. This can translate into inappropriate biting and scratching as well as constant "calling" or meowing. The females are really no better. As estrus approaches, they urinate more often, become quite insistent in being petted, assuming provocative poses, and will also "call" out in a rather mournful sound.

Cats communicate with both vocalizations and body language. They have a wide range of meows that can mean a number of things, and it is the astute owner that can differentiate between "I would like to be fed now" and "please let me outside." Of course, we all know that cats purr. The purring sound is often associated with contentment. However, in the veterinary world you will often hear a client say, "but he was purring" about a very sick or injured cat.

A cat's body language can range from a subtle tip of the tail flick (interest or starting to become annoyed) to a full-out, ears-laid-back, hissing and swiping of the paw indicating anger. It takes a bit of practice to determine when to reach into a cage or carrier and be confident that you will not withdraw a bloody hand. Obviously, the latter description is easy to read, but not all cats will be that forthcoming. Paying attention and using gentle persuasion is usually a good way to get a cat to cooperate.

The world is full of great cat breeds, all of which have their own personalities and traits. Some are loving and seek human attention; some are aloof and tolerate being around people; most like to play some or a lot. They all have hunting instincts and will gladly catch a mouse or bird to play with, offer to you as a present, or eat. The following information on the purebred cats is to help you as a professional learn about them and some of their traits. This is by no means a comprehensive list with only 32 of 44 breeds recognized by the Cat Fancier's Association (CFA) which are the most commonly seen or have some attribute that may assist you in a veterinary practice.

Breeds

Abyssinian

Origins: The origin of the Abyssinian, named for the supposed country of origin which is now Ethiopia, is shrouded in history. It is known that it is one of the oldest recognized breeds of cat, and it was introduced to England in the 1800s. They were imported to the U.S. in the early 1900s but didn't have a breed registry until 1930. World War II had a devastating effect on the breed as there were very few pure bred Abyssinians left in England. Luckily, the American breeders were well established, and the cat is well represented in world. The Cat Fancier's Association (CFA) recognized them in 1934.

FIGURE 10.00 Abyssinian, Ruddy

Characteristics: Playful, energetic, intelligent, and kid-friendly are all qualities you can expect from the Abyssinian. "Abys" are great climbers and prefer to be either overlooking the activities going on in the home or in the mix of things. They are inquisitive and will investigate sounds and sights that catch their eye. Abys have been known to learn tricks easily including fetch. They do not do well as solitary cats if their humans are gone a lot. They will welcome a second Aby or another cat that is equally inquisitive and energetic. Some will be more shy and timid than the usual outgoing majority. They are moderate shedders, and a quick weekly brushing will keep them in good shape. They will need to have their claws trimmed regularly and an appropriate scratching post available. They are slow to mature and have small litters.

QUICK FACTS 10.00 Abyssinian

Life span	Weight	Pattern & Color	Features
9–15 years	9–16 pounds 4–7.5 kilograms	Agouti pattern **Ruddy:** reddish brown body, black ticking **Red:** copper body, cinnamon, or red-chocolate ticking. **Blue:** beige body, gray ticking. **Fawn:** cream body, beige ticking. Chocolate, lilac, and silver (new colors not recognized by CFA). Longhair versions of the Abyssinian are called Somali.	Standard body with a lean, muscular build. **Head:** slightly rounded, moderately wedge shape. **Ears:** large moderately pointed. **Eyes:** almond shaped. **Tail:** thick base, long, and tapers.

FIGURE 10.01 Abyssinian Kittens, Ruddy, Fawn, and Blue (Left to Right)

Genetic Diseases and Conditions: Gingivitis, periodontal disease, renal amyloidosis, progressive retinal atrophy, hyperesthesia syndrome, patellar luxation, pyruvate kinase deficiency, and anemias. About 84% of them have a type A blood, and the rest are a combination of B and AB blood type.

American Bobtail

Origins: Urban legend would have us believe that this cat started out as the result of a domestic shorthair mating with a wild Bobcat. Actually the Bobtails are the result of a spontaneous genetic mutation that turns out to be a dominant gene and possibly related to the same gene as found in the Manx. The CFA recognized this breed in 2000.

Characteristics: American Bobtails are intelligent and playful with moderate energy. They bond with their owners, becoming almost "dog-like" in their devotion. They require attention and will ask for it by meowing or by hopping onto your lap. Once there they will trill, chirp, or purr with delight. They make good traveling companions if introduced to car rides early in life. Bobtails are easily trained to the leash and get along well with everybody. They are keenly sensitive and are used by mental health care professionals to help people in distress. Of course, in keeping with their name, their tails are a natural bob, and no two are the same length. It can be straight, curved, with or without knots or bumps, and can be 1–4 inches in length. Both the short and medium-long hair versions have a double hair coat which makes them look a bit shaggy in the medium-length version. American Bobtails are slow to mature taking up to two to three years to reach full maturity.

FIGURE 10.02 American Bobtail, Brown Tabby

QUICK FACTS 10.01 American Bobtail

Life span	Weight	Pattern & Color	Features
10–15 years	7–15 pounds 3–7 kilograms	All patterns and colors. **Short haired:** double coat of guard and down. **Medium-length haired:** double coat of guard and down.	**Cobby**: moderately long body, almost rectangular. **Head:** modified wedge, slight concave between nose and brow. **Eyes:** large, almond shape. **Ears:** medium size, round tips, small tufts. **Tail:** short, slightly curved, kinked, or bumps.

Genetic Diseases and Conditions: They tend to be very healthy cats with no known or reported genetic predispositions to health issues.

FIGURE 10.03 American Bobtail, Kitten

American Curl

Origins: In 1981 two stray kittens were found in California whose ears were curled back so the front of the pinna faced the back of the head. Both kittens were longhaired, and one of them, called "Shulamith," was bred and gave birth to six kittens with most of them having the curled ears. Breeders started selectively breeding for the curled ears in 1983, and they were recognized in 1992 by The International Cat Association (TICA), and in 1999 the CFA admitted both the short- and long-haired varieties.

FIGURE 10.04 American Curl, Silver Tabby

Characteristics: American Curls are great pets that love being with their people, children, and other pets. They are affectionate and attentive, often making cooing sounds when being petted. They are smart and curious and keep a kitten-like attitude throughout adulthood. They are born with straight ears that start to curl at about 3–5 days and take up to two and a half months to curl into a tight rosebud, after which the curl starts to relax until about four months of age where the curl is set into its permanent position. The cartilage becomes hard and stiff to the touch, and care must be taken when cleaning the ears so as not to damage the pinna. Cats that are to be shown must have a 90–180 degree arc in the curl. American Curls have both long- and short-haired coats that are soft, silky, and flat and require moderate grooming. They are very healthy cats because of their large gene pool with non-pedigreed cats. This gene was a dominant naturally occurring spontaneous mutation. However, a genetically homozygous Curl does not show any structural defects as associated with the Scottish Fold gene.

QUICK FACTS 10.02 American Curl

Life span	Weight	Pattern & Color	Features
14–15 years	9–11 pounds 4–5 kilograms	All patterns and colors. Medium long and short hair varieties both have a minimal undercoat.	Standard body that is very close to an oriental type. **Head:** modified wedge with a straight nose. **Eyes:** walnut oval. **Ears:** minimum of a 90-degree curl not to exceed a 180-degree curl. **Tail:** equal to body length and plumed.

Genetic Diseases and Conditions: Some reports of narrowed ear canals, but other than that they seem to be a healthy and hardy breed with no known genetic defects or predispositions to diseases.

American Shorthair

Origins: Jane Cathcart is credited with developing the American Shorthair. In 1900 she imported two male British Shorthairs and crossed them with "native" non-pedigreed cats. The non-pedigree cats were "working" cats that controlled vermin on ships and jumped ashore as the ships docked. In 1904 she registered the first American Shorthair with the Cat Fancier's Association. This breed went through several name changes over the years from Domestic Shorthair to American Domestic Shorthair to finally American Shorthair in 1966.

FIGURE 10.05 American Curl, Cream Tabby

Characteristics: American Shorthairs are friendly and affectionate, calm, yet playful, quiet voiced and competent hunters. This cat should look like an "everyday" cat, and their short-hair coats require minimal brushing. The American Shorthair takes two to three years to fully mature, and the males tend to be quite a bit larger than the females. They come in a variety of colors which are listed in the Quick Facts table below. The tabby color will have mascara or penciling on the face that is often in the shape of an M on the forehead. The blotched or mackerel tabby pattern will have what looks like the wings of a butterfly across the shoulders.

FIGURE 10.06 American Shorthair, Silver Blotched Tabby

QUICK FACTS 10.03 American Shorthair

Life span	Weight	Pattern & Color	Features
15–16 years, some up to 20	Males: 14 pounds 6.4 kilograms Females: 10 pounds 4.5 kilograms	Patterns: **Tabbies:** blotched or mackerel in silver, brown, red, blue, cream, cameo, patched. **Tortoiseshell and Calicos:** chinchilla, shaded silvers, smokes. Bi-colors All colors	Standard body with a muscular build. **Head:** oblong shape with full cheeks. **Eyes:** round. **Ears:** medium size with rounded tips. **Tail:** thick base tapering to blunt end and medium length.

FIGURE 10.07 American Shorthair, White

Genetic Diseases and Conditions: Hypertrophic cardiomyopathy. Blood type 100% A.

Balinese

Origins: The Balinese is a "longhaired" Siamese. The gene responsible for the medium length hair is recessive and has shown up in purebred Siamese litters for years. Helen Smith, an early breeder of the "longhaired Siamese," dubbed them Balinese after the graceful Balinese dancers. Sylvia Holland is credited with refining and establishing the breed standard in the 1960–1970s. She refused to recognize the reds, creams, lynx (tabby), and tortoiseshell points, as she felt they were most likely crossed with other cats. Those colored cats are recognized as Javanese by the CFA; in 2008, breeders merged the two, and now all "longhaired Siamese" are called Balinese.

FIGURE 10.08 Balinese, Chocolate Tortie Point

Characteristics: The Balinese are sociable and love to be in the thick of things with their people. They get lonely if owners are gone for long periods of time and may get into mischief. If this is the case, it is suggested that a second cat be added to the family. Balinese, like their Siamese cousins, are vocal and will "talk" to their owners, sometimes persistently. Luckily they reportedly have a quieter voice than the Siamese. Balinese are loving, playful, acrobatic, and extremely intelligent. They are a color-pointed cat with bright blue eyes, a medium-length coat and a plumed tail. You may still see them in the "old-style" with the apple-shaped head and standard body style. However, most have been bred to the oriental body style with the wedge-shaped head and almond-shaped eyes. A weekly brushing is all that is needed to keep the coat looking nice and to catch any of the shedding hair. Kittens are born without the points which start to come in by four weeks of age.

QUICK FACTS 10.04 Balinese

Life span	Weight	Pattern & Color	Features
19–20 years	7–9 pounds 3–4 kilograms	Color Point pattern. **Balinese colors:** seal, blue, chocolate, lilac, and red. Medium-length hair version, without an undercoat so fur lays flat with a silky feel. **Javanese are solid color points:** red (smoke and silver), cream, fawn, cinnamon, and lynx point.	Oriental body type, fine bones, and firm muscles. **Head:** long, tapering wedge in proportion to the body. **Eyes:** almond shaped. **Ears:** large, pointed, with a wide base. **Tail:** long, thin whip tail tapers to a fine point.

Genetic Diseases and Conditions: Hereditary liver amyloidosis, asthma, bronchial disease, aortic stenosis, dilated cardiomyopathy, megaesophagus, hyperesthesia syndrome, lymphoma, lysosomal storage disease, feline acromelanism, nystagmus, and progressive retinal atrophy.

FIGURE 10.09 Balinese, Lilac Point

Bengals

Origins: Bengals are hybrids of domestic cats crossed with the Asian Leopard Cat (ALC) of Malaya. Confirmed crosses were written about in 1934 and again in 1941. In the 1970s, Dr. William Centerwall used them to aid in his genetic studies in the natural resistance of the ALCs to feline leukemia. These cats were given to Jean Mill who is credited with refining the breed further and getting them accepted by the TICA in 1983. Greg and Elizabeth Kent were also successful breeders, and many of the modern-day Bengals carry their bloodlines. Most of the other cat fancier's organizations do not recognize hybrid cats.

FIGURE 10.10 Bengal

Characteristics: Only fourth generation cats from the original crossing possess a domestic-cat temperament. All generations before this require special handling and often special permits as they tend to retain the "wild" side of their ALC parentage. Bengals are considered to be highly intelligent and often dog-like in their behavior.

Playing fetch and walking on a leash are favorite pastimes with their owners. Bengals are beautiful cats that resemble their ALC parentage with large spots and/or rosettes, which are spots with black rings around them, along the back and sides of the body and a light or white belly. Stripes appear on the face and around the legs and tail. Their hair coat is easy to care for with a weekly brushing to capture the shedding hair being all that is necessary. There is a long-haired variant from the domestic cats used in the hybrid matings that seems to be a recessive gene.

QUICK FACTS 10.05 Bengal

Life span	Weight	Pattern & Color	Features
11–12 years	14–15 pounds 6.3–6.8 kilograms	Spotted Tabby with short, soft and silky, thick fur. Colors include brown, seal mink, black silver, seal silver lynx point.	Standard, longer, leaner but muscular. **Head:** rounded modified wedge, longer than wide. **Eyes:** oval to almost round **Ears:** medium sized, short. **Tail:** medium length with black tip.

FIGURE 10.11 Bengal, Rosette Pattern

Genetic Conditions and Diseases: Distal neuropathy, autosomal recessive disorder that results in an early onset of primary photoreceptor disorder leading to blindness, flat-chested kitty syndrome, hypertrophic cardiomyopathy, hip dysplasia, patellar luxation, and progressive retinal atrophy. They have a higher incidence of AB blood types.

Birman

Origins: The "Sacred Cat of Burma" has a history shrouded in legend, and though it cannot be disproved, it is a great story. In Burma, many of the territories were ruled by the temples. The temple cats were white and thought to be the souls of departed priests. Apparently, these temples were often raided, which happened as one monk was praying in the temple. The raiders killed the monk, and as he lay dying, his favorite cat jumped upon his chest and looked into his eyes. As the monk died, the cat's fur turned brown on the ears, face, legs, and tail, his body turned a golden cream color from the reflection of the idol in the temple, and his eyes turned blue, but his paws stayed white. The other monks arrived on the scene and declared that the cat had absorbed his master's sins until he was pure as indicated by the white on all four feet. As the monk was laid to rest, the other cats in the temple began to turn the same color, and so they became sacred to the monks of Burma. Whether the story is true or not, France is credited with the first cats recognized as Birmans in 1925. How they came to France is also shrouded in legend and mystery. They were almost lost during World War II with only two cats surviving. They had to be heavily outcrossed with Persians and Siamese to rebuild the breed. The early 1950s saw them breeding true once again, and so the breed was restored. England recognized them in 1966 and the CFA in 1967 with TICA following in 1979.

FIGURE 10.12 Birman, Tortie and Seal Point

Characteristics: Birmans have a quiet dignity about them that endears them to their owners. They are gentle and unfazed with new surroundings or disruptions. They are not known to be great hunters, although they like to stroll about the garden. Birmans are unhappy if kept in confinement, so boarding them is not an option for their mental health. Birmans have a cobby body that is much leaner than a Persian. They have a broad face with a distinctly Roman nose, blue eyes, and a plume tail. The U.S. breeders like a long mane around the neck, called a ruff, that stands out from their medium-long silky coat. They do not have an undercoat and so do not mat. Their coats are always color-pointed, and they should always have white feet. The "gloves" are symmetrical on each paw, with the white stopping at the articulation of toes to metacarpals on the front feet and metatarsals on the rear feet. The white should extend further up the back of the rear legs to the hock ending in a "V-shape," which is referred to as "laces." Kittens are born white and start to develop their color points in a week; it often takes a Birman two years to fully develop the coat color.

QUICK FACTS 10.06 Birman

Life span	Weight	Pattern & Color	Features
14–15 years	10–11 pounds 4.5–5 kilograms	Color-point pattern. Long to medium-long hair with a "golden mist" which is a faint golden beige cast on the back and sides. **Colors:** seal, chocolate, blue, lilac, red, cream, tabby, tortie, parti. **Gloves:** white front paws that end in an even line across the paw at or between the metacarpal pad. White on the back paws extend from the tips of the toes up into a "V" on the dorsal surface of the hock, called laces.	Cobby, leaner than usual. **Head:** rounded with flat spots between ears and in front of each ear, Roman nose. **Eyes:** almost round. **Ears:** medium length, rounded point. **Tail:** medium length in proportion.

FIGURE 10.13 Birman, Blue Point

Genetic Diseases and Conditions: Hypertrophic cardiomyopathy that can result in heart failure, paralysis of the hind legs due to clot embolization, congenital hypotrichosis, tail-tip necrosis, hemophilia B, corneal sequestration, corneal dermoid, congenital cataract, spongiform degeneration, distal polyneuropathy, shaking and trembling kittens, atypical granulation of neutrophils in about 46% of 100 cells, and high concentrations of urea and/or creatinine in the blood. Blood types are 82% A, with the rest being B and AB blood types.

Bombay

Origins: In 1953 Nikki Horner started to develop this hybrid by breeding Burmese cats to American Shorthair cats both with sable- or black-hair coats respectively. They were called Bombay in recognition of India's native Black Panther, and they received CFA championship status in 1976.

FIGURE 10.14 Bombay

Characteristics: It is said that the Bombay is a dog and a monkey rolled into a cat's body. They love to play and will invent games to amuse themselves and their family. This is a very social breed that loves to be in the company of others, although they will select a favorite from the family members. They will seek out attention, and once the perfect snuggle position is achieved they have a very distinct purr that emanates from them. It is said that they do get along well with other cats; however, there may be some issues as a hierarchy is established. The Bombay is a glossy black, short-haired cat with a muscular standard body and large, bright golden to copper eyes in a rounded head.

QUICK FACTS 10.07 Bombay

Life span	Weight	Pattern & Color	Features
15–20 years	6–10 pounds 2.7–4.5 kilograms	**Solid pattern:** shorthair – fine, satin-like, shimmering. Black	**Standard:** medium sized. **Head:** rounded, with slight indentation at bridge of nose. **Eyes:** rounded, set far apart. **Ears:** medium size, set well apart. **Tail:** straight, medium length.

Genetic Diseases and Conditions: Craniofacial abnormality occasionally but otherwise a fairly healthy breed of cat.

FIGURE 10.15 Bombay, Kitten

British Shorthair (and Longhair)

Origins: The British Shorthair is one of the most ancient identifiable cat breeds in the world dating back to the first century AD. The prevailing theory is when the Roman Empire invaded Britain, the invaders brought cats from Egypt with them. These cats bred with the local wildcat population, and over the centuries, because of their isolation on the island, they developed into a distinct breed. In the early nineteenth century, selective breeding began especially with the blue-gray variant that came to be called the "British Blue or English type." Late in the century they fell out of favor because of the importation of Persians and other longhaired varieties. Breeding stock became increasingly rare, so Persians were crossed with the British Shorthairs bringing back the breed and eventually developing the British Longhair.

In an attempt to save the British Shorthair the GCCF decided to accept only third-generation Persian/British Shorthair crosses. This lead to another shortage of breeding stock which lasted until well after World War II. After the war, breeders worked very hard to reestablish the breed and achieved formal recognition from both British and American organizations in the late 1970s.

FIGURE 10.16 British Longhair, Black Silver Shaded

Characteristics: British Shorthairs are described as easy-going, dignified, and loyal. They tolerate children and other pets well. They are a favorite among animal trainers because of their sweet nature. They tend to prefer living indoors and don't care to be too far off the ground. They also tend to be "easy keepers," so their food should be monitored to prevent obesity. They are one of the few breeds that demonstrate sexually dimorphic characteristics. Adult males will be noticeably heavier and will develop jowls. It usually takes a British Shorthair three to four years to fully mature. These robust cats have a very thick, short hair coat that is easy to keep neat with occasional brushings as there is no undercoat. The eyes can be any color, but in the favored "British Blue" they are a deep copper-orange.

FIGURE 10.17 British Shorthair, Tortie

QUICK FACTS 10.08 British Shorthair

Life span	Weight	Pattern & Color	Features
12–13 years	22–24 pounds 10–11 kilograms	All patterns and all colors allowed; the "British Blue" is the most popular. Coat is short and thick and not double coated.	Cobby body, muscular yet compact. **Head:** round, very large. **Eyes:** large and round. **Ears:** medium sized with broad base and rounded tips. **Tail:** medium length, thick base tapering slightly.

FIGURE 10.18 British Shorthair, Gray and White or "Bi-color"

Genetic Diseases and Conditions: Hypertrophic cardiomyopathy, polycystic kidney disease, and gingivitis. About 64% of British Shorthairs have type A blood and have a higher incidence of the AB blood type.

Burmese

Origins: Burmese were developed in America, and most Burmese can trace their roots back to one female cat called Wong Mau. Wong Mau may have been a descendant of cats from Thailand as similar cats are seen in historical illustrations. It is thought they were brought to Burma when Thailand was invaded in the 1800s. Wong Mau was brought from Burma to the U.S. by Dr. Joseph Cheesman Thompson in 1930 and bred to a seal point Siamese cat and then bred to her son to produce dark brown kittens. The Cat Fanciers Association first recognized the Burmese in 1936 but suspended the registrations in 1947 because Siamese were still being used by breeders to increase genetic diversity. The breed standard was reinstated in 1958 as this practice was stopped. American and British breeders developed their own standards, and those cat registries that do not recognize them as one breed will refer to the British type as the European Burmese.

Characteristics: The Burmese is strongly people-oriented with many dog-like characteristics, often forming strong bonds with their people. They want to be involved in any action going on in the home. They learn to play games such as fetch, love to go for walks on a leash, and if introduced early go for car rides. Kittens have a high energy level which lasts well into adulthood. Once they are adults, the males tend to chill out while the females tend to be in charge and rule the household. Both will tell you what they want with a quieter and sweeter tone than a Siamese. Burmese do not do well if left alone for long periods of time, so another cat or even a dog will be a good companion for this breed. If left to their own devices, they sometimes get into mischief including opening cupboards or refrigerator doors. Interactive toys will help keep this kitty from being bored. They do not do well outside. Their silky short hair coats are easy to maintain with weekly brushings. Although there is a wide range in body lengths, all Burmese will fool you regarding how much they weigh. It is said they are a "brick wrapped in silk." Burmese can have faintly visible color points but not any barring or spots.

FIGURE 10.19 Burmese, Champagne

QUICK FACTS 10.09 Burmese

Life span	Weight	Pattern & Color	Features
15–16 years	8.8–15.4 pounds 4–7 kilograms	Solid pattern in sable (dark brown), champagne (warm beige), blue (medium gray), and platinum (pale gray) both with fawn undertones. Coat is very short and satin-like.	Standard with a compact appearance. **Head:** rounded with a short muzzle. **Eyes:** rounded, set far apart. **Ears:** medium sized, broad base, rounded tips. **Tail:** medium length.

Genetic Diseases and Conditions: Lipemia of the aqueous humor, corneal dermoid, orofacial pain syndrome, congenital peripheral vestibular disease, Burmese head defect, hypokalemic polymyopathy, flat-chested kitten syndrome, kinked tail, elbow osteoarthritis, endocardial fibroelastosis, dilated cardiomyopathy, and diabetes mellitus. Blood type: 100% A. In 2008 University of California at Davis discovered that the Burmese has the second-lowest level of genetic diversity of all the breeds studied. This is of great concern to breeders as the Burmese does have a fairly impressive list of genetic diseases and conditions. For this reason, the Burmese breeder's council is allowing outcrossing using Bombay, Tonkinese, and Burmese-type cats imported from Southeast Asia to improve genetic diversity.

FIGURE 10.20 Burmese, Sable

Cornish Rex

Origins: In 1950 in Cornwall, England, a wavy-coated kitten was born to a flat-coated mother. This kitten known as Kallibunker became the foundation of the Cornish Rex breed. Tissue samples showed this to be a recessive gene. Kallibunker was bred back to his mother, and two kittens in the litter had curly coats. When one was imported to the U.S. it was bred to a Siamese, and the resulting cats further refined the slender, fine bones, long legs, narrow heads with huge ears, and a long whip-like tail that was seen on Kallibunker. Burmese, Siamese, and British Shorthaired cats were thrown into the mix to increase genetic diversity. The term Rex means "king" in Latin and was said to first be applied to curly haired rabbits that were entered into a show by King Albert I of Belgium. They were not a recognized breed, but the show officials didn't want to offend the king, so they wrote "Rex" beside their names, and that was the start of using the term Rex to describe curly or unusual fur.

FIGURE 10.21 Cornish Rex, Calico

Characteristics: The Cornish Rex may look delicate, but they are far from it. This people-oriented cat loves to play and will quickly learn fetch and catch and will even play "ball" by itself by tossing toys into the air and catching them with their paws. The Cornish Rex is extremely curious and will explore everywhere it can. Care must be taken as this can mean looking inside the dishwasher or clothes dryer, which doesn't usually end well if the cat is trapped inside. They are playful throughout their lifetime and love to run, jump, and play. The Cornish Rex hair coat is only the downy undercoat, making it soft and wavy. It is only about a centimeter long, and so it is understandable that the Cornish Rex is a heat seeker and will often insist on being under the covers with their people. Some believe because of its short coat, the cat must be hypoallergenic. This is not necessarily true as most people are allergic to the proteins in the cat's saliva that gets onto its coat from self-grooming. The profile of the Cornish Rex is very distinctive with a pronounced Roman nose, large bat-like ears on an egg-shaped head with high, hollow cheekbones. They have a curved back and are often called the "Greyhound" cat because of their arched back and thin long legs.

QUICK FACTS 10.10 Cornish Rex

Life span	Weight	Pattern & Color	Features
11–12 years	7.7–15.4 pounds 3.5–7 kilograms	**Pattern:** tabby, bi-colors, calico, tortie, smoke, and color-points. All solid coat colors. Coat is a tight, uniform wave, very soft, and silky feeling, with no guard hairs.	**Oriental:** long, slender, and tubular, naturally arched back and long slender legs. **Head:** fairly small, Roman nose. **Eyes:** medium to large, oval shaped, and slightly slanted. **Ears:** large, full based, set high on head. **Tail:** long and slender.

Genetic Diseases and Conditions: Cornish Rex are sensitive to anesthetics, and care must be provided when they are to be anesthetized. Other ailments include: congenital hypotrichosis, umbilical hernia, and hypertrophic cardiomyopathy. About 33 % of Cornish Rex have type B blood.

FIGURE 10.22 – Cornish Rex, White and Gray Tabby

Devon Rex

Origins: A kitten by the name of Kirlee was discovered by Beryl Cox of Devon, Britain, in 1960. Kirlee was one in a litter of stray cats. Initially, it was thought that he was a Cornish Rex, but test matings with a Cornish Rex produced flat-coated kittens, proving that he had a different gene mutation. Crosses between Devon and Cornish Rexes are not permitted; however, they are carefully crossbred to American and British Shorthairs in order to enlarge and strengthen the gene pool. CFA recognized the breed in 1979.

Characteristics: This distinctive looking cat has a personality to match. They are extremely loyal to their family including children. They are vocal cats that will trill, coo, and chirp all the while they are playing and exploring with their people. They love to snuggle and will often perch on their person's shoulders. The Devon Rex is an intelligent cat and will easily learn tricks and play fetch or tag, often tirelessly. Because of their social nature, they do not do well alone and should have a companion. Luckily, they are not picky about their friends and will do well with a dog, ferret, or even a rabbit. Devons love

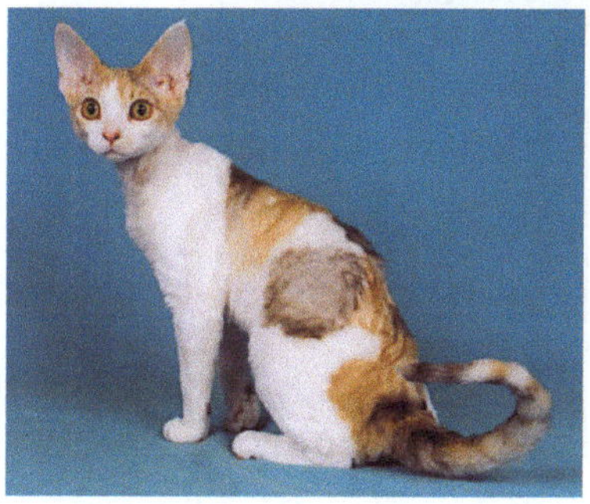

FIGURE 10.23 Devon Rex, Calico

to be high up so they can survey their surroundings. They are great climbers and often surprise their owners with their acrobatics in getting into nooks and crannies. Their short-hair coat has guard and down hair only, so they tend to be "heat seekers" like their Cornish Rex cousins. They will happily share your bed under the covers. The guard hairs do shed, so they will need an occasional soft brushing to capture the loose hair. Kittens often have a heavier coat that molts as they age. Their "pixie" or elfin-like face in what appears to be an overly large head, ears, and round eyes adds to their charm and often explains their winsome personality. Their wavy hair coat includes their whiskers that are often short and curled so that they almost don't seem to have any. One last bit of information about Devons is they are foodaholics. They will help themselves to food left out or on your plate during dinner, or they may open a cupboard to help themselves.

QUICK FACTS 10.11 Devon Rex

Life span	Weight	Pattern & Color	Features
11–12 years	6–9 pounds 2.7–4 kilo-grams	Tabby, bi-colors, calico, tortie, smoke, and color-point patterns. All solid coat colors. Coat is short, without guard hairs, soft, fine, and "rexed," which is a rippled wave.	Oriental: lithe, broad chest. **Head:** modified wedge, three distinct convex curves (outer edge of ear lobes, cheekbones, and whisker pads), full cheeks. **Eyes:** large, oval shape, wide set, sloping towards outer edges of ears. **Ears:** extremely large, set low, taper-ing round tips. **Tail:** long, fine and tapering.

Genetic Diseases and Conditions: Congenital hypotrichosis, hypertrophic cardiomyopathy, malassezia dermatitis, urticarial pigmentosa, "Devon Rex myopathy," and hip dysplasia. Approximately 41% of Devons have B type blood.

FIGURE 10.24 Devon Rex, Gray and White or "Bi-color"

Egyptian Mau

Origins: This relatively rare cat is said to have descended from cats that exiled Russian Princess Natalie Trubetskaya met in Italy in 1952 while visiting the Egyptian ambassador. She was given two kittens which she brought back with her to the U.S. and started a breeding program. The word "mau" means "cat" in middle Egyptian. Genetic tests show that she must have used various cats from the U.S. for breeding as these cats are closely related to the Maine Coon,

Korat, and American Turkish Angoras. However, there are native cats in Egypt that have the same distinct coat pattern, and frescos from 1550 BC depict this same spotted cat. Some of the cat fancier groups recognized the Egyptian Mau in 1968.

FIGURE 10.25 Egyptian Mau, Silver

Characteristics: Egyptian Maus are family oriented and will be gregarious in their territory. They will select a favorite on which they dote but will certainly interact and be happy with others within the family. With strangers, however, they tend to be more guarded and often will not respond well to being handled by strangers. They have a very musical voice that they use when they are happy, want to be fed, or are at play. They will also wiggle their tails in a very unique way when they are pleased. It is very similar to the wiggle that most cats do when they are marking their territory with urine. However the Mau is just demonstrating its happiness and does not mark when the "wiggle-tail" starts. The Egyptian Mau has a striking hair-coat pattern that has naturally occurring spots with either the "M" pencil mark, found on tabbys, or a "scarab beetle" like pattern on their foreheads. A uniquely identifying characteristic of their coat pattern includes a dark dorsal stripe running the entire length of their bodies. They are a medium sized, slender yet muscular cat with forelegs shorter than the hind legs and a unique loose flap of skin from flank to hind leg that allows them to sprint faster and turn quicker than other domestic cats. They have a "wild cat" shaped head and brilliant green eyes. Maus seem to be sensitive to cool weather and will often be found curled up on heat sources including their owner's lap or bed.

QUICK FACTS 10.12 Egyptian Mau

Life span	Weight	Pattern & Color	Features
18–20 years	10–12 pounds 4.5–5.4 kilograms	"Naturally spotted" pattern. **Colors:** silver, bronze, smoke, black and blue/pewter, and dilute: blue, silver, spotted, smoke. **Coat:** varies with color; smokes have silky, fine-textured fur and in the silver and bronze colors, the fur is dense and resilient; all have "gooseberry green" eyes.	**Standard:** muscular. **Head:** slightly rounded wedge, slight rise from bridge of nose to forehead. **Eyes:** large, almond shape. **Ears:** medium large, moderately pointed, often tufted. **Tail:** medium long, thick with a slight taper.

FIGURE 10.26 Egyptian Mau, Bronze

FIGURE 10.27 Exotic Shorthair, Dilute Tortie

Genetic Diseases and Conditions: Sensitive to anesthetics and have a gestation period that can last up to 73 days but otherwise a healthy breed of cat.

Exotic Shorthair

Origins: Exotic Shorthair cats were created by outcrossing American Shorthair cats with Persians as a means to improve the American Shorthair. However, other breeders of American Shorthair cats pushed the cat-fancier organizations to disallow such practices, but one breeder saw the benefits of a shorthaired Persian and continued to develop the breed using both Russian Blues and Burmese cats in the mix. In 1966 the CFA recognized them as a new breed and in 1987 allowed outcrossing only to Persian cats. Because of this some Exotics will carry the recessive longhair gene, so one of four offspring from mating two such cats will have long hair. The CFA does not recognize these as Persian but as Exotic Longhairs, whereas other organizations accept them as Persians.

Characteristics: The Exotics are calm, affectionate, quiet, sweet, peaceful, and loyal companions that are playful, but not destructive. They enjoy home life and easily accept other animals into the family. They adapt well to indoor life and will protect homes from invading rodents thanks to their American Shorthair genes. Males are, in general, more affectionate than females; however, expect a hug from both as you give them attention. They do not like being left alone for long lengths of time and will do well with another cat or even a dog for company. Exotics may or may not sleep with you as their thick coats often keep them quite warm, so they'll seek out cooler parts of the house in which to nap. Grooming isn't as demanding as with Persians, but weekly brushing will help to remove dead hair and thus control the hair flying around the house. Their dense, plush hair coat does stick straight out giving the Exotic a "teddy bear" like appearance. It takes an Exotic two years to fully mature.

QUICK FACTS 10.13 Exotic Shorthair

Life span	Weight	Pattern & Color	Features
12–13 years	7.7–13.2 pounds 3.5–6 kilograms	All patterns and colors. **Coat:** short hair dense, fluffy, standing straight out.	Cobby. **Head:** round, massive head, brachiocephalic. **Ears:** small, rounded tips, set low. **Eyes:** large, full round. **Tail:** short but proportional to body.

Genetic Diseases and Conditions: The brachiocephalic faces of the Exotic can cause difficulties in breathing or at the least cause noisy breathing; the faces have creases that collect tears and mucus which requires cleaning to avoid facial dermatitis. Another detriment of the brachiocephalic face is dental malocclusions, excessive tearing, and heat sensitivity. Cherry eye and entropion are seen, along with feline polycystic kidney disease, a predisposition for ringworm infections, and seborrhea oleosa. Breeding Exotic Shorthairs is very difficult, and low litter numbers are normal as well as ceasarean sections because of their oversized heads. Exotics have a high occurrence of B-type blood.

FIGURE 10.28 Exotic Shorthair, Torbie and White

Havana Brown

Origins: This rare breed started in England by outcrossing chocolate or seal-pointed Siamese cats with black domestic shorthairs in 1950. The goal was to produce a self-colored brown cat that would retain its oriental shape. Two kittens were imported to the U.S. where breeders selectively bred for a moderate version with the same colored Siamese cats. They were successful in achieving a warm brown-colored cat that was called the Havana Brown and accepted by the CFA in 1964. The breeding gene pool for this cat is still very small with only about 260 unaltered registered cats in the U.S. and Europe.

FIGURE 10.29 Havana Brown

Characteristics: Havana Browns are sensitive cats with a real need to "feel" their environment. It is not uncommon to see them reach out a paw to investigate something new or to caress their favorite person. Havanas can be outgoing or reserved in personality, although both will enjoy playing a game with other cats, dogs, or children. They have a charming voice and will often get in the last word. This green eyed beauty has a short, dense coat of chocolate colored hair that is soft and luxurious. They require minimal grooming and will submit to such willingly if introduced at a young age. Havanas are a medium-sized cat that often weighs more than it looks, and males are often heavier than females. They have a distinctive muzzle shape that has a definite "stop" or angle between the eyes and muzzle. Large, forward-tilting ears and brilliant green eyes round out the unique features of this beautiful cat.

QUICK FACTS 10.14 Havana Brown

Life span	Weight	Pattern & Color	Features
10–15 years	6–10 pounds 2.7–4.5 kilograms	Solid pattern. **Color:** chocolate brown. Short, smooth and luxurious.	Standard **Head** – longer than it is wide, rounded somewhat narrow muzzle. **Ears** – large, rounded tips, cupped at base. **Eyes** – medium sized ovals, set wide apart. **Tail** – slender and in proportion to body.

FIGURE 10.30 Havana Brown, Whiskers

Genetic Diseases and Conditions: No known genetic diseases are associated with this breed; however, the young seem to be more susceptible to upper respiratory infections.

Japanese Bobtail

FIGURE 10.31 Japanese Bobtail, Van - Calico

Origins: This naturally occuring breed comes to us from Japan via China or Korea over 1,000 years ago. They were pressed into service by the silk merchants to protect this valuable commodity from rats and mice. The first Japanese Bobtails were imported into the U.S. in 1968. In 1973 the Shorthair Japanese Bobtail was accepted for championship by the CFA with the Longhair Japanese Bobtail following in 1993.

Characteristics: The talkitive Japanese Bobtail has a generous "vocabulary" and will often carry on full conversations with their people. Their people claim that they almost seem to sing when they speak. They are a gregarious cat that will help you meet guests at the door, read the newspaper or emails with you, and help you go for a walk as they are easily accustomed to a harness and leash. They are full of energy and will get along with dogs and children, often playing tirelessly with both. Bobtails are very intelligent and quite enjoy feline agility tracts, often being the cat to beat in trials. Getting to trials or cat shows is also not a problem as they love to travel. The Bobtail's tail is a recessive gene and breeds true. The tail is unique to each cat with none being the same length or having the same curve, angle or kink. It may be flexible or rigid, and the only standard is that it fits the cat's body. Both varieties of Bobtails do not have an undercoat, so grooming is a weekly brushing as neither will mat.

QUICK FACTS 10.15 Japanese Bobtail

Life span	Weight	Pattern & Color	Features
15–16 years	7.7–15.4 pounds 3.5–7 kilograms	Solid colors, calicos, vans, and bi-colors, all tabby patterns. All colors. **Medium to longhair:** no undercoat, ruff around neck, britches, ear and toe tufts. **Shorthair:** no undercoat, soft, and silky.	Standard: long, lean, muscular. **Head:** triangular-shaped head with high cheekbones. **Ears:** large, set wide apart at right angles to head. **Eyes:** large ovals, pronounced slant. **Tail:** each unique to cat, must have any combination of curves, angles or kinks, no longer than 3 inches.

Genetic Diseases and Conditions: They are a genetically diverse pedigreed breed and so tend to have very few common issues. Unlike the dominant Manx gene for short tails, the Bobtail's gene is not associated with skeletal disorders if bred to each other. Bobtails tend to have smaller litters of 3–4 kittens, but they tend to be very large and healthy kittens. The Van-colored cats will often have heterochromic eyes which are two different colored eyes. About 16% of Bobtails will have type B blood.

FIGURE 10.32 Japanese Bobtail, Tabby

Korat

Origins: Allusions to the Korat breed can be found in a book of poems from between 1350 and 1767 that is housed in the National Library in Bangkok. They are considered good luck symbols in Thailand, and a gift of a pair of Korats to the bride ensures a good marriage. They first appeared in Britain between 1889 and 1896 and were called "Blue Siamese," but since they didn't have the same bone structure of the Siamese at the time, they fell out of favor. In 1959 the first imports of Korats came into the U.S., and they were accepted into championship status in 1966.

FIGURE 10.33 Korat

Characteristics: The Korat is credited with extraordinary senses; their hearing, sight, and scenting abilities are not matched by any other cat. They tend to be quiet and cautious and will start at loud noises or ungentle handling. They form strong bonds with their family and will often select one on which to cuddle as close as possible. The Korat has a semi-cobby body with a tapering waist and heavily muscled forequarters. It is said that the Korat has five hearts. The first three are associated with the head. Looking at the Korat straight on you will see that it has a "Valentine-like" heart shaped head. The second heart is seen by looking down over the top of the head, and the third is associated with the head and nose. The fourth heart shape can be seen from in front of the shoulders while the cat is sitting and, of course, the fifth is inside its chest. Is it any wonder these "heart-filled" cats form deep bonds with their owners? Their unique gray coat is lighter at the roots and darkens to a deeper blue as it moves up the shaft until it reaches the tips which are silver colored. This color pattern gives the Korat a light-absorbing "halo" effect. Kittens will be born with blue eyes that change to an amber or golden-green color that eventually changes to peridot green as they mature at two to four years of age.

QUICK FACTS 10.16 Korat

Life span	Weight	Pattern & Color	Features
14–15 years	10–11 pounds 4.5–5 kilograms	Solid pattern in silver tipped blue. **Coat:** single layer, short, glossy, and fine.	**Standard:** almost cobby, heavier than they appear, with a broad chest. **Head:** heart-shaped, nose has a downward curve just before the leather. **Ears:** large, rounded tips, flared at base, set high on head. **Eyes:** large, round. **Tail:** medium length, heavy base tapering to round tip.

FIGURE 10.34 – Korat, Family

Genetic Diseases and Conditions: Genetic neuromuscular degenerative disease, low body fat which is an anesthetic concern, and gangliosidosis.

LaPerm

Origins: This spontaneous genetic mutation occurred in the early 1980s on the Koehl's farm in Dalles, Oregon. A kitten named Curly was born and is the founding father to all LaPerms. A formal breeding program wasn't initiated for another ten years when cat fanicer's were contacted. The breed was named LaPerms because of their medium length shaggy, curly coat. The LaPerm, considered a rex breed, gained recognition from CFA in May 2008.

FIGURE 10.35 – LaPerm, Calico

Characteristics: LaPerms are intelligent and active cats being happy to play any game, including fetch, with their people. They are affectionate often seeking out attention from their people, and once gained they return the affection with a loud purr and petting with their paws. In addition running your fingers through their amazing coats is a pleasurable experience. The dominant gene is carried by both males and females. LaPerms can be long or short haired and are either born bald or with short straight or wavy hair. At about three to four months of age they either start to grow soft curly coats, or the old hair falls out and starts to come back in waves or ringlets. The longhairs will have either tight ringlets or corkscrew curls, and the shorthairs will have textured coats that part down the midline of the back. LaPerms do not have an undercoat which makes the coat springy and loose, with the tightest curls on the underbody and at the base of the ears. The longhaired variety will have a curly ruff and plume tail, and the shorthair will have a bottle-brush tail. Both varieties molt throughout their lives and so can look rather motley at times.

QUICK FACTS 10.17 LaPerm

Life span	Weight	Pattern & Color	Features
10–14 years	5–8 pounds 2.3–3.6 kilograms	All patterns and colors. **Longhair:** springy, light, curly or wavy curly, ruff around neck (tight ringlets or long corkscrew curls), plumed tail. **Shorthaired:** actually more medium length with shaggy waves or curls, bottle-brush tail.	**Oriental:** fine to medium bones, hips slightly higher than shoulders. **Head:** modified wedge, slightly rounded. **Ears:** slightly flared, medium to large, set wide. **Eyes:** medium, large almond shape. **Tail:** proportionate to body.

Genetic Diseases and Conditions: Generally these cats are healthy and few problems have been reported.

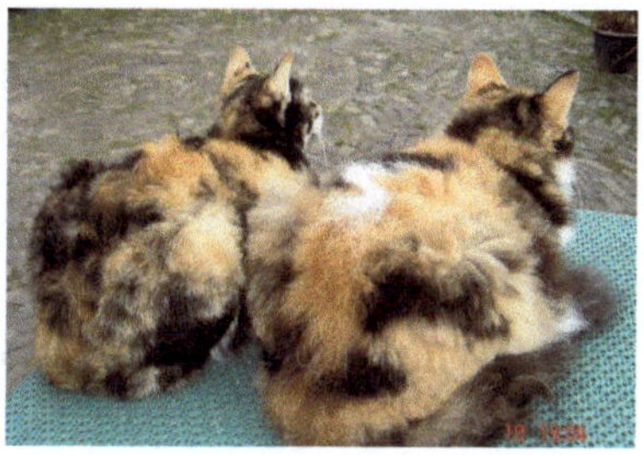
FIGURE 10.36 – LaPerm, Calico

Maine Coon Cat

Origins: The Maine Coon Cat originated in Maine although its start is shrouded in mystery and conjecture. The most acceptable explanation is

FIGURE 10.37 – Maine Coon Cat, Blotched Tabby

FIGURE 10.38 – Maine Coon Cat, Brown Tabby with "M" Mark on Forehead

long-haired cats that traveled with early explorers left their ships to breed with the local cats in the various ports. In 1895 the first Maine Coon was shown at a cat show in New York, and it won best in show. Their popularity declined when Persians and other longhairs were introduced to the U.S.; however, in the 1950s the Central Maine Cat Club was created to save this breed from extinction. CFA gave it championship status in 1976, and today it is the third-most-popular breed registered. They are Maine's state cat.

Characteristics: The Maine Coon is a family-oriented cat that will get along with two- and four-legged family members alike. They are known as the "gentle giants" of the cat world; however, they don't tend to be clingy and are a bit cautious around strangers. Male cats tend to be more playful and fun loving than the females. Both will "talk" to their people using a variety of meows, chirps, chatter, and yowls. Maine Coon cats tend to like to play in, wash their food in, or bathe in water. They are renowned mousers that can survive quite well without humans and often are very unhappy if made to spend all of their time indoors. Maine Coons are giant cats that weigh well over 20 pounds without being overweight. They are 10–16 inches (25–41 cm) at the withers and from end to end up to 48 inches in length. Their double hair coats are medium length, somewhat oily, shaggy, and rugged looking. They have a thick ruff around their necks, shaggy pantaloons on their back legs, tufts of hair on ears and between toes, and a gloriously long bottle-brush tail. Maine Coons have a very large head with a strong square jaw and high cheekbones. Many of them are polydactyl, which means they have one or more extra toes on each foot. Maine Coon cats are slow to mature taking up to four or five years to reach their full growth.

QUICK FACTS 10.18 Maine Coon Cat

Life span	Weight	Pattern & Color	Features
12–13 years	Males: 15–25 pounds 6.8–11.8 kilograms Females: 10–15 pounds 4.5–6.8 kilograms	Solids **Tabby:** striped and mackerel, Calico, Tortie, Chinchilla, Smoke. Pointed cats are not recognized. **Colors:** White, black, blue, red, cream. **Coat:** silky but heavy and shaggy; a long ruff, britches, and tufts on ears and between toes.	**Giant:** large, long, muscular body with broad chest, long legs. **Head:** longer than wide, square muzzle, high cheekbones. **Ears:** large ears tapering to a point. **Eyes:** large, wide set ovals. **Tail:** long, wide base.

Genetic Diseases and Conditions: Hypertrophic cardiomyopathy, spinal muscular atrophy, hip dysplasia, and polycystic kidney disease.

Manx

Origins: The Isle of Man in the Irish Sea is the ancestral home of the Manx. The gene responsible for taillessness is a spontaneous mutant, dominant gene that was naturally occurring among the island's cats. The original tailless cats were shorthairs, but it's thought that the ancestors of the Norwegian Forest River cats introduced longhair among the native cats. Today the longhaired Manx are known as the Cymric. The CFA recognized this breed in 1906.

FIGURE 10.39 Manx, Black and White or "Bi-color"

Characteristics: The Manx has not strayed far from its working roots and is considered a great hunter that is extremely agile despite not having a tail or much of a tail. Their powerful hind quarters allow for great acceleration, quick turns, and high leaps. Mothers are often seen teaching their kittens to hunt using whatever is available, whether it be a toy or a mouse. Manx are known to be fun loving and jokesters, often pranking their people. They are hardy, "dog-like" cats that are extremely devoted to their families. They get along great with children and other pets, and there are many stories of them actually being protective of those they think of as theirs. It is reported that once a Manx has bonded with a family, it is difficult for the cat to adjust to being re-homed. The gene responsible for taillessness is an incomplete dominant gene which can result in variable tail lengths in a single litter, which means that even a long-tailed kitten born of one Manx parent will carry the tailless gene. The only tail lengths allowed in the show ring are a "rumpy" with no hint of a tail or a "rumpy riser" with a slight rise of cartilage at the tail head. A "stumpy" is a partial tail of fused vertebrae while a "stubby" has non-fused vertebrae that can be kinked or otherwise deformed and about half the normal length, and a "longie" is a half- to full-length tail. Manx are "round" cats meaning that they have a round head and eyes, their ears should remind you of the rockers on a cradle and the back is rounded because their hind legs are longer than their front legs. Because of this feature, they have a "bunny-hop" gait when they run. Both the shorthair and longhair varieties have dense double coats that are very thick and soft.

QUICK FACTS 10.19 Manx

Life span	Weight	Pattern & Color	Features
12–13 years	Males: 11–12 pounds 5–5.4 kilograms Females: 8–11 pounds 3.6–5 kilograms	All patterns and Manx colors and tabby patterns are recognized. Color/patterns showing hybridization, chocolate or lavender ticked tabby, pointed with or without white are not allowed. **Shorthair:** double coat, longer outer coat, and close dense undercoat, hard, yet glossy coat. **Longhair (called Cymric):** medium length, double coat, britches and ruff present, hair on abdomen is longer than main coat with toe and ear tufts, soft, and silky texture.	**Cobby:** compact and muscular, with sturdy bone structure, rear legs longer than front. **Head:** rounded with prominent cheeks and jowls, slight dip between forehead and nose. **Ears:** wide base with gently rounded tip. **Eyes:** rounded. **Tail:** different lengths, including tailless, rumpy, rumpy riser, stubby, or stumpy.

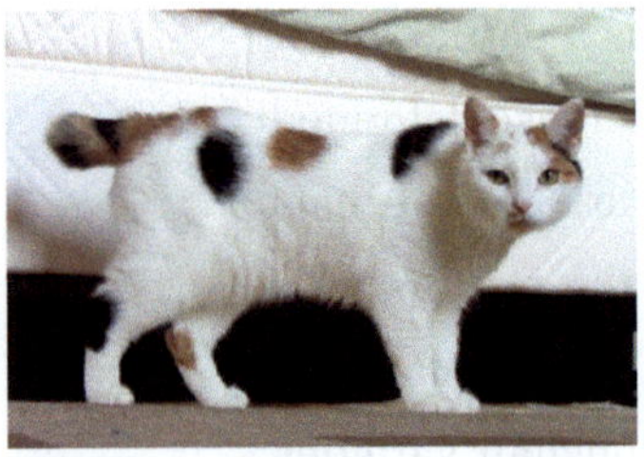

FIGURE 10.40 Manx, Stumpy - Calico

Genetic Diseases and Conditions: Tailless Manx are usually not bred to other tailless Manx as this can result in miscarriages or other serious genetic abnormalities such as intertrigo, megacolon and constipation, rectal prolapse, sacro caudal dysgenesis (incontinence), and spina bifida. These usually show in "rumpies" and will manifest in kittens at four months of age. Other issues include ocular dystrophy and arthritis.

Norwegian Forest Cat

Origins: A very popular and indigenous breed of Norway, the Norwegian Forest Cat has been around since the time of the Vikings. The cat fanciers of Norway established an organization in 1938; however, their efforts were stymied during World War II, and the breed was almost lost. The club reorganized and reestablished breeding programs. King Olaf designated the cat as the official cat of Norway, but it wasn't until the 1970s that the cats were exported to other countries, and in 1994 the CFA recognized them as a breed.

FIGURE 10.41 Norwegian Forest Cat, Black and White – "Tuxedo"

Characteristics: Norwegian Forest Cats are sensitive yet friendly cats that usually have a quiet voice. They will pick the time and place for affection and can be quite demanding about getting what they want. They do well inside as long as they have lots of toys which should include a tall "cat tree" on which to perch and scratch. They still retain their hunting prowess and will gladly keep the home free of vermin. These heavily boned cats are taller than most other breeds and require more food to maintain their weight than other cats. They have a double coat of long, thick guard hairs that is water repellent and a woolly undercoat that is thicker on the legs, chest, and head. They have a triangular-shaped head with a strong chin, almond eyes, large tufted ears, and a bottle-brush tail. In the spring they shed the undercoat and so require frequent brushings to keep the shed hair under control. They are slow to mature with many needing up to five years to reach their full growth.

QUICK FACTS 10.20 Norwegian Forest Cat

Life span	Weight	Pattern & Color	Features
14–16 years	Males: 12–16 pounds 5.4–7.3 kilograms Females: 9–12 pounds 4–5.4 kilograms	All coat patterns and colors with the most common being brown tabby with white markings; the exception is no color points. **Coat:** double, dense undercoat with long, smooth water-resistant guard hairs; bib has three separate sections: short collar at neck, side mutton chops, and frontal ruff; britches are full.	**Giant:** well balanced, sturdy bone structure, rear legs longer than front. **Head:** triangular shape, flat forehead. **Ears:** medium-large, rounded tips, broad base. **Eyes:** large, almond shaped. **Tail:** long and bushy.

Genetic Diseases and Conditions: Hypertrophic cardiomyopathy, polycystic kidney disease, retinal dysplasia, glycogen storage disease type IV causing late-juvenile-onset neuromuscular degeneration, hip dysplasia, and complex rearrangements of glycogen branching enzyme causing perinatal hypoglycemia.

FIGURE 10.42 Norwegian Forest Cat, Orange and White Tabby

Ocicat

Origins: The Ocicat came about by accident. Virginia Daly of Michigan was attempting to develop an Abyssinian-pointed Siamese. The first attempt looked like Abyssinians, but the second generation resulted in a spotted kitten being born in the litter. A second mating of the same parents resulted in more spotted kittens. Other breeders had the same success, and the offspring were registered as the Ocicat which is what Mrs. Daly's daughter nicknamed the first spotted kitten because it looked like an Ocelot. American Shorthairs were added, which gave them a larger frame and added silver to the original six colors. This is the only spotted cat that is 100% domestic. The CFA accepted them in 1966, but they didn't gain championship status until 1987.

FIGURE 10.43 Ocicat

Characteristics: The Ocicat is an overtly friendly cat with an amazing personality. They enjoy being around people and other species of animals. They are easily trained to walk on a leash, fetch, come when called, speak, and sit on command. They like water and will sometimes join their owners in the shower. They do extremely well in feline agility as they are toy-driven. Their beautiful coats are not only spotted but also come in four other patterns and 12 colors. They are large, well-muscled cats with short, tight, shiny coats. Their heads are wedge shaped and are longer than they are wide with tilted ears.

QUICK FACTS 10.21 Ocicat

Life span	Weight	Pattern & Color	Features
12–13 years	12–13 pounds 5.4–6 kilograms	Spotted, ticked, classic tabby, solid, and pointed patterns. **Colors:** tawny, chocolate and cinnamon, their dilutes, blue, lavender and fawn, and all of them with silver: black silver (ebony silver), chocolate silver, cinnamon silver, blue silver, lavender silver, and fawn silver. **Coat:** short, satin texture, close-lying and sleek, all hair shafts are ticked.	**Standard:** long, medium to large body, and athletic. **Head:** modified wedge, broad muzzle. **Ears:** moderately large, if tufted they extend vertically. **Eyes:** large, almond shape, angling toward ears. **Tail:** long, somewhat slim with slight taper and dark tip.

Genetic Diseases and Conditions: Progressive retinal atrophy, hypertrophic cardiomyopathy, renal or liver amyloidosis, and early periodontal disease.

FIGURE 10.44 Ocicat

Oriental Shorthair

Origins: Developed in the U.S. the Oriental Shorthair maintains the shape of its ancestor the Siamese, but it can be any color and pattern other than a color point. The shorthair was accepted by the CFA in 1977, and the longhair variety followed in 1995.

FIGURE 10.45 Oriental Shorthair, Silver Tabby - Classic

Characteristics: The Oriental Shorthair is people-oriented and loves to be in the mix of things. They do not do well if left alone for long periods of time or if ignored once their person gets home. They will get along with other cats and do better with a friend if their family is gone for long periods during the day. Once you are home, the Oriental will let you know if you are late and then proceed to "tell" you about their day. They are very playful and love nothing better than "finding" toys that may be in cupboards or bags or closets. This cat retains the Siamese build: long, lean, muscular bodies with long legs and tail. They have a triangular head with almond-shaped eyes that are usually green and large triangular ears. The shorthair requires little grooming, and a weekly comb out will keep the longhairs looking neat.

QUICK FACTS 10.22 Oriental Shorthair/Longhair

Life span	Weight	Pattern & Color	Features
13–14 years	8.8–14.3 pounds 4–6.5 kilograms	All patterns and colors (300 different color and pattern combinations) with shorthair and longhair varieties. **Shorthair:** fine textured, glossy, close. **Longhair:** medium length, fine, silky, no undercoat.	**Oriental:** long, svelte, fine-boned, and good muscular shape. **Head:** long tapering wedge. **Ears:** extremely large, wide base. **Eyes:** almond-shaped, medium sized. **Tail:** long, thin at base, tapers to fine point; longhair has a plume tail.

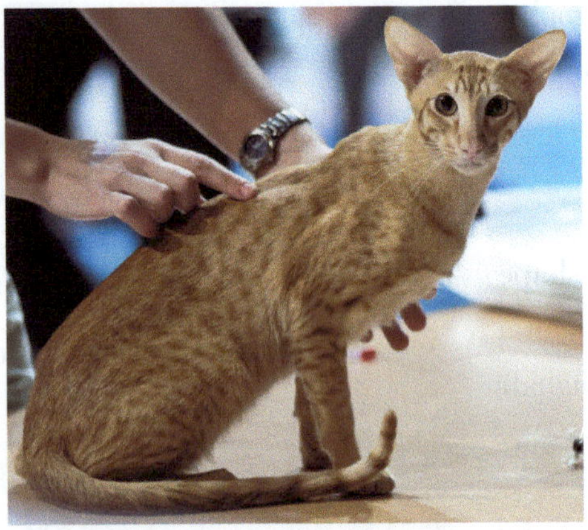

FIGURE 10.46 Oriental Shorthair, Red Spotted Tabby

Genetic Diseases and Conditions: Hereditary liver amyloidosis, dilated cardiomyopathy, aortic stenosis, crossed eyes, nystagmus, progressive retinal atrophy, megaesophagus, hyperesthesia syndrome, lysosomal storage disease, asthma, uroliths, lymphoma, and mast cell cancer. All Oriental Shorthairs have type A blood.

Persian

Origins: Longhaired cats were imported from Persia, Turkey, Afghanistan, Burma, China, and Russia to Western Europe in the 1600s. The interbreeding of Persians with Angoras was very common during those days but is no longer practiced with modern breeders of today. The first breed standard was issued in 1889, but not all fanciers were in agreement with the standard. The "traditional or doll-faced" Persians are considered to be the original version. They have a more pronounced muzzle which is more popular with the public and seems to have fewer health issues. In 1950 the "Peke-faced or ultra-typing" Persian was developed and registered with the CFA. The 2007 breed standard calls for this version to have an almost vertical alignment of forehead, nose, and chin. Himalayan cats are hybrids from Siamese mated with Persians and until 1984 they were a separate breed until the CFA merged it. The Exotic Shorthair is a shorthaired Persian that was developed in the 1950s and were recognized by the CFA in 1966. Because of their remarkable natures and classic looks, the Persian is the most popular cat in the U.S.

FIGURE 10.47 Persian, Orange Tabby - Peke-Faced

Characteristics: Persians are quiet and composed cats that like a quiet and orderly household. They are playful but will not demand play like other breeds. They are content to snuggle and nap either on their person or draped across the back of the couch. Their sweetness and quiet voices are treasured by their people. Persians come in a wide variety of colors and coat patterns, and because of this they are shown in seven color divisions. See the Quick Facts table for a list of these divisions. Their luxurious double haircoat requires daily combing with a metal comb to keep mats, tangles, and hairballs to a minimum. Regular bathing and careful drying is important for fur health. The alternative that many owners resort to is shaving the body short.

Because of their flat faces the tear ducts often malfunction and cause staining or inflammation of the skin if tears get into the skin folds on the face. A daily wash and dry of the face is important to keep their faces from becoming infected. The Persian has a rounded head, large round eyes, and small ears on a very cobby body.

FIGURE 10.48 Persian, Himalayan or Color-poin

QUICK FACTS 10.23 Persian – Doll-face

Life span	Weight	Pattern & Color	Features
14–15 years, some up to 20	7.7–15.4 pounds 3.5–7 kilograms	Variety of patterns and colors divided into various divisions: • Solid color division includes white with copper, deep blue, or odd eye combinations. Blues, blacks, creams, chocolates, and lilacs, all with copper eyes. • Silver and golden division includes chinchilla, shaded silvers and goldens. • Shaded and golden division includes shell and shaded cameos, shell and shaded torties, shell and shaded blue-creams. • Smoke include six colors: black, blue, cream, cameo (red), smoke tortie, and blue-cream smoke. • Tabby: classic, mackerel and patched patterns in silver, blue-silver, red, brown, blue, cream, cameo, and cream-cameo. • Particolor: tortoiseshell and blue-cream. • Bicolor and calico: calicos, bi-colors, smokes and whites, tabby, and whites. • Himalayan or color-point: all point colors as seen in the Siamese, all must have blue eyes. **Coat:** long, thick, stands away from body, ruff is deep, britches.	**Cobby:** broad, short stature, good muscle tone. **Head:** massive, round, brachiocephalic. **Ears:** small, rounded tips, set low on head. **Eyes:** large, round, and brilliant in color. **Tail:** short, in proportion to body.

FIGURE 10.49 – Persian, "Doll-face" - Shaded

Genetic Diseases and Conditions: Polycystic liver and kidney disease, dystocia, hypertrophic cardiomyopathy, dermatophytosis (some have reactions to Griseofulvin which is used to treat ringworm), idiopathic periocular crusting, facial fold pyoderma, idiopathic facial dermatitis, seborrhea oleosa, congenital portosystemic shunts, basal cell tumor, sebaceous gland tumor, primary seborrhea, idiopathic periocular crusting, multiple epitrichial cysts, coloboma, lacrimal punctal aplasia, entropian, trichiasis, corneal sequestration, congential cataract, retinal degeneration, epiphora, lysosomal storage disease, Chédiak-Higashi syndrome, hip dysplasia, malocclusion, calcium oxalate urolithiasis, cryptorchidism, peritoneopericardial diaphragmatic hernia, systemic lupus erythematosus, alpha-mannosidosis, and heat sensitivity. About 14% of all Persians have type B blood.

FIGURE 10.50 Ragdoll – Bi-color

Ragdolls

Origins: During the 1960s, Ann Baker of California purchased kittens from a neighbor that were docile, affectionate, and had a tendency to go limp when picked up, thus the Ragdoll name. She bred these cats to others with similar temperaments with a mind to enlarge their size, calm their demeanor, and fix their color point patterns. She was not interested in petitioning the cat fancier groups for recognition, starting her own organization, the International Ragdoll Cat Association (IRCA), with strict rules as to how the cats were bred or sold. No Ragdoll were allowed to be registered in any other cat association. In 1975 a husband-and-wife team, Denny and Laura Dayton, started the process to gain CFA recognition. The name "Ragdoll" was trademarked and could not be used until 2005 when it wasn't renewed. An offshoot of the IRCA established the Ragamuffin breed in 1994.

FIGURE 10.51 Ragdoll, Pointed

Characteristics: Ragdolls are very loving cats and crave attention from their humans. They'll greet you at the door when you come home, play a rousing game of fetch, and come when they are called. They are known for their gentleness and often play without extending their claws. If you sit, be prepared for the Ragdoll to sit on you or at least with you. They like nothing more than snuggling with you on the couch or in your bed. The Ragdoll's coat is semi-long, full, and silky, although it requires only a weekly grooming to keep it looking neat because it has only guard hairs. They shed seasonally, and a steel comb is good for use on the Ragdoll's coat. Kittens are born white with the color coming in at about 8–10 weeks of age with full color and coat coming in at two years and full height and weight at four years of age.

QUICK FACTS 10.24 Ragdoll

Life span	Weight	Pattern & Color	Features
18–20 years	Males: 12–20 pounds 5.4- 9.1 kilograms Females: 8–15 pounds 3.6–6.8 kilograms	Patterns include: • Pointed: One color darkening on the nose, ears, tail, and paws. • Mitted: same as pointed with white paws and white from chin to abdomen, with or without blaze. • Bi - color: white legs, white inverted V on face, white abdomen, and sometimes white patches on back. • Lynx: variant of Bi-color with tabby markings **Colors:** red, seal, chocolate, blue, lilac, cream, and tortoiseshell pattern in all colors and three patterns. Coat is abundant with long guard hairs and minimal undercoat; ruff and feathery britches are preferred.	**Cobby:** large, long, solid, heavy bones, muscular, rectangular shape, hind legs longer than front. **Head:** triangular, proportional to body. **Ears:** medium size, rounded tips, set wide. **Eyes:** large ovals, vivid blue. **Tail:** long, plumed.

Genetic Diseases and Conditions: Hypertrophic cardiomyopathy, calcium oxalate uroliths, and predisposition to feline infectious peritonitis. Ragdolls tend to have a high incidence of type B blood.

Russian Blue

Origins: This is naturally occurring breed that may have come from a port city in Russia or from the Archangel isles of Great Britain and Northern Europe. The first appearance of the cat outside of Russia was in 1875 in Great Britain where it competed with other blue cats until 1912 when it was finally given its own class. After WWII their numbers were almost obliterated so breeders used the Siamese to salvage the breed. Eventually the Siamese was bred out of the line. Americans had specimens as well and worked to increase the blood lines by using Scandinavian and British Russian Blues. The CFA recognized the breed in 1947.

FIGURE 10.52 Russian Blue

FIGURE 10.53 Russian Blue, Kittens

Characteristics: Russian Blues are known as sensitive and loyal cats that bond with their people. It is reported that they will pick up on how you are feeling and do their best to cheer you up. They tend to be shy around strangers but will play with dogs, ferrets, other cats, and children. They are excellent hunters, so small pocket pets and pet birds should be protected from them. Blues are intelligent and curious, often finding things to play with that their owners never expected to become toys. Some are quite talkative while others will only meow occasionally. These beautiful cats have a short, thick double hair coat with both guard and downy hairs of equal lengths tipped with silver. This makes them shed a bit less than other cats. They have brilliant green eyes in a broad, wedge-shaped head with a straight nose and large, wide-based ears.

QUICK FACTS 10.25 Russian Blue

Life span	Weight	Pattern & Color	Features
10–16 years	8–15 pounds 3.5–7 kilo-grams	Solid pattern in silver-blue color, vivid green eyes. **Coat:** double coat, thick, short, and plush.	**Standard:** fine boned, long, muscular. **Head:** medium wedge, blunt muzzle. **Ears:** very large, wide base, pointed tips, set on sides of head. **Eyes:** round, set wide apart. **Tail:** in proportion to body, moderately thick base.

Genetic Diseases and Conditions: This is a generally healthy breed with few reported problems; however, they are prone to obesity. Their gestation period is approximately 65 days.

Scottish Fold

Origins: In 1961 a kitten by the name of Susie with folded-forward ear flaps was born on a farm outside of Perthshire, Scotland. She had kittens, and two of them had the same folded-forward ears. One of the kittens was purchased by William Ross who wanted to start trying to breed them. He had help from a geneticist by the name of Pat Turner. He registered his cats with the British cat fancier's group in 1966. In 1971 the cat fanciers group withdrew their recognition because of genetic deformities of limbs and tails as well as ear problems such as infections, mites, and deafness. However, some Folds had made

FIGURE 10.54 Scottish Fold, Blue

their way to the U.S. where breeders used British and American Shorthairs to stabilize the breed. The CFA recognized them in 1978. Genetic studies have found that the gene responsible for the fold can be either one (heterozygous) or two (homozygous). Mating homozygous folds to homozygous folds results in severe health issues and is considered unethical. Mating a homozygous fold to a normal-eared cat will result in heterozygous folds; however, this should not happen in ethical breeding programs since there shouldn't be homozygous folds available. Ethical breeding would be heterozygous folds to normal eared cats which will result in 50% of the kittens born having folded ears and the other 50% having normal ears.

FIGURE 10.55 Scottish Fold, Red Cream Tabby

Characteristics: Scottish Folds are as sweet as their expressions lead you to believe. They are very good natured and can adjust to a single household or to a house full of children and other pets. They are undemanding and will only occasionally let you know they are around with a tiny meow. Folds get very attached to their people, and if left alone for long periods of time they can become depressed. They enjoy playing and especially like the outdoors. They have been seen sleeping on their backs and sitting with their legs stretched out and front paws on their belly in what looks very similar to a sitting Buddha. When purchasing a Fold, it is important to check the flexibility of the tail by GENTLY moving it in a slightly upward-arch; this should not cause any pain. The feet should be checked for splayed toes, and the legs should not appear course, shortened, or lack mobility. Scottish Folds have either short or long hair coats and a rounded body appearance; their heads, cheeks and eyes are round with their folded ears really accentuating the round appearance. Kittens are all born with straight ears that don't start to fold forward until they are about 21 days old. Remember that about half of the litter will not have folded ears. The folds range from one fold to double or triple creases that cause the ear to lie totally flat against the head.

QUICK FACTS 10.26 Scottish Fold

Life span	Weight	Pattern & Color	Features
10–15 years	Males: 9–13 pounds 4–6 kilograms Females: 6–9 pounds 2.7–4 kilograms	All patterns and colors except color-point. **Shorthair:** short to medium-short dense, plush, soft in texture. **Longhair:** medium-long to long hair, britches, tail plume, toe tufts, and ear furnishings are desirable.	**Cobby:** medium-sized, rounded body that is well padded. **Head:** rounded, prominent cheeks with jowly appearance. **Ears:** folded forward and downward, small, rounded tips. **Eyes:** large, rounded. **Tail:** medium to long, tapering to a round tip.

Genetic Diseases and Conditions: Arthropathy; polycystic kidney disease; hypertrophic cardiomyopathy; osteochondrodysplasia; degenerative joint disease especially in tail, ankle, and knee joints; and progressive arthritis. About 14% of Scottish Folds will have a B blood type, and they have a high incidence of AB blood as well.

Siamese

Origins: Siamese are descended from native cats found in Thailand which used to be called Siam. They made their appearance in the United States in 1878 when President Hayes received "Siam" as a gift from the American Consul in Bangkok. In 1884 a breeding pair of Siamese cats was brought to the UK by the British Consul-General in Bangkok as a gift to his sister. The breed soon spread to other parts of Europe and Asia. They started appearing in English cat shows almost immediately and in American shows by the early twentieth century. Officially recognized by CFA in 1906, they are one of the first breeds of pedigreed cats shown. In the 1950s judges started favoring longer, leaner cats, and the breeders started selecting for that preference, and thus the "oriental" body type was born. By the mid-1980s the only Siamese shown were this type of body style. Fortunately some breeders continued the "old style" lines of Siamese that has a more standard body with an "apple" or round head. See Figure 10.58.

FIGURE 10.56 Siamese, Seal Point

Characteristics: The vocal Siamese is considered the king/queen of extroverts. They form strong bonds with a particular family member, although they will snuggle up with anyone else if their person is not available. They do get lonely and will actually get depressed if left alone for long periods of time. A cat or dog friend will be accepted, and the two will usually become firm friends. Their vocal range is legendary from low wheezes to sharp cries that can sound like a human baby. They are playful, persistent, and loving. They have a unique hair coat pattern that is used as a basis for all cats that have a "pointed" hair coat. This pattern is usually a darker color of hair on the face, ears, legs, and tail than on the body. The gene responsible for this causes a malfunction in the tyrosinase enzyme making it heat sensitive. It fails to work at normal body temperatures but activates in the cooler areas of the skin such as the head (cooled by the sinus cavity), ears, legs, and tail. It also makes a difference as to where the cat lives. In cooler climates the cats will be darker, and in warmer climates the cats are lighter. There are four natural coat colors (which are the only ones recognized by the CFA), and because of crosses with other cats we are seeing more colors available. Kittens are born either all white or cream colored with the

FIGURE 10.57 Siamese, Lilac point

points coming in at about four weeks of age. Siamese tend to darken as they age in part because they don't regulate their body heat as well and in part due to the aging process. The coat is short, shiny, and sleek without a downy layer. They have a long (some say tubular), lean, muscular body with long legs, a triangular head, blue almond-shaped eyes, large triangular ears, and a long whip tail.

FIGURE 10.58 Siamese, "old style" Seal point

QUICK FACTS 10.27 Siamese

Life span	Weight	Pattern & Color	Features
12–20 years	Males: 9–14 pounds 4–6.3 kilograms Females: 6–10 pounds 2.7–4.5 kilograms	Pattern is color-point. **Colors:** seal (dark-almost black points fawn body), blue (cool gray or slate points with bluish white body), chocolate (light brown with creamy white body), and lilac (pale, warm gray with white body); other colors include red, lynx (tabby), tortie, and cream. **Coat:** short, fine, and glossy.	**Oriental:** medium-sized, long, lithe, tubular, well-muscled body. **Head:** long, tapering wedge, good proportion to body. **Ears:** very large, pointed, wide base. **Eyes:** medium-sized, almond shape, slanted toward nose. **Tail:** long, thin, tapering to fine point.

Genetic Diseases and Conditions: Endocardial fibraelastosis, paten ductus arteriosus, aortic stenosis, progressive retinal atrophy, food hypersensitivity, junctional epidermolysis bullosa (claw shedding), congenital hypotrichosis, vitiligo periocular leukotrichia, unilateral periocular depigmentation, eyelid coloboma, Grisofulvin drug reaction, cleft palate, megaoesophagus, pyloric dysfunction, small intestine adenocarcinoma, portosystemic shunts, amyloidosis, haemophilia B, porphyria, hip dysplasia, mast cell tumors, lipoma, sweat gland tumors, nasal cavity tumor, mammary tumors, vestibular disease, deafness, and hydrocephalus. Siamese are always type A blood.

Sphynx

Origin: In the 1970s hairless kittens were spontaneously born in Minnesota, U.S.A., and in Toronto, Canada. Both groups were used to start the Sphynx breed, although the various breeders had a difficult time producing cats that were interested in breeding, could carry a litter to term, or even live very long after birth. Crosses with the Devon Rex (which were later halted because of different health issues), American Shorthairs, and Domestic Shorthairs (which are the only allowed out-crossings by

FIGURE 10.59 Sphynx, Red Tabby

FIGURE 10.60 Sphynx, Black and White

the CFA) were used to strengthen the bloodlines and even out the issues faced by the early breeds. The gene responsible for the hairlessness causes a malfunction of the structure of the hair, so it is easily damaged and falls out. Both the Devon Rex and Selkirk Rex have this same gene, but the keratinization of the hair follicle still produces the necessary proteins to prevent the damage to the hair structure. The CFA accepted the Sphynx for competition in 2002. They are among the rarest of cats.

Characteristics: The Sphynx are loving cats that like to snuggle and play. They tend to be extroverts, seeming to crave attention and will not shy away from strangers. Sphynx will take up "dog-like" activities like walking on a leash and playing fetch. Sphynx also enjoy other cats and get along well with dogs, often snuggling up with them during nap time. Though they are billed as hairless cats, they are not truly completely hairless. They have a very fine layer of down, and the skin tends to feel warm and quite soft, like the texture of a tanned chamois cloth. They may or may not have whiskers. If they do have whiskers, they can be whole or broken. They tend to have "extra" or loose skin, so wrinkles occur on various body parts. Because of their near hairless condition, their skin needs a lot of care. They will require sunscreen if going outdoors, lotion for dry times of the year, and weekly baths. Because there isn't enough hair to catch the body oils normally absorbed, oil tends to build up on the skin, under the nails, and in the ears. A weekly or bi-weekly bath with careful attention to cleaning the ears is important to maintain the health of the Sphynx. In essence, the Sphynx requires more "coat" care than other cats with a full coat. In cold climates the Sphynx will need to be kept in a warm home and provided a warm nest to snuggle into or a sweater. The Sphynx is often touted as a non-allergenic cat because it doesn't have hair. This is usually not the case as most allergy sufferers are allergic to the saliva proteins that get deposited on the cat's skin (in this case) when they clean themselves. Sphynx have a strange body shape with a somewhat "pot belly," a wedge-shaped head, large lemon-shaped eyes, very large ears, and a long whip tail. The skin on the Sphynx is the color that the coat would be if present, so you will see solid, point, van, tabby, and tortie "printed" skin!

QUICK FACTS 10.28 Sphynx

Life span	Weight	Pattern & Color	Features
13–14 years	7.7–15.4 pounds 3.5–7 kilograms	**Hairless:** possibly short, soft, downy peach-fuzz over body or a bit longer hair on feet, outer edges of ears, tail and scrotum. Skin is wrinkled around muzzle, between ears, and across shoulders. If whiskers are present, they should be short and sparse. Skin is pigmented to what the fur would have shown if not hairless!	**Standard:** medium length, muscular, broad chest, and well-rounded abdomen. **Head:** modified wedge, longer than wide with prominent cheekbones. **Ears:** large to extremely large. **Eyes:** large, lemon-shaped, set wide. **Tail:** whip-like in proportion to body.

Genetic Diseases and Conditions: Kittens often suffer from respiratory infections, and many breeders will not release kittens until they are 12 weeks or older to ensure they are healthy before moving to a new home. Sunburn and skin cancer from sun exposure, dry skin from cold or dry air, urticaria pigmentosa, hypertrophic cardiomyopathy, mitral valve dysplasia, and muscular dystrophy-like disease are other problems. Sphynx have a high incidence of type AB blood.

Tonkinese

Origins: In the late 1960s and early 1970s, breeders in Canada bred Siamese and Burmese cats together to try to create mink-colored cats with aqua-colored eyes. They were also aiming to get the best personality traits from the parent breeds. "Tonks" were moved from the crossbred classification to full breed status by the CFA in 1984. However, the gene for the Tonkinese is recessive, so about half of the kittens in a litter will be mink, a fourth will be pointed, and the last fourth will be solid colored.

FIGURE 10.61 Tonkinese, Mink

Characteristics: "Tonks" are great cats with great personalities. They are fun loving often inventing games to play by themselves, with their person, or with another cat, preferably another Tonk. They will insist on being in

FIGURE 10.62 Tonkinese, Lilac Mink

the thick of things and must be included; otherwise, they will get bored and find mischief. Tonkinese are easy to train to a leash and will happily go for walks to explore the neighborhood. Tonks will talk and expect to be listened to as well. Many owners report having full conversations with their Tonkinese. They are very loving cats and will often pick a person they like the most and will be insistent that the person take time to give them attention. Tonkinese have a standard body shape that is heavier than they seem with proportional long legs and tail. They have a slightly rounded, wedge-shaped head and blunt muzzle, their eyes are almond shaped, and their ears are moderately sized and positioned to the outside of the head. Their hair coats have all three layers, giving them a full, satin-like fur that is quite a beautiful color. Weekly brushing and an occasional bath is all that is needed to keep them neat.

QUICK FACTS 10.32 Tonkinese

Life span	Weight	Pattern & Color	Features
15–16 years	14–15 pounds 6.3–6.8 kilograms	**Patterns:** solid (gold or green eyes), pointed (blue eyes), and mink (aquamarine eyes) all with points that are lighter to darker for each pattern. **Colors:** natural (medium brown), blue (pale gray), champagne (pale buff-beige), platinum (creamy gray); colors darken with age. **Coat:** medium length, close-lying, soft, and silky.	**Standard:** medium length, muscular, well balanced. **Head:** gently rounded, modified wedge, longer than wide, slight convex curve on forehead. **Ears:** medium, oval tips, set on sides of head. **Eyes:** almond shape in proportion to head. **Tail:** in proportion to body.

Genetic Disease and Conditions: These cats are generally healthy with no reported specific diseases or conditions. They can be prone to gingivitis, amyloidosis, asthma, aortic stenosis, crossed eyes, nystagmus, progressive retinal atrophy, megaesophagus, hyperesthesia syndrome, lymphoma, and are sensitive to anesthesia. Tonkinese always have type A blood.

FIGURE 10.63 Turkish Angora, Tortie Smoke

Turkish Angora

Origins: The Turkish Angora, sometimes just called Angora cats, are a distinct breed of cat that hails from Turkey with documentation showing they were present in the early 1600s. They were among the longhaired cats imported to Europe and were used to create or refine the Persian cat. This was done to such an extent that the breed was almost lost because of the popularity of the Persian cat, and they were placed in the same category at shows. In the early 20th century the Ankara Zoo in Turkey started a breeding program to preserve the Turkish Angora and in particular to breed the white with odd eyes, which is still the choice of breeders today. In 1963 a pair were exported to Canada and were accepted by the CFA in 1972. At that time only the white cats could be shown as Turkish Angoras; however, in 1978 that standard was changed, and now all colors and patterns are accepted in the show ring. The show cats of today are fine boned and not all that representative of the Turkish version which are a much larger and sturdier breed.

FIGURE 10.64 Turkish Angora, White – Odd Eye

Characteristics: These beautiful cats are full of life and energy. They often bond with one member of the family in particular but will be friendly towards others in the family as well. Angoras love to be in the air and so will climb to the highest points they can in order to observe and "direct" the activities in the home. They get along with other cats, dogs, and children but can become quite assertive and rule the pack. "Turks" are problem-solving and intelligent and get bored easily, which often leads them into mischief. The odd-eyed and white-coated cats are often deaf, but this doesn't seem to slow them down. They learn quickly and are great observers of the daily activities. Turks often love to be in the thick of the household activities, and if there is a party going on they will act as the gracious host. They have long, lean bodies, almond-shaped eyes, a plume tail, and great ears that are wide at the bottom and come to a nice point. The Turkish Angora has soft, medium-length hair that is kept neat with weekly brushings. They tend to shed more heavily in the summer, so bi-weekly brushings are important to reduce hairballs. For years breeders concentrated on producing only white, odd-eyed cats, but today all of the colors are accepted.

QUICK FACTS 10.30 – Turkish Angora

Life span	Weight	Pattern & Color	Features
12–18 years	5–9 pounds 2.3–4 kilograms	**Patterns:** Solid, tabby, and smoke in all colors. White cats often have one blue eye and one amber eye. Pointed, chocolate, and lavender colors are not accepted in the show ring. **Coat:** single coat, various lengths, usually medium long, ruff and britches should be long, tufts between toes, bottle-brush tail.	**Standard:** medium sized, long, lean, fine-boned. **Head:** small to medium size in proportion to body, long smooth wedge. **Ears:** large, wide base, set close together and high on head. **Eyes:** large, almond shaped. **Tail:** long, tapering from wide base.

Genetic Diseases and Conditions: Deafness in one or both ears because of the gene responsible for the white coat and blue eyes, hereditary ataxia, and hypertrophic cardiomyopathy. Turkish Angoras have a high incidence of type B blood.

FIGURE 10.65 Turkish Van, White and Red

Turkish Van

Origins: Lake Van, Turkey, is home to these great cats. In 1955 local cats were given to two visiting British women who took them home to the UK The first "Vans" were imported to the U.S. in 1982 and were accepted by the CFA in 1994. They are a relatively rare breed in the U.S. but are quite common in Turkey. The blood lines remain pure as no other cats can be bred to the Turkish Vans except other Turkish Vans, so imports from Turkey are common. Genetics have been done on both the Turkish Angora and the Turkish Van proving they are distinctly different breeds. The piebald color pattern was thought to have come from the Van, so the term "van-pattern" is used to acknowledge its origins. However, the piebald spotting gene is a common one showing up in domestic cats and other animals like horses and ball pythons.

FIGURE 10.66 Turkish Van

Characteristics: The Turkish Van are excellent hunters and are social towards people, often developing strong bonds with a particular person. They like to initiate play and will often bring a toy over to their person to play fetch. Vans are supposedly water-loving cats because they seem to have an intense fascination with water. Owners report them playing with the water in their dish and following them into the shower or bathtub. This behavior may or may not be any more intense in Vans than other breeds of cats. Vans are large cats with muscular shoulders, back legs that are longer than front, and large paws. They take a long time to mature, some up to three to five years of age before they have their full height, weight, and length. The Turkish Van is a semi-long haired cat with no undercoat, so the coat lays very flat and sleek. It is waterproof, so bathing them can be a challenge and really should be done only if necessary.

QUICK FACTS 10.31 – Turkish Van

Life span	Weight	Pattern & Color	Features
15–16 years	Males: 16–17 pounds 7.2–7.7 kilograms Females: 12–14 pounds 5.4–6.4 kilograms Can reach lengths of 3 feet (1 meter) long from nose to tip of tail.	"Van" color pattern: body color bright white with colored fur restricted to head and tail, one or more random patch on body is allowed. **Colors:** red, cream, black, blue, red tabby, cream tabby, brown tabby, blue tabby, tortie, dilute tortie, brown, or blue patched tabby. Color-points are not accepted. Eyes can be blue, amber, or "odd-eyed." **Coat:** semi-long, soft, no undercoat, summer coat is short, winter coat is substantially longer and thicker; feathering on ears, legs, feet, and belly. Facial fur is short; ruff and brush-tail become more pronounced with age.	**Giant:** sturdy, broad, muscular, long body, deep chest. **Head:** broad wedge, medium-length nose, prominent cheek bones. **Ears:** moderately large, wide based, set high and well apart. **Eyes:** moderately large, rounded, set at a slant. **Tail:** proportionate to body, bottle brush.

Genetic Diseases and Conditions: The breed sometimes has hypertrophic cardiomyopathy but is otherwise generally healthy. Turkish Vans have a high incidence of type B blood.

CHAPTER 3 Pocket Pets

Species of Pocket Pets

Pocket pets typically include rodents such as Chinchillas, Gerbils, Guinea Pigs, Hamsters, Fancy Mice, and Rats. Hedgehogs, Ferrets, Sugar Gliders, and Rabbits which are not rodents but are often placed in this category in veterinary medicine. Pocket pets have been kept as pets for centuries. Early records indicate that the Japanese kept pet mice, and in Victorian times the gentry kept ferrets for hunting. Guinea Pigs and Rabbits have been and continue to be a food source for many people around the world.

One aspect that set the rodents apart from the other pocket pets is their diets. They and rabbits are herbivores, meaning that they derive their nutrients from plant materials. There are formulated pellets for each species supplemented with Timothy hay or alfalfa. Treats can be small amounts of fruits and vegetables, but care must be taken to not overfeed these animals as they can become obese. The only rodent that requires something different is the Guinea Pig, and it requires Vitamin C. They cannot synthesize vitamin C from their food and so must be given it as a supplement. Rats and Mice will consume foods we consider protein-based but do not require it as a part of their nutritional needs. Hedgehogs and Sugar Gliders are omnivorous and do require vegetable matter as well as proteins such as insects. Ferrets are strictly carnivores and as such require diets high in protein. Each species' diet is discussed further on in the chapter.

A number of the pocket pets are coprophagic, which is the ingestion of feces. In the case of rabbits, it is an important part of their nutritional health. They are hindgut fermenter which means that when food reaches their cecum, it starts to ferment the fiber that is necessary to maintain their GI tract. As the bacterium breaks the fiber down into a metabolically manageable substance, it is formed into a cecotrope, which is a soft,

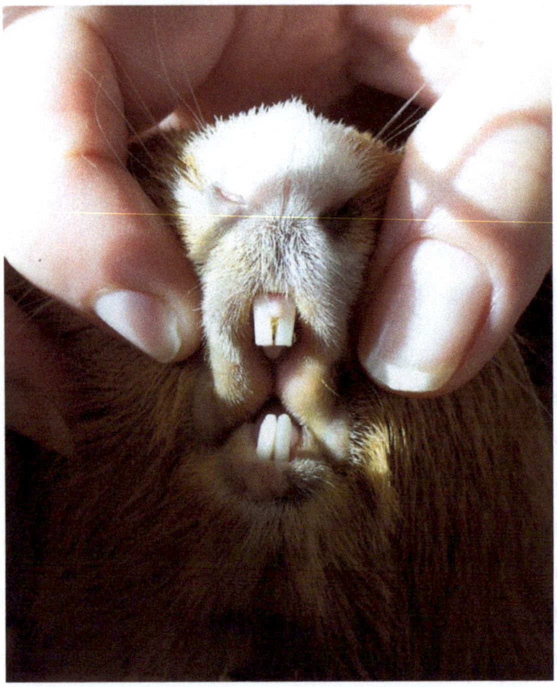
FIGURE 11.00 Guinea Pig, Incisors – Normal Occlusion

mucus-covered fecal pellet. This "grape bunch-like" pellet is passed, and the rabbit ingests it whole to keep it protected from the stomach acids. Once it reaches the small intestine, it is absorbed, releasing nutrients and, most importantly, vitamin B. If the pellets become sticky, it is usually an indication of a diet too high in carbohydrates and not enough fiber. Reducing the amount of fresh vegetables and fruits and increasing the hay will usually fix this imbalance. Chinchillas, Gerbils, Guinea Pigs, Hamsters, Mice, Rabbits, and Rats also produce cecotropes.

The pocket pet's dentition differs from species to species. One common feature of rodents and rabbits are their incisors or front teeth.

They are open-rooted teeth and grow throughout their lives, so they require foods or chew toys that will wear them down. However, rabbits have four top incisors versus two in the rodents. There are two "peg teeth" or small incisors that sit behind the upper incisors. Rabbits have white enamel, and rodents have yellow—or bright orange-colored enamel that is directly related to their nutritional health. A lighter color indicates a lack of iron which is influenced by the amount of calcium the animal is getting in its diet.

Rabbits also have more premolars and molars than other rodents. Another difference between rabbits and rodents is that rabbits have diphyodont (two sets of teeth); the first set or deciduous teeth are replaced by adult teeth soon after they are born. Rabbits, Chinchillas, and Guinea Pigs have aradicular hypsodont teeth which have an open root, so they grow throughout the animal's lifespan. This is one of the reasons why these animals require unlimited access to grass hays to help them keep their teeth ground to a proper length. The other rodents have closed roots that stop growing because their food of choice is softer. The following are the dental formulas for each of the species in this chapter:

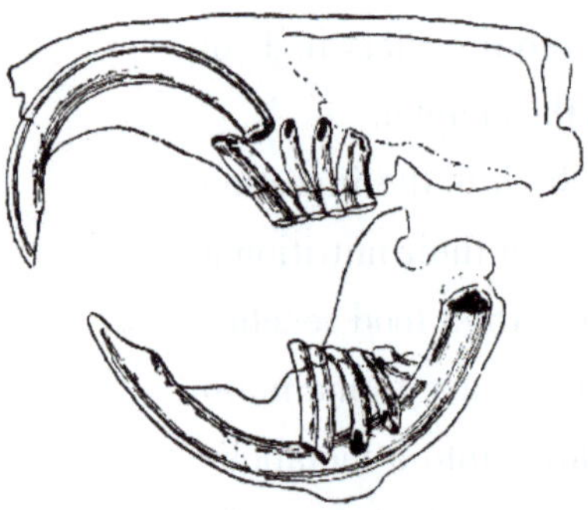
FIGURE 11.01 Guinea Pig Dentition

Gerbils/Hamsters/Mice/Rats: 1/1 incisors, 0/0 canines, 0/1 premolars, 3/3 molars x 2 =16

Chinchillas/Guinea Pigs: 1/1 incisors, 0/0 canines, 1/1 premolar, 3/3 molars x 2 = 20

Rabbits: 2/1 incisors, 0/0 canines, 3/2 premolars, 3/3 molars x 2 = 28
Hedgehogs: 2/2 incisors, 1/1 canines, 3–4/2–3 premolars, 3/3 molars x 2 = 34–40
Sugar Gliders: 3/3 incisors, 1/0 canines, 3/3 premolars, 4/4 molars x 2 = 40
Ferrets: 3/3 incisors, 1/1 canines, 3/3 premolars, 1/2molars x 2 = 34

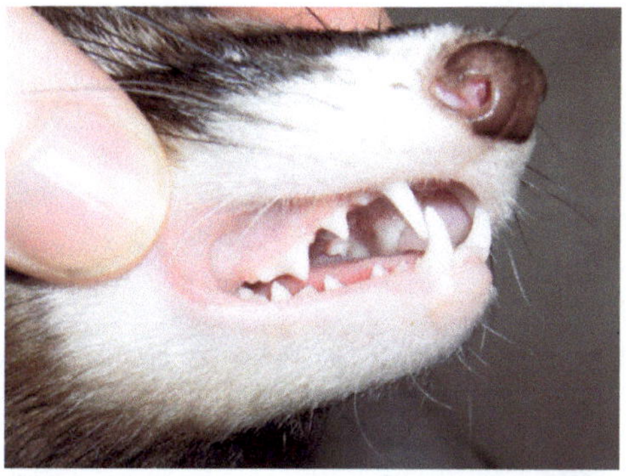

FIGURE 11.02 Ferret Dentition

If the rodents and rabbits do not get the proper foods and chew toys, their teeth can overgrow or become worn irregularly. This causes the teeth to not meet in the proper manner and is called malocclusion. The improper wear can also cause hooks that can cause injuries to the cheeks. If it is a case of overgrowth or irregular wear, the teeth can be treated by a veterinarian. If the malocclusion is allowed to continue, the animal can starve to death.

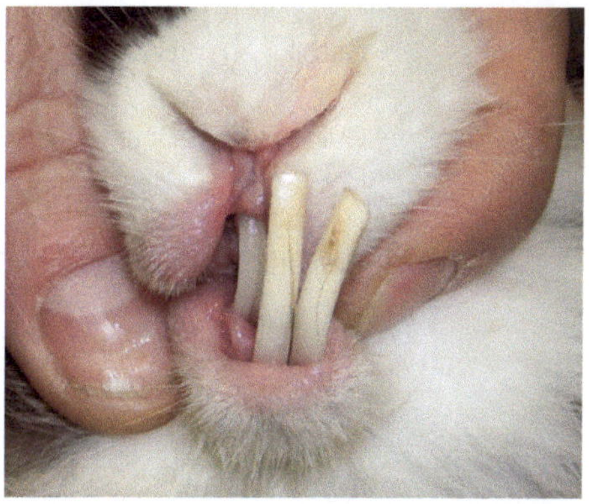

FIGURE 11.03 Rabbit Malocclusion

Another aspect of rodents that sets them apart from other animals is that their reproduction cycles tend to be very short, so babies (pups) are born naked, blind, and deaf, which is referred to as altricial.

However, they do grow very quickly, and many are sexually mature in four to six weeks.

Guinea Pigs are different in that their babies are precocial which means they are born with their eyes open, they have hair, and can eat adult food, although they do suckle. Sugar Gliders are marsupials, and their babies (joeys) are born extremely small, and they crawl from the vagina to the pouch where they latch on to a teat and stay there. Ferret babies (kits) are born blind and deaf, but they do have hair so they are semi-precocial.

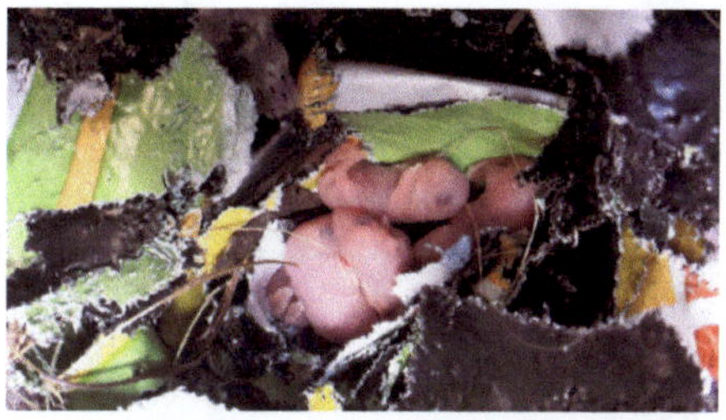

FIGURE 11.04 Mouse Pups

In the laboratory settings, individuals are identified in a number of ways. Ear tags, fur dying, tattooing tails or toes, microchips, and ear notching are a number of examples. There are a number of schemes for ear notching, and it depends on the placement of the notches on the edge of the ear, about 3 mm inside the ear or more to the middle of the ear. The numbers from left to right ear can indicate 1–6 left ear or 10–60 on the right ear.

FIGURE 11.05 Ear Notched Mouse

TABLE 11.00 Gender Terms for Pocket Pets

Species	Males	Females	Young	Altered Sexes Neutered (NM) Spayed (SF)	Parturition	Group
Chinchilla	Boar/ Bull	Sow/ Velvet	Kit		Whelping	Litter-young
Ferret	Hob	Jill	Kit	Gib – NM Sprite – SF	Kindling	Busyness Business
Gerbil	Buck	Doe	Pup			Horde
Guinea Pig	Boar	Sow	Pup			
Hamster	Buck	Doe	Pup			Horde
Hedge-hog	Boar	Sow	Hoglet, Piglet, Pup			Array
Mouse/ Mice	Sire	Dam	Pup/ Kitten			Colony/ Nest
Rats	Sire/ Bull Buck	Dam/Doe Cow	Pup			Colony Horde
Rabbits	Buck/ Jack	Doe/Jill	Kit/ Kitten Bunny Nestling	Lapin – NM	Kindling	Herd Litter Wrack Nest
Sugar Gliders						

Many of the larger pocket pets appreciate being out of their cages and interacting with humans, Guinea Pigs, Rabbits, Rats, Ferrets, and Sugar Gliders among them. However, it is important that owners understand the dangers of allowing these pets freedom to roam. Inside electrical cords, toilet bowls, other pets, small children, open flames, spaces between walls, and backsides of appliances can be dangerous. Some plants are toxic or poisonous to the herbivores, such as lily of the valley, foxglove, buttercups, ragwort, rhubarb, tulip bulbs, onions, ivies, and oak tree leaves. Animals should always be supervised when out of their cages and watched to make sure they don't get into trouble.

Each of the animals in this chapter requires specific caging and should be given the biggest cage possible. See the Quick Facts Tables for minimum cage dimensions. Many appreciate "two stories" for climbing and a nest or sleeping box so they can get out of the daylight as most are either nocturnal (active at dark) or crepuscular (active at twilight). Many will also need a solid-floor exercise wheel.

FIGURE 11.06 Solid Floor Exercise Wheel

Cages should be placed in areas that are quiet during the day and out of direct sunlight for these animals. Cages also need to be kept clean. Bedding or substrate material is often used on the floor of the cage to soak up urine and should be recycled newspaper, corn cob or woodchips from non-toxic wood. Rabbits, Ferrets, and Chinchillas will use a litter box, and regular, non-clumping cat litter is appropriate for their use. The entire cage should be cleaned weekly with hot water and a dish detergent or other disinfectant. Water bottles and food dishes should be cleaned daily.

The animals in this chapter make delightful pets for people that don't have a lot of room or have pet restrictions for dogs and cats. The previous table is representative of the species discussed further on in the chapter.

Species of Pocket Pets

Chinchilla

Origins: Originally from Western South America, Chile, and Peru in the Andean mountains, chinchillas live in social groups that are called herds. Their natural home is burrows or crevices in rocks; they breed year-round and can have two to six **precocial** kits. As defensive maneuvers, they will spray urine, slip their fur (which means it will come out in big bunches, which ruins their pelt), and jump up to six feet (1.8 meters). They are omnivores in the wild, eating plants and insects. Chinchillas in the wild are critically endangered due to overhunting for their fur.

FIGURE 12.00 Chinchilla, Gray

Characteristics: Chinchillas have amazingly soft fur that is very thick because of over 60 hairs sprouting from one hair follicle. Because of their thick fur, they do become hyperthermic if the temperatures go above 75°F (23.9°C). They must be kept in an air-conditioned area during the hot seasons. If for some reason the heat can't be controlled, placing a bucket filled with ice and covered with a lid that they can sit upon or curl up against will help them cool off. They have extremely long whiskers or vibrissae and very large, bat-like ears. They disperse heat from their ears, so bright red ears can indicate overheating. Their natural coloring is silver, and there are several man-made colors which can be found in the Quick Facts table below.

Chinchillas are used in the fur trade and are raised for that purpose on farms. They are also used in research for behavioral studies and auditory research because they have very large tympanic bullae. They are also used in studies on Chagas disease, gastrointestinal disease, pneumonia, listeriosis, and *Yersinia* and *Pseudomonas* infections. As pets, chinchillas are affectionate and fairly easy to care for; however, they do require a large cage so that they can get exercise. Ideally it should be a two-story cage with a sleeping box where they can be in the dark during daylight hours and a large, solid-floored exercise wheel. They will use a litter box to urinate; however, they will defecate wherever. Appropriate cage substrates are the same as those recommended in the introduction to pocket pets. They are social animals and require interaction with their people; otherwise, they will become lonely and sad. If owners are unable to provide this interaction, it is recommended that two chinchillas live together.

FIGURE 12.01 Chinchilla, Silver Mosaic – Note the Long Vibrissae

Chinchillas need to be offered a daily pumice dust bath which absorbs oils and dirt in order to keep their fur clean. The dust container must be taken out after the dust bath, so it does not defecate in the dust, thus contaminating the dust bath. Chinchillas require chew toys to keep their front teeth worn down as they grow continuously throughout their life span. Natural sticks from apple, birch, or kiln-dried pine are safe for them to chew, and pumice stone and purchased wooden sticks are also appropriate. Chinchillas should be provided fresh water daily in a sipper bottle to keep it clean, 1–2 tablespoons of a good pelleted diet, and free-choice Timothy hay. Do not feed chinchillas fruits and vegetables as they can cause bloat which can lead to death. A salt block as a mineral supplement and vitamin C drops added to the water bottle are also important for chinchillas. Dried fruits, raisins, or seeds can be given as a treat but should be limited as they can cause diarrhea. Chinchillas will consume their own, what looks like, feces which is called **coprophagia**; refer to the introduction to pocket pets for an explanation of this behavior. It is completely normal and necessary for them to maintain their digestive health.

Individual chinchillas are identified by ear tags or tattoos. Male chinchillas have a longer ano-genital distance than females. If breeding chinchillas is a goal, the females have to be bred early; otherwise, their pelvis will fuse making birth difficult. They reach sexual maturity at 7–8 months of age.

QUICK FACTS 12.00 Chinchilla

Life Span	Length/ Weight	Color	Genus species	Gestation Period & Litter Size	Activity & Caging
9–17 years Sexual maturity 8 months of age	8–10 inches 20–25 centimeters 1–2 pounds 450–900 gram Males are smaller than females	Silver, black velvet, violet, sapphire, beige, white and blends.	*Chinchilla laniger*	105–116 days Litter size 1–6 kits - precocial	Nocturnal Minimum Caging: 3'x 2'x 2' 1 x 0.6 x 0.6 m

Genetic Diseases and Conditions: Septicemia, respiratory and gastrointestinal infections caused by *Listeria, Pseudomonas, Pasteurella, Proteus, Salmonella* and *E. coli.*, parasite infections from Toxoplasmosis, Giardia, Eimeria, and Cryptosporidium. Stress or rough handling will cause them to "slip their fur" or they will "barber" their fur when nervous or bored. Dental problems include overgrown molars resulting in watery eyes and gum infections. Chinchillas may also have fractures, seizures, ringworm, diarrhea, constipation, rectal prolapse, gastric ulcers, bloat, penile hair rings, choking, fungal infections, heat stroke, heart problems, diabetes, bites from fellow chinchillas, ear infection, and hairballs.

FIGURE 12.02 Chinchilla, Gray Female

Ferrets

Origins: The domesticated ferret descends from the European polecat which likely came from North Africa about 2,500 years ago. It is thought that Romans used ferrets to control rabbit populations, and they were the responsible for spreading ferrets throughout their empire. England is also responsible for introducing ferrets to New Zealand for rabbit control (also introduced), which quickly turned

FIGURE 12.03 Ferret, Fitch or Polecat Color

into an ecological disaster for the native bird populations when the rabbit populations were reduced. Ferreting is the term used for when ferrets are used to hunt rabbits. The U.S. saw imports of ferrets as early as the 1600s, and they were used to protect grain stores well into the 1800s. Ferrets were not commonly seen as pets until the 1980s. They are banned in many countries and states in the U.S. because of their potential impact on ecological niches; if they are allowed, many municipalities require licenses, micro-chipping, and sterilization before one can be acquired as a pet.

Characteristics: Ferrets have long, muscular bodies, short legs, and a thick, greasy-feeling hair coat. The natural color of ferrets is a sable or Fitch color which is predominantly brown and yellow. There are more colors being developed, and the American Ferret Association has attempted to classify these variations in their showing standards. Ferrets have four types of teeth: incisors, canines, premolars, and molars, totaling 34 teeth. They must be taught not to bite much like a kitten or puppy is taught. They spend 14–18 hours a day asleep and will happily romp and play during the hours they are awake, either with their humans or with other ferrets. They love to explore and examine things they find on their explorations. Ferret proofing their play room is very important to avoid electrical shocks and accidental poisonings from house plants. It is a good idea to have a small bell attached to their collars as they have a tendency to burrow into couches and other hidey holes, and with the bell on you can at least locate them if they go missing. Ferrets have an anal scent gland that is often removed before they are sold as pets. If it is not removed, they have a musky odor that can become quite strong as they reach sexual maturity. They can also express these glands when excited or angry. This is often accompanied by a behavior called a "war dance or weasel war dance" which is a series of hops, leaps, and bumping into objects or beings. If accompanied by a clucking or dooking sound, it is an invitation to play. If accompanied by a hissing noise or squeaks it can indicate that the ferret is upset, frightened, or angry.

Ferrets are used in research as a human influenza model as they can get the flu from humans and vice versa. They have also been used in studies on cardiovascular disease, nutritional and respiratory diseases, virology, reproductive physiology, anatomy, endocrinology, and neuroscience. Ferrets are also used to pull cables through walls and tight spaces like aircrafts. A pull line is attached to a harness, and the ferret is released into the space; when it comes out the other end, the line is attached to the cables allowing the humans to pull them through the tight space. Of course, the ferret is well rewarded with special food treats for doing such a job.

Ferrets are high-maintenance pets and require a great deal of time spent with them as they do bond to their owners. They require toys to play with and can be quite entertaining while at play. They will destroy toys quickly, and small pieces can become a choking hazard. Hammocks, corrugated tubes, dryer vent tubing, and large PVC pipes for tunneling are also entertaining to a ferret. It is recommended that families with children under six not get a ferret because they do bite if held incorrectly or if accidently hurt. Ferrets are natural hunters, so pet rodents, rabbits, and birds are not safe around them. They can get along with dogs and cats but should be supervised so that nobody gets hurt. Introducing them to dogs and cats should go at a slow pace, and once they are allowed to be together, an escape route for the ferret is a must in case there is an unexpected altercation.

FIGURE 12.04 Ferret, Albino

They are carnivores and require a protein-rich (30% or more) diet. Commercially prepared foods for ferrets is preferred although they can do well on cat food diets. They can also be fed either pre-killed or live rodents or rabbits to mimic their natural diet choices. Ferrets will imprint on the type of food fed to them, making it really hard to switch their diets, so it is recommended that young ferrets (kits) be exposed to as many different types and flavors of food choices as possible.

Ferrets will use a litterbox but must be trained to use one regularly. Place a litterbox in one corner of the cage and bedding and food/water dishes in the other corners. When they use the litter box without fail, give them a bit more space, increasing the space if they continue to use the box. Be ready for accidents as they will happen, and keep the litterbox clean at all times. While outside of their cage a litterbox in every corner of a ferret-proofed room is a must. They do prefer to sleep in a nest box or something closed in, and if you have more than one ferret, a hammock to accommodate them all is perfect. If owners prefer to have the ferrets locked up while they are gone, an open-sided cage with a nesting box or hammock works fine, with enough room for a litterbox, water, and food dishes in addition to the required space. Wood chips are fine for a substrate if needed. They can be kept together but should be separated during breeding season as the males will fight.

Regular grooming chores are nail trims, ear cleaning, teeth brushing, and flea prevention and/or treatment. Bathing is not usually recommended as it strips the oils from their coats, and then their body tries to compensate by producing more oils, making the ferret smellier and itchy. If a ferret has to have a bath, it should be no more often than monthly. Yearly vaccinations include rabies, and modified live canine hepatitis/distemper vaccine. Identification is easiest with small collars and identification disks; tattooing is possible, but the ears are quite small and furry. Fur dyes can be used temporarily, especially on the albinos.

QUICK FACTS 12.01 Ferrets

Life Span	Length/ Weight	Color	Genus species	Gestation Period	Activity & Caging
7–10 years Sexual maturity 4–8 months of age	20 inches 50 centimeters Including tail 1.5–4 pounds 0.7–2 kilograms Males larger than females	Sable, albino, dark eye white (DEW), and silver. Other color variations are from these four colors.	*Mustela putorius furo*	42 days Litter Size 3–7 kits-altricial	Crepuscular Minimum 36" x 24" x 24" per ferret – bigger is better, two-tiered for more than one ferret a must

FIGURE 12.05 Ferret, Male

Genetic Diseases and Conditions: Heatstroke, cancers (adrenal glands, pancreas, and lymphatic systems), insulinoma, compacted anal glands, human influenza virus, infectious canine distemper and hepatitis, Aleutian disease, mink enteritis virus, gastroenteritis, epizootic catarrhal enteritis, coccidia, foreign bodies, hairballs, dental issues, diseases caused by *Bordetella* and *Pseudomonas sp.* bacteria, gastrointestinal obstructions, cardiac disease, urolithiasis, cystitis, renal disease, sarcoptic mange, ear mites, heartworms, flea allergy dermatitis, back injuries, and aplastic anemia in unspayed females. Some man-made colors can have a genetic defect known as Waardenburg syndrome.

FIGURE 12.06 Gerbils, Showing Long Tail

Gerbils

Origins: Gerbils hail from Africa, Asia, and India. Once known as a desert rat, there are almost 110 species of gerbils all found in arid habitats. In the wild they are social animals living in large groups or clans in elaborate underground tunnel systems and burrows. Gerbils are territorial and do not get along with other gerbils they don't recognize as part of their clan. Gerbils were first brought to Paris in the 1800s and quickly became popular as pets. They were brought to the U.S. in 1954 to be used in research by a Dr. Schwentker and were introduced to the pet market in 1964. California bans gerbils because of their threat to the ecosystem and agricultural operations.

Characteristics: The gerbil has a medium-length hair coat that has a greasy appearance and doesn't lie flat on the body. They have a long tail that is tufted at the end that can slip off if grasped by predator or child. They have small ears, long vibrissae, and powerful hind legs that allow them to jump or stand up to look around. In research they are used to study filarial worms, neurological conditions that include epileptiform convulsions, endocrinology studies because of their ability to concentrate urine, in cholesterol studies as they are prone to increased serum and hepatic cholesterol levels on low-fat diets, and in stroke and radiation-poisoning studies.

As pets, gerbils are not typically aggressive; however, they do require frequent handling to keep them tame. If frightened or cornered, they may bite out of fear. They enjoy interacting with people, especially if living alone. It is recommended that if little human interaction is going to be available, a second gerbil should be purchased. Same-sex gerbils from the same litter can be housed together. If added later as adults, they will fight, and the weakest can be killed although if introduced slowly by dividing the cage or placing the cages next to each other, they may learn to recognize each other as clan mates and not fight. Foot stomping is a sign of aggression, and if this should occur during a face-to-face introduction, a quick separation is called for. Use a small wash cloth or grasp them by the scruff of the neck to pick them up. Scooping them up and holding them in the palm of the hand is not recommended because they jump. Many a gerbil has been injured by jumping out of hands or off tables and landing on the floor. Do not pick them up by the tail as the skin will slip off, leaving just the vertebrae behind. The bones will eventually fall off, and the skin will heal around the stump. It is estimated that 20–50% of all pet gerbils will suffer a seizure at some point during their life. These are thought to be caused by fright,

FIGURE 12.07 Gerbil

rough handling, or stress. The convulsions can be mild to severe and do not typically appear to have any long-term effects; however, in rare cases death has resulted from a very severe seizure.

The best housing for a gerbil is ten-gallon tank designed for reptiles. They are escape proof and easy to clean. Gerbils should not be housed in plastic cages as they can gnaw through them in a short period of time and escape. In addition, if they swallow any of the plastic, it can cause a gastric blockage. Gerbils are diggers and re-arrangers and can become quite upset if not given materials that can accommodate this behavior. They require either a deeper depth of substrate material than other pocket pets to facilitate this behavior or, even better, placing small cardboard boxes inside their cage for them to chew to bits and then arrange to their liking. An added bonus of the cardboard is that it can be used as part of or all of their substrate material. Because of their desert heritage, they produce little urine and so have little urine smell. Appropriate cage substrates are the same as those recommended in the introduction to pocket pets. Gerbils appreciate a sleeping box, a solid-floored exercise wheel and enjoy being in an exercise ball that can be rolled around on the floor. Gerbils will also require wood materials to keep their incisors trimmed; otherwise, they will become overgrown. A twelve-hour light/dark cycle is good for them with an ideal temperature between 60–70°F with 30–50% humidity.

A pelleted diet with small offerings of fruits and vegetables is a good diet choice for gerbils. Alfalfa is a good source of calcium, and a small amount can be given daily. Reserve sunflower seeds as special treats given occasionally or as a reward during training. If given free choice, they will eat only these seeds which can cause them to develop alopecia, other skin problems, and obesity. They require a water source, and a sipping tube works perfectly to keep the water from becoming contaminated.

Ear notching is a satisfactory method of identification and is often used in research facilities. Male gerbils have an obvious scrotum and a longer anogenital distance than females. Breeding pairs tend to be monogamous and maintain lifelong relationships. It is not uncommon for gerbils to mate while the female is giving birth or immediately afterward, so if future litters are not wished for, the male should be removed from the cage as the mother goes into labor. The male can be reintroduced after 48 hours. Gerbils are altricial; however, within ten days the pups will begin to grow fur, at about 19 days of age the eyes will open, and at four to five weeks the pups can be weaned. Start handling the pups at about ten days of age to familiarize them to humans by taste, touch, and smell. Sexing a pup can be done when they are one week old; nipples indicate females, no nipples indicate males. This must be done before the coat of belly fur forms as they are really hard to find with a full coat.

QUICK FACTS 12.02 Gerbils

Life Span	Length/ Weight	Color	Genus species	Gestation Period	Activity & Caging
3–5 years Sexual maturity 70–90 days	3–6 inches 7–15 centimeters not including tail Males 80–130 grams Females 50–80 grams	Agouti (natural), black, albino, white, cinnamon, lilac, blue, cream, silver, sable, and spotted. New mutations: fat-tailed gerbils, "duprasi," long-haired, bushy-tailed	Meriones unguiculatus	23–26 days Litter size 3–7- altricial	Diurnal Minimum Caging: 36 square inches (232 cm squared), solid floor

Genetic Diseases and Conditions: Bald nose ("sore nose," from staphylococcus, stress, lost mate, or overcrowding), Salmonellosis, ***Demodex sp.***, basophilic stippling, ventricular septal defects, epileptiform seizures, ovarian cysts, Tyzzer's disease, gastritis, overgrown incisors, tumors, fractures or spinal injuries from being dropped, obesity, tapeworms, inner ear infections, deafness, and tail chewing. Antibiotic toxicity: a dose of penicillin-streptomycin will kill most gerbils within an hour.

FIGURE 12.08 Gerbil, Multiple Colors

Guinea Pig

Origins: Guinea pigs originated in the Andes Mountains of South America where they have been domesticated since 5000 BC and were used as a food source, given as gifts, and used in medical research and religious ceremonies. In the wild they live on plains in small groups. They do not store food or burrow but will take shelter in burrows of other animals and crevices. Introduced to Europe in the 1500s, they quickly became popular as pets among the upper class and royalty. There are numerous

FIGURE 12.09 Guinea Pig, English

FIGURE 12.10 Guinea Pig, Abyssinian

organizations devoted to competitive breeding and showing of Guinea pigs, and many man-made breeds have been developed because of this. "Cavy," another name for the Guinea pig, is short for its genus species name, and it is easy enough to figure out where it came from; however, the name "Guinea pig" is not. They are rodents and not related to pigs at all, and they did not originate in Guinea. There is all kinds of speculation as to where the name came from, but we may never know the true origins of the name.

Characteristics: Guinea pigs have a solid, oval-shaped body with large, thin ears. They are the most vocal of all pocket pets. Whistles express excitement when they hear their owner come home or the refrigerator door opening. Bubbling or purring is a sign of enjoyment or contentment. Rumbling is used to establish dominance in a group, or it can be a response to something fearful. If it is cornered, one may hear a chattering sound made by gnashing their teeth together. When chasing each other you may hear a chutting and whining sound from pursuer and pursued respectively. Squealing, shrieking, and chirping are all sounds of discontent, pain, or hunger. Guinea pigs make wonderful pets as they rarely bite or scratch intentionally. They tend to be a bit panicky and may screech or chirp when initially picked up but usually calm down quickly if handled gently. Guinea pigs are social creatures, and though they will bond with their people, they do better with a Guinea pig friend or two. Same sex pairings work well, and, of course, a male (boar) and a female (sow) get along great but expect little ones (pups) within about two months of putting them together. If housed in a group, they do establish a pecking order that is usually male dominated. Sometimes if crowding occurs, the weakest member may show signs of barbering. Barbering is where one Guinea pig chews on the fur of another sometimes escalating into damaging the skin. If seen, it is important to remove the victim and eliminate whatever is causing the stress, which is often overcrowding.

In research, Guinea pigs are used in studies for juvenile diabetes because they have an unusual insulin mutation and in studies on scurvy because they, like humans and primates, cannot synthesize vitamin C and must obtain it in their diet. They are used in tuberculosis research because they are easily infected by the human tuberculosis bacteria. They are used in human pregnancy studies because their placental structure is similar, and they frequently suffer from pregnancy toxemia. The strains used in research are outbred strains; the Hartley and Dunkin-Hartley strains are albinos. Inbred strains are less common and are used for very specific research such as immune system molecular biology and are called Sewall Wright's designated as Strain 2 and 13. A hairless breed of Guinea pig was developed for dermatological studies and came about as a spontaneous genetic mutation.

Cavy fanciers obtained these "naked" Guinea pigs from breeders for research and call them "skinny pigs."

Guinea pigs were originally domesticated for their meat in the Andes. It was usually reserved for ceremonial meals but has become more socially acceptable by many South Americans. They reproduce quickly and need less space and food to reach maturity than other livestock, and they can be raised in urban areas. Guinea pig meat is similar to rabbit in that it is a good quality protein that is low in fat and cholesterol.

FIGURE 12.11 Guinea Pig, Texel

Individuals can be identified by ear notches, ear studs, tattoos, black ink on white ears, green on dark, and fur dying as a temporary marker lasting up to six weeks. Sexing Guinea pigs can be a bit tricky as the anogenital distance is similar. The females have a Y-shaped configuration of the vulva, and the male's testicles may be visible. Males reach sexual maturity at three to five weeks of age and females at four weeks. If females are to be bred, it is important to wait until they are at least 400 grams. If the females are over six to ten months of age, they should not be bred because the pubic symphysis may have calcified due to a metabolic disease which could lead to dystocia. They can rebreed six to ten hours after delivery, so the males should be removed shortly before birth.

There are 13 breeds of Guinea pigs recognized by the American Cavy Breeders Association. The English or American shorthair has a short, smooth coat and comes in solid (self) colors or a variety of colors. The Abyssinian has swirls of hair called rosettes or "cowlicks." Ten symmetrical perfect-centered rosettes are the show standard, and they also come in a variety of colors. The longhaired varieties include Peruvians, Silkys, Shelties, Coronets, and Texels. Peruvians and Silkys have long straight hair up to six inches in length, and the Texels have curly long coats. All three need daily brushing to keep them from matting up with particular care spent around the anal region to avoid fecal buildup. The Teddy has a kinked or bent-hair shaft that causes the coat to stand erect over the entire body. Satins have abnormal hair fibers that produce a sheen, and the Rex has short guard hairs that do not show above the undercoat. The Crested, Dalmatian, and Himalayans round out the thirteen breeds.

As mentioned earlier, Guinea pigs require vitamin C much of which can be obtained from fresh fruits and vegetables such as cabbage, kale, apples, and carrots in small quantities, and unlimited Timothy or Alfalfa hay is also a good source of the other vitamins and minerals they need. Guinea pigs do not appreciate changes in their diets and may not eat if changes are suddenly made. A constant source of food and hay is important to prevent barbering as well has helping them to keep their molars from over growing. Vitamin C can be added to their drinking water at a dose of 10–20 mg/day. Some of the pelleted foods are enriched with vitamin C; however, it stays good in the pellets for only three

FIGURE 12.12 Guinea Pig, Teddy

months after milling, and after that it loses its potency. It is important to check the bags for expiration dates, and if feeding just one Guinea pig, avoid buying huge quantities that will expire before it is consumed. Guinea pigs are coprophagic.

Guinea pigs require a solid floor and enough room to equal four times the size of an adult per pig. Some large-scale Guinea pig breeders will have entire rooms of pigs. They tend to be messy, not picking a corner or area to defecate or urinate but will go all over. When they play a game of tag, the droppings will fly into food and water bowls, or they'll tip the bowls as they run through them. Daily washing of food and water bowls is a must as well as weekly or twice weekly thorough cage cleaning to keep them healthy and reduce odor. If allowed to go longer without cleaning, the urine will often crystallize and harden, making it very difficult to clean. Recycled newspaper, corn cob, or aspen bedding is the best substrate because they will chew on woodchips which can cause constipation.

Guinea pigs have good memories. If a group is startled, they will scatter in all direction, and if threatened with no escape, they will hop or "popcorn" in place similar to a ferret's "war dance." When they self-groom or social groom, they produce a white liquid from their tear ducts that is wiped on their paws and then wiped on their fur. Guinea pigs can swim. There are Cavy clubs and associations dedicated to the showing and breeding of Guinea pigs established worldwide. People can be allergic to Guinea pig urine, saliva, and dander.

QUICK FACTS 12.03 Guinea Pigs

Life Span	Length/ Weight	Color	Genus species	Gestation Period	Activity & Caging
6–8 years, up to 8 is possible Sexual maturity Males 3 months of age Females 2 months of age	8–10 inches 20–25 centimeters 1.5–2.5 pounds 700–1200 grams	Multiple colors and lengths of hair.	*Cavia porcellus*	59–72 days Litter size 1–6 pups - precocial	Crepuscular in nature Domesticated G. pigs active during the day with short sleep-ing periods scattered throughout

FIGURE 12.13 Guinea Pig, Silky

Genetic Diseases and Conditions: Pneumonia from stress, poor nutrition, and hygiene is the most common cause of disease. Other conditions include: LUMPS – cervical lymphadenitis, Coccidiosis, bite wounds, lice (*Gliricola porcelli*) and mites (*Trixacarus caviae*), lymphadenitis, Salmonellosis, gastrointestinal parasites, diarrhea, constipation, abscesses, fungal infections, torticollis, scurvy, ovarian cysts, corneal ulcers, hyperthermia, toxemia of pregnancy, prolapsed uterus, hypocalcaemia, and mastitis. Fractures or ulcerative pododermatitis from wire mesh floors is very common. Congenital issues include eye disorders, palsy, tremors, and "waltzing disease."

The Guinea pig's intestinal flora is predominantly gram positive, and any treatment that drastically alters this will lead to a fatal enterotoxemia. Crystalline potassium penicillin will kill some animals in three to eight days; penicillin will kill 80%–100%; erythromycin, bacitracin, and lincomycin will also kill some. Streptomycin is directly toxic. Neomycin and polymixin B will protect them if administered soon after a dose of penicillin. Antibiotics that are safe to give include tetracycline, chloramphenicol, and sulfonamides; feed ½ to 1 teaspoons of plain yogurt daily while on antibiotics to boost the intestinal flora.

Hamsters

FIGURE 12.14 Hamster, Golden

Origins: The Syrian Hamster or Golden Hamster was first described in 1839, but it took until 1939 for researchers to successfully breed and domesticate hamsters. In the wild they are crepuscular and feed on seeds, fruits, vegetables, and an occasional insect. They have immense cheek pouches that are about 4–8 mm wide and 35–40 mm long, reaching back to the shoulder region. They use these to carry food back to their burrows. When full, the cheeks

FIGURE 12.15 Hamster, Campbell's or Russian Dwarf

can make their heads double or even triple in size. Many an inexperienced hamster owner has called in to the veterinary clinic with concerns of their hamster suddenly developing tumors over the shoulder area. In the wild they can hibernate, allowing their body temperature to fall to ambient temperature. Tame hamsters do not hibernate but can reduce their breathing and heart rate for short periods of time, usually if food is scarce and the ambient temperature is in the 20°C (68°F) range.

Characteristics: Hamsters are stout-bodied with very short tails, small ears, short, stocky legs, and wide feet. They have thick soft fur that can be short or long and in a number of colors depending on the species. Because the cheek pouches are immunologically privileged sites due to the absence of intact lymphatic drainage pathways, researchers utilize hamsters for transplant studies. They are also the animal of choice for Leptospira isolation and in caries research.

As pets, they have to be handled frequently in order to keep them tame and happy. However, they often will not tolerate being held still for long. Hamsters may bite or show aggression in response to rough handling or sudden disturbances, especially if touched while they are sleeping, being around other hamsters, or when they are ill or in pain. Males tend to be less aggressive than females.

Hamsters can be scooped up with hands cupped around them or they can be scruffed, grasping the loose skin over the shoulders. Remember that with their extensive cheek pouches a person has to get as much skin as possible so that the hamster is not able to turn and bite while being scruffed. Other techniques include placing a small cloth over the top of them or placing an empty can in front of them since this looks like a good place to hide, and they will run right into the can. Individuals are permanently identified by ear notching; fur dyes or fur clipping can be used for short-term identification. Male hamsters have an anogenital distance greater than females. and if gentle pressure is applied to the abdomen, the testicles will descend into the scrotum. Hamsters are strictly solitary, and if housed together they may fight fiercely, sometimes with fatalities occurring. Some dwarf hamster species may tolerate another hamster. Russian Hamsters form close, monogamous bonds with their mates, and if separated the males will often become very depressed. However, Chinese and Syrian Hamster females are often aggressive toward the male if kept together for too long after mating. Females are also sensitive to disturbances while giving birth, and may eat their young if they think they are in danger, although sometimes they will carry the pups in their cheek pouches to hide them in a different location. Some females will cannibalize their young if left with them too long. They should be separated as soon as the young can feed and drink on their own, which is usually three weeks of age.

Hamsters have long sharp incisors that grow continuously throughout their lifetime, and they must be given food stuffs regularly to wear them down. Wood chew sticks or non-toxic tree branches work well to help them keep their teeth short. A solid tank with a tight fitting lid is the best cage option because hamsters are escape artists. Plastic cages are a breeze for them to chew through, and the plastic tunnels sold as enrichment devices should only be accessible when their human is around to watch; otherwise, they will chew their way out of those in an instant. They are extremely susceptible to rapid temperature changes; both extreme heat or cold will stress them, and they don't appreciate long periods of time in direct sunlight. They will utilize a solid floored exercise wheel and enjoy being in an exercise ball that they can roll about the room.

Hamsters should be fed commercial pellets specifically designed for small rodents. Treats may include such items as tiny bits of carrot, apple (no seeds or skin), raisins, acorns, bananas, green beans, broccoli, cabbage, carrots, cauliflower, celery, corn, zucchini, cucumbers, grapes, kale, oranges, peas, spinach, sweet potatoes, turnips, and walnuts. Timothy hay, alfalfa, or a mix of the two, scrambled or hard-boiled eggs, dog biscuits, sugar-free cereal, whole wheat bread, and uncooked pasta, mealworms, or crickets are also nice additions to the diet. Drinking water is provided via sipper tubes. The water should be changed daily, and the tube cleaned thoroughly to prevent it from becoming blocked with food and growing bacteria. Hamsters are coprophagic.

QUICK FACTS 12.04 Hamsters

Life Span	Length/ Weight	Color	Genus species	Gestation Period	Activity & Caging
2–3 years For all species Sexual maturity 7–8 weeks of age	7.1 inches 18 centimeters Males 80–130 grams Female 95–150 grams	Black, grey, honey, white, brown, yellow, red, or a mixed in short or long hair versions.	*Mesocricetus auratus* – Golden or Syrian referred to as: "Honey-Panda-Black-European-Polar-Teddy Bears" and "Dalmatian"	16 to 19 days Litter size 2–16 pups - altricial	Nocturnal Minimum caging is 150 square inches (967 cm^2) of floor space, height 6–10 inches (15–25 cm), nesting or hide box lined with undyed, unscented toilet tissue, and deep substrate for burrowing
Sexual Maturity Russian 4–5 weeks of age	2.2–4.1 inches 5.5–10.5 cm	*Phodopus* hamsters have a black stripe running down their body from head to tail.	*Phodopus campbelli* – Campbell's or Russian Dwarf Hamsters *P. sungorus* – Djungarian or Russian White Coated Hamster.	Russian = 18 to 21 days Litter size 4–8 pups up to 13	
Chinese hamsters 2–3 months of age	C. *cricetus* 13.4 inches 34 centimeters not including the 2.4 inch (6 cm) prehensile tail	*Cricetulus* hamsters have a black stripe running down their body from head to tail.	*Cricetulus barabensis* – Chinese Striped Hamster Cricetus griseus - Chinese Hamster	Chinese = 21 to 23 days Litter size up to 7 pups	

Genetic Diseases and Conditions: Proliferative ileitis ("wet-tail" adenocarcinoma) is the number-one killer. Other conditions include Lymphocytic choriomeningitis, gastrointestinal worms, obesity, malnutrition, weight loss, dehydration, hairballs, foreign bodies, diarrhea, diabetes mellitus, leprosy, brucellosis, skin wounds and infections, hypothermia, hyperthermia, cardiac diseases, malocclusion, dental caries, cheek pouch impaction, and gallstones.

FIGURE 12.16 Hamster, Dominant Spot Chinese

The hamster is like the Guinea pig in antibiotic toxicity; all the drugs plus a penicillin-dihydrostreptomycin combination will kill a hamster in less than an hour. Tetracycline and Neomycin are safe to use. Hibernation can occur if room temperature is sustained at 55°F for a number of days; if suspected dead, warm them up gradually and see what happens.

Hedgehog

Origins: There are 17 species of Hedgehogs. The species that are kept as pets are hybrids of the Four-toed Hedgehog and the North African or Algerian Hedgehog, which is often referred to as the African Pygmy Hedgehog because it is smaller than the European Hedgehog. Two other species kept as pets include the Long-eared Hedgehog and the Indian Long-eared Hedgehog. The African Hedgehog was introduced and has established stable populations in France, Spain, Malta, Morocco, Algeria, and Libya. The European Hedgehog is native to almost all of Europe, and has been introduced to Mediterranean and Scottish islands and New Zealand where they are an invasive species. Wild hedgehogs can live in scrublands, forests, rocky grasslands, and arid areas. They eat insects and ground-nesting bird eggs, and they nest in scrub brush and rock burrows. It is illegal to own hedgehogs in some U.S. and Canadian communities.

FIGURE 12.17 Hedgehog, European with baby

Characteristics: Hedgehogs are lively pets that are easy to take care of but may not be appropriate for families with small children because of their spiny coats and the natural ***Salmonella*** bacteria flora on their coats. The spines cover their entire exterior except for their underbelly which has a soft fur. African hedgehogs are distinguishable from their cousins by the lack of spines on the crown of the head or a lack of a "widow's peak." The spines are made of keratin and do not easily come out of the skin unless under extreme stress or illness. Baby hedgehogs are born with their quills under the skin, which erupt quickly. There is a sheath around their spines which dries and falls off within hours. Baby spines are shed as they mature and are sharper than adult spines. The spines are a defense mechanism where when threatened they will often curly into a tight ball which makes the spines protrude straight out. Male hedgehogs have a mid-ventral penis.

To pick up a hedgehog, you gently grasp it around the belly. If it should roll into a ball, just hold it gently until it relaxes. They are very curious animals, and once it figures out it isn't in danger, it will unroll and begin to sniff its surroundings. Hedgehogs are solitary animals and don't seem to require interaction with humans but also don't seem to mind being handled. They should be handled daily to keep them tame.

FIGURE 12.18 Hedgehog, African Pygmy

Hedgehogs are naturally clean animals and will often groom themselves. One characteristic called "anointing" is when they smell something interesting they will often lick it and then proceed to hyper-salivate. They spit the foamy saliva onto their bodies and rub it around with their paws and nose. It is thought to be a scent-camouflaging technique. Many owners will bathe the hedgehog to remove the scent. To bathe a hedgehog, gather the materials: towel, toothbrush, and if desired a cat shampoo. Run about an inch of warm water into the sink and add a few drops of shampoo. Place the hedgehog into the water and use a cup to dip water to pour over the back. Avoid getting the face and ears wet. Gently scrub with a toothbrush over the back then use your hands to wash the belly. Set the hedgehog onto a towel and drain the sink, then refill and rinse the shampoo off with clean water. Once rinsed, place the hedgehog onto a towel and gently dry. It should be completely dry before going back into the cage. Hedgehogs can get chilled, so it is important to provide a warm area when they are drying. This is a good time to trim or check the toenail length, and if in need of a trimming, a human nail clipper will quickly get the job done.

Hedgehogs prefer quiet environments, so cage placement is important. Keep them out of the high activity areas of the home and away from loud noises. They need to be out of direct sunlight but still where they can get the day-and-night sequence. The cage should be equipped with a litter box using recycled newsprint (they will use it), a nest box, something to climb upon like a tree limb or a rock, and the substrate can be newspaper or recycled paper bedding at least several inches deep to facilitate burrowing. They also enjoy a "swim," so a shallow pan of warm water can be offered up a couple of times a week. A shallow "kiddy" swimming pool makes a great area for this activity. Place the pan in the center of the pool and allow the hedgehog to exercise; they often run laps around the circumference of the pool and then take a dip. A solid-floored exercise wheel, pipes and tubes to crawl into, balls, and chew toys inside the cage are also appreciated. Scoop or replace the litter in the litterbox as needed, and thoroughly clean the tank and toys weekly to keep everything odor free and dry. A wet and/or cold hedgehog can become ill, so ambient temperature should be kept around 75–80°F.

Hedgehogs are omnivores and should be feed a high-protein, low-fat diet in small portions in the evening. Start with 2–3 tablespoons of hedgehog diet or light cat food, 1–2 tablespoons of mixed fruit and vegetables, 3–5 mealworms or crickets. Moist dog or cat food, eggs or cooked hamburger or chicken should be offered about 3–4 times a week. The portion should be gone in the morning; if not, reduce the portion a bit at a time until the quantity is correct. Do not provide food ad-lib as this often leads to obesity. Fresh water from a sipper bottle or a heavy crock should be provided at all times.

Tidbits about hedgehogs: they may have some natural immunity to snake bites, they are used as food and medicine in some cultures, and they do vocalize using grunts and snuffles, only occasionally screaming if very frightened.

QUICK FACTS 12.05 Hedgehogs

Life Span	Length/ Weight	Color	Genus species	Gestation Period	Activity & Caging
Larger species 4–7 years Smaller Species 4–7 years Some up to 8–10 with good diet African sexual maturity 8–10 weeks European Sexual maturity 12 weeks	African 6–9 inches centimeters male 500–600 grams female 250–400 grams European 10 inches 260 mm Male 800–1200 grams Female 400–800 grams	Quills are white tipped with brown bands, color variations; brown, black, cream, gray, and albino.	North African *Atelerix algirus* Four-toed *Atelerix albiventris* Long-eared Hemiechinus auritus Indian long-eared H. collaris European *Erinaceus europaeus*	African 30–40 days 3–10 / litter European 31 to 35 days 3–4 larger species 5–6 for smaller species altricial	Nocturnal Minimum 4'L x 2'W and two tiers, a 20 gallon tank is appropriate with logs set at an angle for climbing and a nest box for sleeping

Genetic Diseases and Conditions: Fatty liver and cardiovascular disease, squamous cell carcinoma, ringworm, obesity, mange mites, internal parasites, pneumonia, enteritis, foreign body, ear infections, respiratory infections, dental problems, foot and nose sores from exploring or rough areas in cages, and potential carrier of Salmonella.

Mouse

Origins: Mice that are kept for pets are a domesticated form of the house mouse and fill multiple rolls; "fancy mice" are those that are selected for color, coat patterns, and size for exhibition, feeder mice are of the same genus except they are used to feed carnivorous animals such as snakes, and laboratory mice are also of the ***Mus*** genus, but they, too, have a number of different varieties that are required for different research parameters.

FIGURE 12.19 Mouse, Brown Fancy Mouse

Characteristics: Mice have small sleek bodies, a smooth hair coat, long hairless tail, large erect rounded ears, and eyeballs that protrude if restrained too aggressively. Mice have been used in research since the 1500s and continue to be the laboratory animal that is used in the greatest numbers because of their homology with humans. Most laboratory mice are hybrids of the subspecies ***Mus musculus domesticus*** and ***Mus musculus musculus*** and are very different in behavior and physiology. There are three different types of laboratory mice: outbred, inbred, and knockout mice, resulting in over 500 distinct strains. Outbred laboratory mice are raised by a large number of commercial breeding farms on a random mating system. Inbred mice are bred to close relatives which increases consanguinity. The ***isogenic*** mouse has been produced by at least 20 generations of brother-sister mating. These mice will accept skin and organ grafts from any animal of the same strain. Mice are also used in tumor and cancer research, organ transplants, toxicity testing, diabetes and renal disease, behavior, giardiasis, obesity, and autoimmune disease research.

Besides the number of strains available there are also four separate ecological classifications that are based on the microbiological status of the animals. Germfree or Axenic have no microorganisms detectable by current methods. These mice have to be maintained in plastic or steel isolators. Defined-flora animals carry one or more specified microorganism and are also maintained in isolators. Specific pathogen-free animals (SPF) have a microflora with a backward definition—that is, the organisms that are absent are defined. These animals are maintained behind air lock double doors, and the food and equipment are autoclaved to keep the reappearance of pathogens to a low level.

As children's pets, they are relatively cheap, hardy, not too long lived and are generally very docile if handled often. A major disadvantage is their urine has a strong ammonia smell often referred to as a mousy odor. To reduce the smell, the substrate material should be cleaned every two days and the cage disinfected every two weeks to prevent infection and disease. They tend to scatter their bedding, so a cage with high solid sides is recommended. It is important to know that mice can overwhelm owners with their reproductive capacity. A buck and a doe can have a litter every three weeks. A mouse is sexually mature at four to six weeks of age, and many are often pregnant when purchased. Adolescent mice and/or mature females can be kept in large groups, but males will fight, and often the stronger will kill the weaker. Enrichment items can be any non-toxic thing they can chew up. Toilet paper rolls, light cardboard boxes like cereal boxes (which also work well as nest boxes), tissues for nesting materials, sections of PCV pipes, and wood chips that can be purchased at pet stores provide enrichment items for mice. A solid-floored wheel is also appreciated for exercise.

FIGURE 12.20 Mouse, Agouti Even Mismark

FIGURE 12.21 Mice, Multiple Colors

Newborn males are distinguished from newborn females by the greater anogenital distance. Ear notching is the most common way of identifying mice in the laboratory, fur dying or fiber-tipped pen marks on the tail can be used, and tattooing is feasible but not common. Mice have a harderian gland connected to their lacrimal glands that produces "red tears" (porphyrin). This is normal, although they often increase during times of stress or illness. Their small body makes it difficult for them to regulate their body temperature, so they should be kept away from drafts and fluctuations in temperature.

Mouse pellets are formulated to provide a complete diet. However, they do appreciate fresh vegetables like carrots, lettuce, and spinach, plus apples, bread, and low-sugar cereal. Other foods like sunflower seeds can be used as treats but remember to keep portions small as they can make mice obese. If fed a powdered meal, often used in research diets, the teeth should be checked frequently for overgrowth of the incisors. Water provided in a sipping bottle is appropriate for mice. Mice are coprophagic.

QUICK FACTS 12.06 Mice

Life Span	Length/Weight	Color	Genus species	Gestation Period	Activity & Caging
1–2 years Sexually maturity – Males 6–8 weeks of age	6–7 inches 15–17.5 centimeters 1–1.6 ounces 29–44 grams Some show mice can get to 4.6 ounces 130 grams	See below	*Mus musculus*	19–20 days 10–20 pups average litter - altricial	Nocturnal Minimum caging: 2 square feet
Colors & Color Patterns	Black, beige, chocolate, blue, white, cream, coffee, gold, ivory, orange, silver, lilac, champagne, red, fawn, and dove. Coat types: short, satin, long hair, long hair satin, frizzy, frizzy satin, hairless. Tan & Fox – any color on top, Tans have a golden-red underside, and Fox have a nearly white underside. Marked mice are any color but there are 9 patterns: banded, broken marked, broken merle, Dutch, even marked, Hereford, rump white, spotted Tans, variegated, ideal markings of mice. Any Other Color: cinnamon, chinchilla, pearl, blue agouti, black, blue, chocolate and grey silver. Any Other Color Pattern: Blue Point Himalayan or Siamese, Brindle, Himalayan, Merle, Reverse Siamese, Roan, Seal Point Siamese, Siamese Sable, Splashed.				

FIGURE 12.22 Mouse, Fancy Black and White

Genetic Diseases and Conditions: Internal and external parasites, pinworms, mites, ticks, rickettsial parasites of the blood, respiratory infection (pneumonia), Salmonellosis, tapeworms, tumors, obesity, cardiovascular disease, diabetes, malocclusion, diarrhea, and Tyzzer's disease. Chronic ulcerative skin lesions particularly in the C57Bl, A/J, and BALB/c inbred mice. Zoonotic disease found in 100% of wild mice populations is Lymphocytic choriomeningitis.

Rabbits

FIGURE 12.23 Flemish Giant Rabbit with Shetland Sheepdog

Origins: Rabbits have been domesticated for years and have been used for fur, wool, food, companionship, research, and exhibition. In the 16th century they were used for food and fur with a number of breeds being developed at that time. In the 19th century rabbit fanciers developed more breeds so that now there are 48 breeds of rabbits recognized by the American Rabbit Breeders Association (ARBA) and 50 breeds recognized by the British Rabbit Council. They range in size from 20 pounds, as is the Flemish Giant in Figure 12.23, to as small as the Dwarf Hotot at 2 pounds in Figure 12.30.

FIGURE 12.24 Rex Rabbit

Characteristics: Rabbits make great pets in that they rarely bite and usually don't scratch unless alarmed or accidently if trying to escape. They will need to be handled daily to keep them tame and will often bond with an attentive owner. Care must be taken when picking up a rabbit, and unlike the cartoon rabbits they should never be picked up by their ears. Grasping them at the scruff of the neck with one hand and carefully lifting and tucking the back legs towards the belly with the other hand is the most appropriate way to pick up a rabbit. If rabbits are allowed to dangle they may kick out with the back legs with enough force to actually fracture their vertebrae. Rabbits will demonstrate fear by becoming very still, a bit twitchy; their eyes may bulge, and they may thump their back foot, and if very frightened, they emit a very high pitched scream.

Rabbits have multiple uses as mentioned earlier. In research they have been used for male reproductive system toxicology, the production of antibodies to develop vaccines, and for studies on asthma, stroke-prevention treatments, cystic fibrosis, diabetes, and cancer. The New Zealand

FIGURE 12.25 Rabbit, New Zealand White

FIGURE 12.26 Rabbit, Jersey Wooly

FIGURE 12.27 Rabbit, Satin

White rabbit is an albino rabbit that weighs about 5–6 pounds and is the most common breed used in the laboratory setting.

Rabbits are a good source of lean protein, and the giant breeds such as the New Zealand and Californian are preferred because they are a white rabbit, so the carcass is pink, unlike other rabbits with dark fur having a dark-colored meat. Rabbits raised for meat are slaughtered at various ages depending upon the type of meat required. Fryers are butchered at 70–90 days and weigh between three and five pounds (1.3–2.3 kilograms). Roasters are three to six months of age and weigh five to eight pounds (2.3–3.6 kilograms). Stewers are over six months old and weigh over eight pounds (3.6 kilograms).

There are a number of breeds of rabbits that provide wool that can be spun into yarn. The Angora (English, Satin, Giant, and French), American Fuzzy Lop, and the Jersey Wooly are all producers of wool. The fur is sheared, combed, or plucked from the hair coat during the molting period and then spun.

Fur from rabbit hides is used to make coats that are warm and quite stunning. The breeds most commonly used for fur production are the Satin, Rex, Palomino, and Chinchilla rabbits.

Rabbit exhibitions are quite popular and are a means in which anyone can participate in either county fairs or other fancier shows throughout the year. Purebred rabbits are shown, and a registration with the ARBA is required. The ARBA holds an annual convention where over 25,000 rabbits are in attendance. Some of the most popular breeds are lops (see the English Lop, Figure 12.28 below), which are rabbits whose ears fall to either side of their head and come in either a standard or miniature size. Rabbits are also shown in jumping competitions.

Bucks and does are territorial and should not be housed together. Does have been known to castrate or eviscerate bucks using their back feet. If the goal is to raise rabbits, the doe is brought to the buck's cage where copulation takes place. The doe is then removed and brought back a few hours later to increase litter size. Then she is placed back into her own cage. Does are induced ovulators and can be bred at almost any time, although fertility is reduced in the winter months. Sexual maturity, gestation periods, and weaning ages all depend upon the size of the rabbit. See the Quick Facts Table below. Baby rabbits are called "kits," and an old term for an adult rabbit which is often used in reference to their meat is "Coney." Neutering or spaying a rabbit before puberty helps reduce reproductive-related health issues.

FIGURE 12.28 Rabbit, English Lop

Pet rabbits can be kept outdoors or indoors and should be housed separately, even if neutered. An outdoor cage is referred to as a hutch and usually consists of a wood or metal frame with wire caging material all around it, even the floor, which allows feces and urine to fall out of the cage. The cage should also have a solid floor space for the rabbit to rest upon, usually with a nest box to get out of the wind. Shade should be provided to prevent heat stroke, and they must be taken into a cooler space if temperatures go over 90°F. If wintered outside, deep hay or straw bedding should be placed inside their nesting box to keep them warm. Indoor cages with a solid floor work well, and if provided a litterbox, they will use that most of the time. Litter made of paper, alfalfa or oats works well and should be cleaned at least twice weekly if not more often to control odor. Most of the diseases associated with rabbits are caused by dirty cages. Keeping a cage clean, providing regular brushing, trimming toenails, and a quality diet is all a bunny needs to remain healthy. If rabbits are allowed to roam about the home, a litterbox should be placed for easy access, and the room should be rabbit proofed. They will chew on electrical cords and furniture. Toys are appreciated and can range from cardboard boxes, to paper towel or toilet paper tubes. Purchased wooden toys will help keep their teeth worn down as well.

Diet depends on how the rabbit is going to be used. A pet rabbit can be fed pellets but should be supplemented with grass hay (alfalfa and pellets are fed ad-lib up to six months of age), dark leafy greens (not iceberg lettuce), vegetables, and small pieces of fruit. An adult rabbit should be fed 2–3 tablespoons of pellets per five pounds (2 kg) of body weight daily. Unlimited access to grass hay is fine for the adult. The stereotypical carrot should be fed only occasionally because carrots are high in sugar and can cause obesity in rabbits. Meat and fur rabbits are usually fed a commercial pelleted diet that is alfalfa based with other grains to provide the carbohydrate requirements. Grass hay is also offered but is usually fed in hoppers to prevent fecal contamination. Fresh drinking water

FIGURE 12.29 Rabbit, Giant Checker

should be provided at all times from a sipper bottle. If housed outside, a rabbit will need to drink more often, especially in the summer months, so the bottle should be checked twice a day. Rabbits practice coprophagia.

QUICK FACTS 12.07 Rabbit

Life Span	Length/Weight	Color	Genus species	Gestation Period	Activity & Caging
5–15 years Sexual maturity Minis 4–5 months, Medium breeds 5–6 months, and Giant breeds are 6–7 months	6–12 inches 15–30 centimeters Weight (see chart below)	See below	Oryc-tolagus cuniculus	28–32 days 7–8 kits per litter - altricial	Crepus-cular Minimum Caging depends on adult size should allow 5 hops length wise and tall enough to allow them to stand on their back feet and not touch their head to the top.
Colors	Agouti—brown-gray, copper, chocolate, opal Broken—white with any color patches or spots Chinchilla—chocolate, lilac Steel—chocolate, lilac Otter, black, blue Pointed—white with black, blue, chocolate or lilac color on nose, ears, feet, tail Sable—marten, point Self—black, blue, lilac, castor, cinnamon, chocolate, cream, fawn, frosted pearl, gray, light gray, orange, pearl, red, sandy, seal, white with blue or red eyes Shaded: color transitions from dark to light—frosted pearl, sable, sable point, Siamese sable, seal, tortoise Silver—fox, marten Tan Pattern—otter and marten Ticking Tortoiseshell –orange or fawn with black, blue, chocolate, lilac Tri-colored—white with black/orange, lavender blue/fawn, chocolate/orange, gray/fawn.				

Breeds by Size and Color or Color Pattern	**Small - 2 to 6 pounds (0.9—2.7 kilogram)** American Fuzzy Lop—variety of coat colors, dense, coarse coat. Britannia Petite—ruby eyed white, black otter, black, chestnut agouti, slender, fine boned build with a sleek, silky coat. Chinchilla—fine boned. Dutch–white band around shoulder area, white on face and back feet, and the rest can be black, blue or brown, chocolate, steel or tortoise. Dwarf Hotot—white with black eye rings. Florida White–Pure white. Rounded body. Havana–black, blue, chocolate, compact, rounded body. Himalayan–black, blue, chocolate, lilac, color on ears, nose, tail, feet, and legs. Holland Lop–agouti, broken, pointed white, self, shaded, ticked, or wide band color groups, muscular build, lop ears. Jersey Woolly–wide variety of colors, long wooly coat. Mini Lop–agouti, broken, pointed white, self, shaded, ticked, or wide band color groups, muscular and compact. Mini Rex–black, blue, broken group, castor, chinchilla, chocolate, Himalayan, lilac, lynx, opal, red, seal tortoise, white. Netherland Dwarf–self group, shaded group, and agouti group, tan pattern group, fawn, Himalayan, orange, steel, tortoiseshell. Polish–black, blue, chocolate, blue-eyed white, ruby-eyed white, also have very short ears. Silver–black, brown, fawn with silver/white guard hairs. Tan–black, blue, chocolate, or lilac with tan (eye circles, nostrils, jowls, ears, backs of legs, toes, chest, belly, tail, and neck collar). **Medium—6 to 9 pounds (2.7—4.1 kg)** Alaskan—black, thick coat, originated in Germany American Sable–sepia brown - soft, dense fine coat with coarse guard hairs. American—blue, white Argentes—black, blue, brown, or creamy white. Belgian Hare—reddish tan or chestnut with slate blue under coloring. Slender build, fairly stiff coat. English angora–long silky hair. English Spot–white with black, blue, chocolate, gold, grey, lilac or tortoise. Markings include butterfly mark on nose, colored ears, eye rings, spine marking (which is herring-boned) and a spot on the cheek and a chain of spots along the body. Long arched body (like a hare). French angora—long silky hair Harlequin—black, blue, chocolate, lilac - alternating bands of color and white, with half the face white and the other half colored, and the ear on the white side colored, and vice versa. Lilac—compact body and dense coat. Rex—black, black otter, blue, broken group, Californian, castor, chinchilla, chocolate, lilac, lynx, opal, red, sable, seal, white. Rhinelander - white with black and bright golden orange markings (6–8 round markings on each side of the back part of the body, as well as a spine marking, butterfly mark on the nose, eye circles, colored ears, and round cheek spots). Satin Angora–very fine wool. Silver Marten–black, blue, chocolate, sable with silver-tipped guard hairs. Standard Chinchilla–chinchilla coloration. Rounded body.

QUICK FACTS 12.07 Rabbit (Continued)

	Large–9 to 11 pounds (4.1–5 kg) American Chinchill–dense, fine hair that is smooth and glossy. Californian–white, with black nose, ears, feet, and tail - rounded body, medium build, and short smooth coat. Champagne d'Argent–bluish white with black hairs interspersed and slate blue under color. Cinnamon–rust or cinnamon color with grey ticking on back and grey on belly. Rust colored spots inside hind legs as well as butterfly mark on nose and eye rings. Creme d'Argent–creamy white with orange undercoat, and butterfly marking on nose. English Lop–agouti, broken, self, shaded, ticked, or wide band color groups. Very long lop ears. Giant Angora–soft fine undercoat (wool), straight stiff guard hairs, and a wavy fluff with a guard tip in between. Hotot–white with black eye ring. Well rounded body. New Zealand–black, red, white. Long muscular body. Palomino–golden, lynx. Satin–black, blue, broken group, Californian, chinchilla, chocolate, copper, red, Siamese, white. Silver Fox–jet black with silvering. **Giant–11 pounds (5 kg) and over** American Checkered Giant–white with black or blue markings. Checkered Giant–white with black or blue markings - long, hare-like body. Flemish Giant (Patagonian)–black, blue, fawn, light grey, sandy, steel grey, white. Long rabbit, heavy build (but shouldn't be fat). French Lop–agouti, broken, self, shaded, ticked, or wide band groups. Muscular heavy build. Giant Chinchilla–chinchilla coloration, heavy build, rounded body. Giant Papillon–white with markings - similar to English Spot except patches on sides instead of spots.

Genetic Diseases and Conditions: Respiratory infections (snuffles from ***Pasteurella***); sinus infections from environmental pollutions, stress, or high temperatures and/or humidity; runny eyes associated with dental issues or blockage of the tear ducts; corneal disease; entropion; distichiasis or inflammation of the eyes; sore hocks from wire bottomed cages and dirty cages; inner ear infections from mites or protozoan parasites (***E. cuniculi***); stroke; diseases of the brain or inner ear can cause "wry neck;" malocclusion causing teeth not to wear properly and so they develop molar spurs which causes pain and they'll stop eating; also overgrowth of the incisors which can cause pain; gut stasis; hairballs; fractures; fly strike;

FIGURE 12.30 Rabbit, Hotot Dwarf

abscesses; heatstroke; and cancer. They can be vaccinated for West Nile disease. Rabbits are low risk for zoonotic diseases; however, people with compromised immune systems can acquire ***E. coli***, ***Salmonella***, and ***E. cuniculi*** from rabbits.

Rats

FIGURE 12.31 Rat, Fancy Blaze

Origins: The Brown Rat or ***Rattus norvegicus*** is the wild version of the domesticated "fancy rat." They are originally from Norway and have spread worldwide by stowing away on ships. Domesticated rats are different from their wild cousins in every way. They pose no bigger threat of illness to humans than a dog or cat versus their wild cousins that can carry any number of diseases. The fancy rat came about from the blood sport of baiting, which occurred during the 18th and 19th centuries until outlawed. Rats would be released into an enclosure, and a terrier-type dog would be set upon them. Betting took place as to how many rats the dog could kill in a specific space of time. It is thought that two men, Jack Black and Jimmy Shaw, are the forefathers of fancy rat breeding. They would keep and tame the rats that were of unusual color, adorn them with ribbons and sell them as pets. In England the National Mouse Club included rats in 1901 until 1930 when pet rats fell out of fashion. The National Fancy Rat Society was established in 1976 in England and in the U.S. the American Fancy Rat and Mouse Association (AFRMA) was founded in 1983. Importation of rats is illegal in Australia and in Alberta, Canada, unless they are in a school, laboratory, or zoo.

Characteristics: Rats have long bodies, long tails with little or no hair, long pointed noses, poor eyesight but excellent hearing, and a strong sense of smell. They prefer to be clean and are constantly grooming themselves to stay clean. They are very smart and seem to be able to remember. Rats are used for a number of reasons: as pets, in research, or for feeding reptiles and rehabilitating raptors.

Laboratory rats have served and continue to serve as animal models for research studies in psychology, medicine, and other fields. The Wistar Rat is an outbred albino rat that was the first strain of laboratory rat. It is one of the most widely used rats in laboratory research and was used to develop a number of other strains also commonly used in research. The Sprague Dawley is an outbred albino that is used in medical research mainly because it is docile and easily handled. The Lewis Rat was developed in the 1950s and is also known for its docile behavior. They are unique in that they spontaneously suffer from neoplasms including adenomas and adenocarcinomas of the

FIGURE 12.32 Rat, Dumbo

pituitary and adrenal cortex in both sexes, mammary and endometrial carcinomas in females, and C-cell adenomas, and adenocarcinomas of the thyroid gland and tumors of the hemopoietin system in males. They are also prone to develop transplantable lymphatic leukemia, and when they are old, they often develop glomerular sclerosis. Rats are also used in transplantation research, induced arthritis/inflammation, experimental allergic encephalitis, and STZ-induced diabetes. Long-Evans Rats are outbred rats that were developed by crossing Wistar females with wild gray rats. They are white with a black or brown hood and are utilized in many different studies including behavioral and obesity research. The Bio-breeding Diabetes Prone (BBDP) rats spontaneously develop autoimmune Type 1 Diabetes and have led to many breakthroughs in research of T1DM. Zucker Rats were bred for research on obesity and hypertension. These rats can weigh up to 1 kilogram or 2.2 pounds. Hairless rats are utilized in research on compromised immune systems and genetic kidney disease. There are three varieties of "hairless" rats with various amounts of hair: Rowett Nude (rnu) does not have a thymus gland in addition to no hair, Fuzzy (fz) Rats develop progressive kidney failure, Shorn (shn) Rats were developed from Sprague Dawley Rats and also suffer from kidney failure. The Royal College of Surgeons (RCS) Rat is the first to inherit retinal degeneration. Shaking Rat Kawasaki (SRK) lacks the gene for cortex lamination and cerebellum development. Others include Fischer 344, Holtzman albino strains, and Lister black hooded rats. Inbred strains are also available but are not as commonly used as inbred mice.

The reason we discussed laboratory rats first is that many of the pet rats owe their existence to those strains developed for research. Rats make great pets; they are affectionate and love to play with their humans. They are quiet, require little space and enjoy learning tricks. With patience and food treats, they can learn a variety of tricks and seem to enjoy the interaction and stimulation.

Rats are social, so two rats will be happier especially if owners are away for much of the day. Same-sex pairs or neutered male/female pairs work well together. Don't be alarmed if there are tussles which are more often than not play fighting. Rats are expressive and have a wide range of squeaks and squeals that can indicate excitement, fear, or anger which is also accompanied by body language. Teeth grinding is often associated with contentment whereas teeth chattering is associated with anger or feeling threatened. This can be accompanied by the rat freezing in place as the tail shivers and/or slaps the floor rapidly. Rats can be picked up by placing one hand over and around the shoulders and the other scooping the rear end up. Then perched on a shoulder, inside the hood of a jacket or in a pocket, they will happily accompany their human wherever he or she goes.

Males are easily recognized as they have a large scrotum, and only females have nipples. The anogenital distance in males is also farther apart than in females. Ear notching, tattoos, tags, and fur dying are all means of identifying individuals. Rats also have the harderian gland that produces "red tears" and is normal although it tends to become more apparent when the rat is stressed or ill.

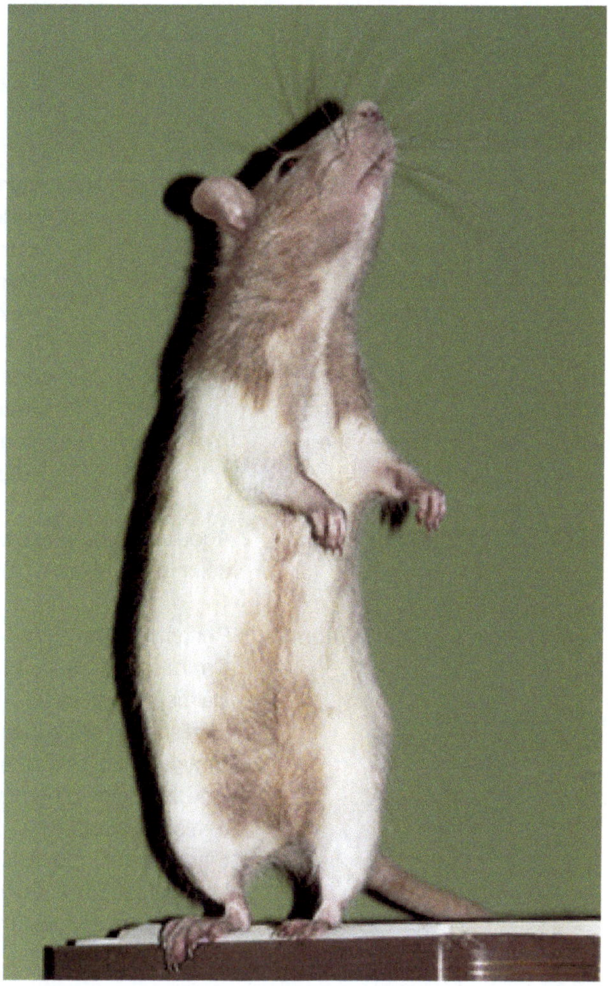

FIGURE 12.33 Rat, Hooded, Down Under Agouti

Appropriate cages for rats can include an aquarium tank or a solid-floored wire cage with multiple tiers in both. They appreciate a solid-floored exercise wheel, climbing apparatus, and toys to gnaw on to keep their teeth in good shape. The position of the cage is important for their mental health; avoid areas with bright nightlights or streetlights shining in on them during the evening hours. Young rats suffer a disease called ringtail which is evidenced by the tail looking scaly and constricted with the tip sometimes drying up and falling off. This occurs when the humidity is below 40%; most rats prefer 40–70% humidity and can tolerate temperatures from 65–80°F. Bedding and nesting-box materials can be recycled newsprint or strips of paper towels. The entire cage should be cleaned weekly, with food and water dishes cleaned daily. Stoppered water bottles and dishes attached to the cage bars work great for feeding and watering rats. Rats are also coprophagic.

Rats are omnivorous, and there are commercially available pelleted diets for rats that provide adequate nutrition. The pellets can be available ad-lib, and a variety of fresh fruits and vegetables can be offered as treats. Watch them with this feeding schedule because rats will eat because they enjoy food, and if they are bored will overeat. If the rat is getting fat, the pellets should be reduced a bit and smaller portions of the fresh foods offered until they are back in shape. Table scraps can be a treat for the rat; however, don't feed greasy or high-sugar foods as this isn't good for you or the rat.

QUICK FACTS 12.08 Rats

Life Span	Length/ Weight	Variety or Type	Genus species	Gestation Period	Activity & Caging
2–3 years Sexual maturity Males 6–10 weeks of age Females 8–12 weeks of age	Males 450–520 grams Females 250–400 grams	Dumbo–large, wide round ears Hairless Manx—tailless Rex—curly whiskers and coat Satin—shinier coat Standard	Rattus norvegius	21–23 days Litter size up to 12 pups - altricial	Nocturnal Minimum cage: be at least 14" x 24" x 12"
Color	Agouti, albino, beige, black, blue, brown, chocolate, cinnamon, fawn, and gold. Eye color can be black, pink, ruby (dark red), and odd-eyed				
Coat Patterns	Berkshire color on back with a white belly Capped color on head only Dalmatia spotted English Irish white that is even or symmetrical anywhere along its belly Himalayan color points on nose, ear, tail, and feet Hooded colored head that runs down back Self solid colors Siamese color points on nose, ears, tail, and sometimes feet Variegated blaze or white strip on forehead with a fully colored head and splotches of colors running down the back				

FIGURE 12.34 Rat, Fancy Female

Genetic Diseases and Conditions: Mycoplasmosis, Coronavirus, Sendai virus, respiratory disease, pneumonia, congestive heart failure, cancer, parasites, hair loss, ringtail, Giardia, pseudotuberculosis, abscesses, ulcerative dermatitis, obesity, malocclusion, pododermatitis, fractures, nephrosis, hyperthermia, dehydration, constipation, diarrhea, ear and eye infections, Tyzzer's disease, vaginal bleeding, paraplegia, and sudden death.

Sugar Glider

FIGURE 12.35 Sugar Glider, Male, showing bald spot where scent gland is located

Origins: Native to Eastern and Northern Australia, New Guinea, and the various islands in the region, including Tasmania where it was introduced. Sugar Gliders breed easily in captivity; however, they are commonly available in the illegal pet trade. They can be kept in the U.S. as pets; however, California, Hawaii, and Alaska do not allow them.

Characteristics: These small marsupials are **arboreal** and nocturnal, having large eyes which enable them to see in the dark. Sugar Gliders have squirrel-like bodies with a membrane (called a **patagium**) that stretches from wrists to ankles that gives them the ability to glide through the air to get to one tree or another. They have a weak prehensile tail that acts as a rudder when gliding and ears that can swivel independently. Their second and third digits are partially fused (called **syndactulous**) on the back feet, which are used for grooming. They have scent glands located on their foreheads (males only, and they have a bald spot in this location), chests, and two near the cloaca that are used for marking territories and each other. Females have a pouch in which their joeys continue to develop after they are born.

FIGURE 12.36 Sugar Glider, Female (left) and Male (right) eating mealworms

Sugar gliders are very social animals and do better with a friend. If kept alone, owners need to spend a great deal of time with them every day to keep them of sound mind. They will happily ride in a pocket and may mark you with their scent as you belong to their colony. They have large front incisors that are designed to gouge trees for sap, so it is important to provide them branches on which they can chew to keep their teeth worn down. Sugar Gliders are vocal and will emit low groans, screams, squeaks, yaps, and hisses.

In the wild Sugar Gliders will cycle through periods of torpor during which their body temperatures will lower as a means to conserve energy. It usually happens during the winter time or when food is scarce. This does happen to them in captivity, but it is usually shorter in duration. They are omnivorous, eating insects, lizards, and small birds, eggs, pollen, fungi, fruits, and, of course, the sap from eucalyptus, acacia, and gum trees. In captivity they require a diet high in calcium balanced with phosphorus (2:1) to prevent hypocalcemia. They need a mixture of foods that include 50% gut-loaded insects (crickets, mealworms, waxworms, moths, lean meat or poultry, hard-boiled eggs, yogurt, cottage cheese, and tofu), 25% fruit (apples, bananas, cantaloupe, figs, grapes, grapefruit, mango, oranges, papaya, peaches, pears, pineapple), and 25% vegetables (avocados, carrots, sweet corn, and potato). There are also commercially prepared diets such as Bourbon's Modified Leadbeaters and High Protein Wombaroo, or there are recipes available on the internet that can be made using human food. Sugar Gliders should not be fed candy, sugar, or chocolate. Nuts can be used as a treat, and a multivitamin and calcium supplement used for captive reptiles works well for Sugar Gliders as well. Use dishes that can be attached up high as they enjoy eating high up in the cage. A stopper bottle for water is important as well.

Caging for Sugar Gliders should be a tall, large bird cage that will accommodate branches for climbing, chew bones, and/or bird toys for chewing. Because of their large eyes it is important to keep them out of direct sunlight and provide them a nesting box that is completely enclosed, so they can sleep during the daylight hours. The cage should be in a quiet area of the home, so they are not disturbed during the day, but they should still be able to tell if it is day or night. Substrate on the bottom of the cage can be shredded paper or recycled paper that can absorb urine.

QUICK FACTS 12.09 Sugar glider

Life Span	Length/ Weight	Color	Genus species	Gestation Period	Activity & Caging
10–14 years Sexual maturity 12 months	11 inches 28 centimeters 90–130 grams	Blue-grey, yellow, tan, or albino, black stripe runs from its nose to mid-back, belly, throat, and chest are cream color.	*Petaurus breviceps*	15–17 days – move to pouch coming out in 10 weeks, wean in about 15–17 weeks of age	Nocturnal Minimum cage 24" x 24" x 48"

Genetic Diseases and Conditions: Malnutrition, calcium deficiency, obesity, fractures, torn claws, self-mutilation, stress (malnutrition, dirty living conditions, overcrowding, boredom, loneliness, over-handling, hyperplasia, polydipsia, coprophagia), parasites, pneumonia, diarrhea, gastroenteritis, constipation, paralysis, blindness, cataracts, and burns.

FIGURE 12.37 Sugar Glider, Hanging From a Tree

CHAPTER 4

Avians

Species of Avians

There are two genera of birds kept as pets. The Passerines are the seed-eating birds, and the most-kept species include the canaries, finches, doves, and pigeons. The Psittacines are the parrot-type birds and includes a number of species from the South Pacific, Afro-Asian region, Central and South America. They are kept to provide companionship and entertainment. However, neither type should ever be adopted without doing complete research on their many needs.

Birds are flock animals, which means they have a very social way of life, with many even mating for life. When removed from a group and living a single life among people, they, especially the Psittacines, demand a great deal of interaction with their people as a surrogate flock. These birds will require specific housing, a balanced, high-quality diet, training, and many, many toys to keep them from biting and chewing on objects when outside of their cages. Some of these birds will live for 80 or more years and have the mentality of a two- or three-year-old human child.

It is recommended that an exotic bird be purchased from a breeder that has hand-raised and socialized the bird. Wild-caught birds are very difficult to tame and can harbor diseases that can be transmitted to people, their age is unknown, they may become forlorn mourning the loss of a mate, and in many countries it is illegal. There are also bird rescues that are a valuable resource for selecting a bird to fit the owner's lifestyle.

Birds are covered in feathers, which is referred to as plumage. Feathers are made of keratin, and each feather has a unique design to keep the bird warm, cool, waterproof, or allow it to fly. The main shaft down the center of the feather is called a rachis. The rachis attaches to a skin follicle with a structure called a calamus which is hollow. Branches

FIGURE 13.00 Parts of a Feather, 1.Vane, 2. Rachis, 3. Barb, 4. Afterfeather, 5 Calamus

or barbs extend from the rachis and collectively make up the vane and give the feather its shape. Each barb has a shaft called a ramus, and from each side of the ramus there are barbules which have barbicels which form a "hook and eye" system. This allows the bird to keep its feathers in place (see Figure 13.00).

There are six types of feathers, known as contour, semiflumes, filoplumes, down, powder down, and bristle. Each serves a different function such as flight, heat conservation, waterproofing, camouflage and display. **Contour feathers** constitute the ordinarily visible plumage (see Figure 13.01 below for the names of the contour feathers). **Down feathers** form an undercoat and are generally not visible. They are soft to the touch because they do not have barbicels. The barbules float free of each other, allowing the down to trap air and provide thermal insulation. **Powder down feathers** are continuously growing feathers where small particles break off the barbules. The resulting powder is used by the birds for cleaning, conditioning, and waterproofing the feathers. It also gives the plumage a characteristic bloom. Semiflumes create the shape of the bird and trap warmth. Semiflumes look like they are totally made of "afterfeathers." Filoplumes are specialized extra-sensory feathers that register vibrations and pressure, and they help keep the other feathers in place during flight and bathing. They have a bit of "afterfeather" at the top of the rachis. Bristle feathers are an extra-sensory feather that stimulates the reflex for snapping up food. The bristle is a straight feather.

Preening is where the bird cleans and reposition feathers by coating them with power down. Preening is a socializing event, and pet parrots will often preen their favorite person. Some birds do not have power down, but instead they have an oil gland that provides cleaning and waterproofing which is spread on the feathers in the same manner as the powder down. Birds enjoy a bath, and it is important to provide the means for them to take a bath to help them maintain their feather condition in addition to preening. For smaller birds a shallow pan of water works well. For the larger birds a few inches of water in a bathtub works better. This will facilitate cleaning up the water that gets flung around in their enthusiasm to get clean. Bathing should be offered several times a week. If the bird doesn't care to bathe, remove the pan so that they don't defecate in it and then later decide to bathe in the contaminated water. Most birds will also appreciate a daily misting with warm water and will get quite enthusiastic, lifting each wing for an extra squirt.

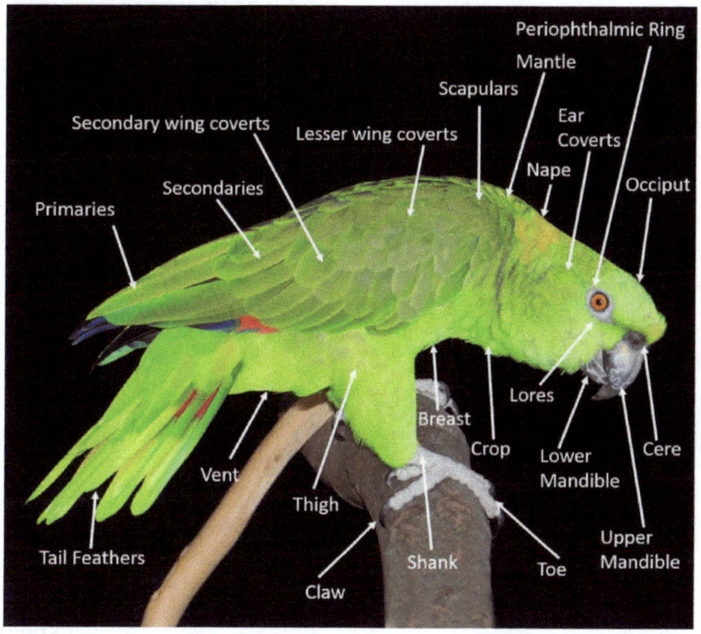

FIGURE 13.01 Names of Feathers and Body Parts

Molt or **Molting** is the periodic renewal of feathers and is a two-step processes. Molt is brought on by hormonal and seasonal changes. The first step is the loss of the old feathers, and the second step is the growth of new feathers. A **pinfeather** or **"blood feather"** is a new feather as it comes out from the skin. It is covered with a sheath and has a bright red shaft. The calamus of the new feather allows the blood to enter the feather until it is fully grown. Once the feather is fully grown, the calamus is sealed off, and the feather is no longer considered alive. Missing feathers are re-grown right away; whereas, damaged feathers are replaced when the bird molts. It is important to feed a quality diet during the molt because it takes a lot of energy to replace feathers. Many birds molt twice a year with some molting more often.

Ectoparasites are parasites that live on the surface of the skin or inside the feathers. Feather mites (red mites) will fill the inside of the calamus, turning it brown. Feather lice will live and chew holes in feathers. Skin mites (***Knemidocoptes mutans***) will dwell in the skin and cause it to become scaly on the cere and/or legs.

It is recommended that wings be clipped on the Psittacine birds. The goal is to keep them from flying long distances and to not get enough lift to fly up. They should still be able to flit to the ground without falling. The primary feathers are clipped short enough for the ends to be covered by the coverts. The number of feathers to clip is different for the different sized birds, and each feather is

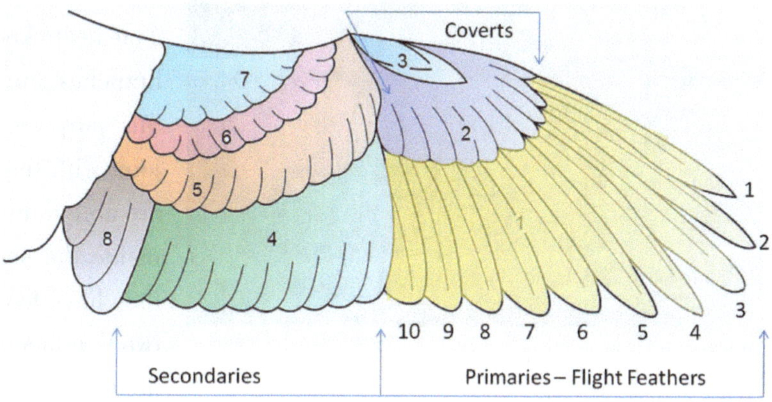

FIGURE 13.02 Outstretched Wing, Showing Primaries and Coverts

FIGURE 13.03 Goffin's Cockatoo in Harness

clipped individually to ensure blood feathers are not cut.

Budgerigars, Cockatiels, Lovebirds, and Conures should have at least five primary feathers cut starting at #1 and cutting to #5. If the Conure is still able to get lift, cut one or two more of the primaries. African Grays, Cockatoos, and Amazons should have at least eight primaries cut, and for Macaws eight to ten primaries. Start with eight, and if they can still lift off, then cut one to two more. Do not leave the first two primaries as they may still provide enough lift to lose your bird out an open door. It is important to check the wings every 8–12 weeks for new feathers, especially if blood feathers were skipped during the clipping. If there are blood feathers, you have to wait until the calamus is sealed and the interior is completely white. A cut or broken blood feather can actually cause a bird to bleed to death. If broken or cut it must be pulled out and pressure applied to the skin follicle to stop the bleeding.

FIGURE 13.04 Zygodactyl Feet

Some owners are opting to not trim their birds' wings and will use a harness to take them outside for exercise.

All Psittacines have a curved beak that is also made of keratin. The beak grows continuously, and it requires a means to keep it trimmed. Chewing toys, non-toxic tree branches, cuttlebones, and mineral blocks work well for this purpose. The mandible has some mobility in the joint with the skull and along with the tongue the birds are able to manipulate nuts and seeds to get at the meat inside the shells. All parrots are zygodactyls, meaning they have four toes on each foot: two that face front and two that face back. Many parrots are actually left-footed, holding food and toys in that foot.

Housing for birds can range from a wire cage for the Passerines to a large perch stand for the large Psittacines. Each bird species discussed in this book will have the minimum cage requirements in the corresponding Quick Facts table. Whatever the housing, it is important that it be cleaned daily, which includes cleaning both the water and feed cups with hot soapy water, changing the papers at the bottom, and wiping off the perches. On a weekly basis the bars and perches should be wiped down with hot soapy water. Every month the entire cage or perch must be disinfected. Without this level of cleanliness, bacteria and parasites multiply which endangers the bird's health.

Parrots should be kept where the center of activity is in the home. A family room that is bird proofed so that the cage door may be left open to allow the bird out whenever someone is at home is important for mental health. This is also a good time to interact with the birds such as talking to or petting, going about daily activities.

Perches made of natural materials of varying diameters must be placed at different levels in the cage to allow for climbing mobility and a choice to select a comfortable roosting spot. Care must be taken when placing the perches so that they do not allow fecal material to fall into the feed and water cups. Birds require plenty of toys, branches from non-toxic trees, swings, tunnels, boxes, and other safe things to chew on and play with. Without them, the birds become very bored, and if their social needs are also not met, they can start to pick at and then pull out their feathers.

The Psittacines should be allowed out of their cages as much as possible so that they may interact with the "flock." This will keep them happy and relieve boredom. Before doing so make sure the home is bird proofed to prevent severe injuries and/or death as follows:

Turn off or close stove tops, especially with pans full of hot water or oil, ceiling fans, stand-alone wood fireplaces, open doors to the outdoors, and open toilet bowls.

Eliminate toxic inhalants such as cigar or cigarette smoke, pesticides, plug-in air fresheners, heavily scented candles, cleaning supply fumes such as oven cleaners, and burned nonstick coated pans.

Other household hazards include easily destroyed toys that break into small parts that can be swallowed, toys with small parts, toxic houseplants, cedar, redwood and other pressure-treated wood, items made of lead or zinc, clear glass windows or walls, electrical cords, curtain cords, lacey or string curtains that can snag a toenail or wrap around a leg, other birds, cats, dogs, and children.

A healthy bird should have dry, open nares; clear, bright eyes; a smooth beak; bright and well colored feathers without breaks or ragged edges; even scale patterns on legs and feet; and a body free of lumps and bumps. They should be alert, bright and responsive, have appropriate posture on the perch, eat and drink well, and move without difficulty.

Signs of illness include lack of appetite; weight loss; undigested seeds or food in the droppings; pasty vent or discolored oral cavity; unusual droppings; ruffled, stained, or discolored feathers; bulges in feathering; watery, closed, or squinty eyes; swelling of eyelids; drooping or hanging wing; shifting from foot to foot; difficulty breathing, tail bobs up and down with respirations; listlessness or mood changes.

Birds that are improperly fed can develop many diseases and conditions that can threaten their lives. Food and water dishes need to be cleaned with hot soapy water to remove bacteria from droppings and food debris. Make sure food dishes are completely dry before adding seeds or other foods to prevent mold or bacteria from multiplying. All birds need fresh water twice a day. If feeding seeds to Psittacines, check the seed cup often as they hull the seeds and drop the hulls back into the cup. It may look full but there is nothing; if you blow on the cup and it empties out, there was nothing but hulls.

A balanced bird diet consists of a mixture of pellets and/or seeds, fruits, vegetables, proteins, and fats. It is best to serve small amounts of the fresh foods either chopped, diced, or shredded. This motivates the bird to give them a try, and they don't need huge quantities. One resource notes that for a cockatiel one grape is equivalent to 49 grapes to a human. Wash all fresh food thoroughly before serving as it may contain bacteria or pesticides that could make your bird sick. Pelleted diets are formulated for each bird species and during specific life events like molting and aging. Select diets without dyes, artificial flavorings, fortification or preservatives. Harrison's, Dr. Harvey's Bird Food Mixes, or Lafeber make quality pellet diets. Avoid feeding large amounts of citrus fruits or vitamin-C-rich foods if feeding pellets as it can increase the amount of iron absorbed from these foods leading to "Iron Overload Disease." Pellet-fed birds also tend to drink more water than usual and will have brown-colored droppings.

Fed in small amounts, most foods are safe to give to birds. However, if variation is not present, and some foods are fed too often, it can cause issues with a bird's health. Some birds have a tendency towards developing gout and should be fed a diet low in protein. Plant-based protein generally does not induce gout, but it can if fed in large amounts. It can also put a pet into breeding or hormonal mode. Please note the following are only suggestions and research should be done to properly feed specific species of birds; also note the percentages foreach type of food appropriate on a daily basis.

Balanced diet for Psittacines consists of 10% fruits, 10% vegetables and greens, 50% pellets (or 25% each of pellets and seeds), 25% seeds, 5% protein, whole grains, and table foods.

Balanced Diet for Passerines consists of 10% seeds, 80% pellets, and 10% fresh fruits and vegetables.

TABLE 13.00 Food Choices for Psittacines

Fruit Choices	Vegetables/ Greens/ Flowers/ Herbs/Grasses	Protein Choices	Whole Grains/ Table Foods/ Herbs/ Grasses	Seeds
Apples, apricots, bananas, berries, cantaloupe, cranberries, figs, grapes, kiwi, mangos, melons, oranges. papayas, peaches, pears, pineapple, plums, pomegranate, baby food. Citrus: small amounts of oranges, nectarines. Raisins (small quantity only): reserved as a treat and for training. Naturally dehydrated fruits and vegetables without sulfur dioxide. (Sulfur dioxide is a preservative known to increase hyperactivity, aggressiveness, feather shredding, or picking).	*Vegetables:* Beets, carrots, cauliflower, chicory, corn on the cob, cucumbers, escarole (endive), green beans, parsnips, peas, peppers, Swiss chard, tomatoes, turnips, zucchini. Cooked: carrots, yams, sweet potato, pumpkin, and squash (roast seeds). *Greens:* Celery (stalks and leaves), collard, dandelion, kelp, leaf lettuce, mustard, seaweed, sprouts (sprouting tips below), turnip, and water cress. *Flowers:* Calendula, carnations, chamomile, chives, dandelion, day lilies, echinacea, eucalyptus, hibiscus, honeysuckle, impatiens, lilacs, nasturtiums, pansies, passion flower, roses, rose hips, sunflowers, tulips, and violets.	Cooked legumes, hard boiled or scrambled eggs, grated cheese yogurt, cottage cheese, sprouts, ripe tomatoes, spinach, plus freshly cooked lean beef, chicken, or fish in small quantities. *Nuts & Oil Seeds:* feed in small amounts as they are high in fat and can be used as a treat and for training. Hard-shelled nuts given whole for birds to crack open themselves are good exercise and entertainment. *Nut choices:* Almonds, Brazil nuts, cashews, hazelnuts, macadamia nuts, pecans, pine nuts, pistachios, sesame seeds, walnuts.	Amaranth, barley, bread, flax, cereal (whole grain, no sugar), corn, couscous, oats, pasta (cooked), quinoa, rice (cooked such as basmati, brown, and jasmine), and wheat. *Herbs:* Basil Leaf, Lavender, Peppermint Leaf, Rosemary Leaf, and Thyme Leaf. *Grasses:* Alfalfa Leaf, Oat Straw, Red Clover Blossoms, Red Raspberry Leaf, Fruit tree blossoms, and Papaya Leaf.	Pumpkin and squash seeds roasted in shell and fennel seeds. Sunflower seeds are used as a treat or reward during training. Some birds love them so much that they stop eating everything else. All birds are particularly fond of millet, and one spray of millet will keep a bird cheerfully occupied for a couple of days. Purchased seeds can be frozen to keep them fresh and free of bugs. Store in vacuum-sealed bags or re-sealable plastic bags to avoid moisture inside the bag.

TABLE 13.01 Food Choices for Passerines

Seeds:	Fruits and Vegetables	Greens
Basic Mix: 40% yellow millet, 18% canary seed, 10% white millet, 9% peeled oats, 5% Japanese millet, 4% safflower, 4% buckwheat, 3% paddy rice, 3% oats, 2% linseed, and 2% hempseed.	Fruits: Apples, carrots, corn on the cob, endives, grapes, peas, spinach, cooked sweet potatoes, baby food, and watercress.	Chickweed, dandelion, nasturtium leaves, or thistle, or salad greens, such as Bok Choy, lettuce or cabbage, should be fed daily if possible.

All birds need a calcium and mineral source. A number of foods contain calcium: oatmeal, bok choy, broccoli, cabbage, carrots, endive, greens, kale, okra, and tahini (sesame seed paste). In addition, egg shells that are baked and crushed can be added to soft foods, or shaving cuttlebones over their food adds calcium. A calcium/mineral block can be purchased and attached to the cage for the bird to chew. Remember that in order to absorb calcium, Vitamin D3 is needed as well. It can be added to the soft food or produced in the skin through ultraviolet rays obtained via daylight sun exposure or indoor full-spectrum lighting. Calcium and minerals are particularly important for maintaining the health of breeding birds.

A crushed stone or shell grit is an important part of nutrition for Passerines but not the Psittacines as they use it in the gut as part of the digestion. A shell grit has the added bonus of calcium which is essential for health and during the breeding season. DO NOT FEED grit or crushed oyster shells with silica sand (crushed glass) to small birds as it can cut the delicate lining of the esophagus, stomach, or intestines.

There are times when a bird requires an enhanced diet to maintain its health. When birds are molting, they benefit from sprouted seeds such as hemp, flax, and niger as well as cucumber. After the molt, cut back on the additional fat and feed your bird niger and hemp seeds as a treat to motivate it to sing.

Sprouted Chinese Mung bean and other seeds are quite good for birds, and they tend to readily eat them. To sprout seeds, place an 1/8th to 1/4th cup of seeds in a clean glass jar. Fill with tap water and let stand at room temperature for twenty-four hours. Rinse and drain completely. Repeat the rinsing and draining daily until the seed has sprouted. Rinse and drain one last time and put them on a paper towel in a bag in the refrigerator. Before feeding check for a foul odor or mold and discard if necessary. Sprouted seeds are lower in fat and are a balanced choice as they supply vegetable proteins, vitamins, minerals, enzymes, and chlorophyll.

If the birds are to be bred, protein is especially important and be should given to hens in the spring prior to the breeding season. Mealworms are a great and inexpensive choice or a high-protein commercially prepared nestling food is acceptable. Shelled hemp seeds have virtually the highest concentrated balance of proteins, fats, vitamins, and enzymes.

TABLE 13.02 Toxic Foods That Are Never Fed To Birds

TOXIC FOODS THAT ARE NEVER FED TO BIRDS			
alcohol	**avocado**	**chocolate**	**mushrooms**
dry beans (black or caster)	caffinated beverages: coffee, tea, sodas	sprouts from the Lima or Navy bean	raw eggs, meats, fish, or chicken
onions	rhubarb leaves	tomato plants	moldy foods
dried peanuts in the shell (potential aflatoxin contamination)	pits or seeds from apples, apricots, cherries, peaches, pears	high-fat, salty, or sugary foods	nutmeg
milk/milk products	poultry feeds	toxic house plants	

Birds that have received antibiotic therapy and are stressed, molting, or sickly would benefit from a dose of probiotics. Probiotics such as Acidophilus and Bifidus regularis are organisms that boost the immune system and help to restore the healthy microbes in the digestive tract.

People who are either inexperienced or unwilling to spend time on training and socializing a parrot should not get a parrot. Remember, they have the mentality of a two- or three-year-old child and will exhibit many of the same behaviors of such child. If there are not boundaries or expectations of behavior, the parrot will eventually develop bad behaviors such as aggressiveness, screaming and screeching, and feather picking. Once these behaviors are established, inexperienced bird owners are not able to easily reverse them. Learning a bit about training and behavior modification is strongly recommended in order to help owners with their birds before problems arise.

For the most part parrots are generally easily trained. This book is NOT designed to be the last or even the first word on training. The learning process for a parrot is not based on hours or days, but in months; time and patience are the keys. There are many resources available that deal with training and socializing birds. Here are a few tips that may help owners get off to the right start:

1. Don't reinforce undesirable sounds. Reacting to screams or screeches reinforces negative sounds. Teaching the birds to talk, sing, or mimic sounds will redirect their noise making, mainly because this takes time, and interaction with the bird is what they need. Never feed or go to the cage when

the bird screams. They quickly realize that even negative attention is better than no attention. These birds can be persistent and will try to wear on your patience by being louder than before. Go for a walk, watch TV, or listen to music, but do not surrender to a bird's demands. The moment they are quiet, pop in and LAVISH them with praise and treats when quiet or making nice sounds. They will quickly learn that being quiet(er) is the way to get attention.

2. PAY ATTENTION to the bird after being out of the house for a length of time, but do so on your terms. Do not speak to the bird or call out to it the minute you walk in the door. Put your stuff down, get settled, and IF the bird is quiet, go to the cage and reward it for patiently waiting. Being allowed out of the cage, a food treat, and some attention is all that is needed to train the bird to wait for you to come to it. Again, don't approach or reward screams. (By the way, this same approach works for reducing anxiety in dogs when you come home.) Also remember that birds are "NOISY." This is their form of communication and to have an absolutely quiet bird is not possible. If a bird isn't making some kind of noise, it should be checked out by a veterinarian as it may be ill.
3. Provide toys and things to do while you are occupied or absent. These birds get very bored especially if alone and unstimulated. Leaving a TV on can help, but remember many of these birds form monogamous relationships, and if there is not another bird in the house, then they will pick a favorite person as a mate.
4. Keep a routine. Moving the cage around the house and unpredictable feeding times can be very upsetting. It can take weeks for a bird to feel safe and comfortable in a new environment. Birds do not mind doing the same thing over and over again. Predictability is safety, and if you aren't safe, then you must be in danger, and that is very scary for a bird.
5. Never force a bird to do something it does not want to do. This can often lead to aggressive behavior. Keep training sessions short (two or three ten-minute sessions daily), in a quiet area, preferably near the cage as the cage's perimeter is its comfort zone. Keep a food treat as a special reward that is given only when the bird does what is asked. It should be something small and easily consumed within a few seconds. If the bird enjoys praise or a head scratch, offer these instead of food. Remember to work in stages; even a partial step in the desired direction should be rewarded. This will take lots of time, repetition, and dedication, so don't give up or quit.
6. Dominant behavior often occurs as a young bird is reaching sexual maturity, and depending on the species, this can start at 3–12 years of age. They quickly learn that their beaks can be used to "discipline" humans. This phase can be very confusing for a bird owner because their wonderful, loving, and cuddly friend has gone berserk. It very important to understand this behavior and guide the bird before this undesirable behavior becomes established. This is where doing research and getting advice from professionals is important and is beyond the scope of this book.

However, please remember when birds go through this nippy phase, it can be hard on children that up to this point were welcomed. It is important to teach children appropriate handling to help prevent painful experiences. Realize this will never entirely eliminate these problems for either children or adults. The tamest bird can become startled and bite; it is a natural reaction. It is important for owners to learn how to read the body language of their bird and hopefully avoid most of these "accidental encounters." The following is a short list of some of the things a bird can convey with body language.

Pinning is when the pupils rapidly contract and expand. If their eyes are narrowed and pinning at the same time as the feathers are fluffed out, the bird is angry, frightened, or bothered and will probably bite. If they are just pinning with a relaxed body and feathers, it usually means the bird is thinking about something.

Beak grinding is when they softly grind their beaks together and is often a sign of contentment.

Growling is often heard when a bird is frightened. If something is bothering the bird, remove it.

Head tucked under wing means the bird is sleeping.

Fluffing or ruffling feathers means the bird is relaxed or is relieving tension. If the feathers remain fluffed for a long period of time, and the bird seems more quiet than usual, it can mean the bird is feeling cold and may be getting sick.

Flattened feathers and body quivering indicates the bird is frightened and usually is accompanied with wide, open eyes, and deep breathing.

Bowing is a way to ask for a pet or scratch.

Wing flapping can be exercising their wings, a way to get attention, or a display of happiness. Sometimes if they hold their wings away from their body, they may be warm and are trying to cool off.

There is never a guarantee that a meaningful relationship between a person/family and a pet parrot develops, and if things don't go well, it may end up that the bird is terribly abused, not intentionally or with malice, just in frustration and ignorance. For this reason, parrots are not recommended as a pet for the uninformed. The adults will have to take on most of the interaction and responsibility as a bird can become dangerous for children to handle. These amazing creatures need to be handled gently and with confidence, and how well the people-parrot interaction goes really depends on the experience of the owner, as well as the socialization-level of the parrot. It will also depend on the amount of time spent with the bird and the family dynamics. Birds pick up on the stress and anger humans display, directly impacting a bird's personality and likelihood for aggressive behavior. They require a calm, enjoyable, and consistent interaction with people. Without that interaction, they become sad, bored, and sometimes downright psychotic, which is not good for the bird or the people.

Species of Avians

African Grey Parrot

Origins: The African Grey Parrot originated in West and Central Africa and some islands located in the Gulf of Guinea. In the wild, monogamous pairs nest in tree cavities and lay three to five eggs. Fledglings leave at 12 weeks of age. They eat fruit, nuts, leaves, bark, flowers, and insects. Imports and habitat loss has put this bird on the vulnerable list. The U.S. has banned importation of these birds for pet purposes.

FIGURE 14.00 African Grey, Whole body

Characteristics: Also known as Congo Greys, they are sensitive birds that will bond to one person in the family. They are extremely intelligent, having the ability to learn words and mimic sounds throughout their life span. In 1977, Dr. Pepperberg proved that Greys were not just mimickers but actually understood and could communicate with their owners. She established that they are capable of reasoning, which explains their ability to figure out puzzles. They remember bad experiences and often associate these experiences with certain people or places. These birds are not recommended for the novice bird owners. They require a huge commitment of time and resources to keep them entertained, busy, and in toys.

FIGURE 14.01 African Grey, Showing off the red tail.

Bored or lonely parrots can start to pick on their feathers or become aggressive. In addition to lots of attention, providing good food, puzzle toys, non-toxic tree branches, and other activities such as playing and snuggling are recommended. Outdoor exposure for sunshine and exercise can be accomplished with a harness to allow exercising of the wings and to absorb Vitamin D. A bath, shower, or misting several times a week will keep their feathers in nice shape and is much appreciated by the bird.

QUICK FACTS 14.00 - African Grey – Congo African Grey

Life Span	Height/Weight	Color	Genus Species	Other Information
40–60 years	10–14 inches 25–36 centimeters 0.83–1.2 pounds 380–554 grams	Adults: gray, black bill, white around eyes, tail and under-tail coverts bright red, grayish-yellow eyes	*Psittacus erithacus* Red-tailed Grey Parrots, Grey Parrot, Congo African Grey Parrot	Sexually monomorphic Minimum Housing: 2' x 3' x 2' durable non-toxic metal with variable perches

FIGURE 14.02 African Grey Parrot in Flying Harness

Genetic Disease and Conditions: Aspergillosis, Chlamydiosis and other bacterial infections pneumonia, hypovitaminosis A, sarcocystosis, PDS Proventricular dilation disease (PDD), Psittacine beak and feather disease (PBFD), Macaw Wasting Disease, Hypocalcaemia, vitamin D deficiency (weakened immune systems, soft bones, abnormal beak development, reproductive problems, malfunctioning uropygial gland), feather picking, self-mutilation, liver disease or kidney problems, hyperkeratotic swellings, malignant tumors, oral abscesses, choanal atresia, tapeworms and blood parasites, toe deformities, and hematuria from lead poisoning.

African Grey Parrot - Timneh

FIGURE 14.03 African Grey, Timneh

Origins: These birds originated on the western coastal regions of Sierra Leone, Liberia, Ghana, and the southwestern parts of the Ivory Coast of Africa with some isolated populations in Guinea. They are smaller and do not have a solid red tail like the Congo Grey but rather have a mottled red chest, abdomen, and tail. They are relatively common; however, their numbers are decreasing due to capture for the pet trade and habitat destruction. Their status in the wild is vulnerable.

Characteristics: Timneh African Greys (TAGs) are a separate species from the African Greys but like them have a wide-ranging vocabulary and the ability to mimic sounds. They tend to speak sooner and will have a different inflection to their voice, almost like a small child's tone. They are not as high-strung as the Congos and will be content to play by themselves if their people are busy or there are others in the home. They don't often resort to feather picking as quickly as the Congo Greys.

QUICK FACTS 14.01 African Grey – Timneh African Grey

Life Span	Height/Weight	Color	Genus Species	Other Information
Up to 50 years	9–12 inches 23–31 centimeters 0.6–0.87 pounds 275–400 grams	Dark charcoal grey, bone-colored on part of the upper beak, maroon mottled tail, scalloped pattern on edges of feathers, white face	*Psittacus timneh*	Sierra Leone Gray Parrot

Genetic Disease and Conditions: Bacterial, viral, and fungal infections; feather picking; allergies; beak malformations in chicks; papillomas; lipomas; kidney disease; toxicity; heavy metal poisoning; and Psittacosis.

FIGURE 14.04 African Grey, Timneh Side View

FIGURE 14.05 Amazon, Yellow Headed

FIGURE 14.06 Amazon, Blue Fronted

Amazon

Origins: There are 27 species of Amazon parrots that can be found from South America to Northern Mexico, and nine species are found in the Caribbean. Because of habitat and wild pet bird trade, they are considered threatened or endangered in the wild. They are an all-green parrot with various colors on the head, nape, shoulders, wings, and tail. They have a short, blunt tail.

Characteristics: Different species of Amazons will have slightly different personalities and characteristics. The Double Yellow-heads, Yellow-naped, and Blue-fronted Amazons tend to have the highest potential for talking. The White-fronted, Blue-fronted, and Mealy Amazons tend to be more affectionate and make great companions. Many do not snuggle or enjoy being touched and so have gained a reputation for being bad-tempered or hyperactive. Overall, they are a lively bird that will talk, sing, and mimic words and sounds. They can be trained to do tricks and are considered very entertaining just with their own antics. Amazons tend to be couch potatoes, so providing them with toys to chew and keeping them active as well as outside activities on a harness are important to prevent obesity. Young, hand-fed Amazons will adapt to new homes quickly if they have been socialized and continued to be socialized throughout their life spans.

Adults will often bond to one individual in the home and may become aggressive towards other people, especially during mating season. Careful training and rewards will sometimes help these birds; however, they are never quite trustworthy because they apparently will bite for no particular reason. These birds are not great for a first-time bird owner. It is said they have the brain and temperament of a two-year-old human. Now think of having a spoiled, rotten "child" for 50-plus years.

QUICK FACTS 14.02 Amazon Parrot

Life Span	Height/Weight	Color	Genus Species	Other Information
50+ years	10–18 inches 25–46 centimeters 0.76–1.3 pounds 350–600 grams	Most have a solid green body with multiple variations of color on head, lores, wings, and tail Blue-fronted Yellow-naped Yellow-headed	*Amazona sp.;* *Blue-fronted, Amazona aestiva;* *Yellow-naped, Amazona auropalliata;* *Yellow-headed, Amazona oratrix*	Sexually monomorphic Minimum Housing: 2' x 3' x 2' durable non-toxic metal, variable perches

FIGURE 14.07 Amazon, Yellow Naped

Genetic Disease and Conditions: Malnutrition, obesity, respiratory infections, vitamin A deficiency, lethargy, weakness, Chlamydiosis, bacterial disease, tumors, foot and mouth sores, blood in droppings, feather picking, cloacal protrusions, and egg-binding.

Budgerigars – "Budgie" or Parakeet

FIGURE 14.08 Budgerigar, Male, Natural Color

Origins: Native to Australia, they are distributed throughout the continent. They favor scrublands, open forests, and grasslands all with a close-by water source. They tend to live in small groups; however, if the food and water source is readily available, they will live in a large flock. Their natural diet is grass seeds, fruits, berries, leaves, buds, and insects (especially during breeding season). They need to drink daily about 5% of their body weight and love to bathe in shallow ponds.

Characteristics: The Budgie is the most popular companion bird in the U.S., and for a reasonable price they are a great "starter bird." They are quieter than other parrot-type birds in that they don't usually shriek; however, they will burble, peep, and hum to themselves or with a friend. They are intelligent, able to learn about

FIGURE 14.09 Budgerigar, Man-made English, Female, Natural Color Male (Left to Right)

300 words, although not as clearly understood as other birds, and whistle tunes. Males, identified by their blue cere, are more "talkative" than females. They are relatively gentle if hand raised and then kept tame by frequent handling, and they have outgoing personalities. The younger a bird is, the more adaptable it is to new experiences. Youngsters are identified by their large black eyes, pink cere, and stripes over the forehead. Budgies will interact with all family members and usually don't have a particular favorite. These birds are "bullet" shaped and have a long tail that is almost half of its entire length. The natural color pattern of budgie is found in the Quick Facts table below. Man has developed numerous colors, feather variations, and sizes of budgies.

There is a color and feather combination for all tastes. However, because of this, they are not as hardy as the originals and suffer from a variety of diseases and conditions. If a Budgie should require medications or vitamin supplements, it should not be given in the water supply. The reason is that in the wild, Budgies can go for days without water, and for a sick bird this is basically a death sentence. Budgies do require a cuttle bone or other chew toys to keep their beak in good shape. Another unfortunate issue with Budgies is that they breed readily in captivity and thus are very inexpensive. This means they are bought on a whim, and after the novelty has worn off or they don't perform as expected, they are cast aside. Many a Budgie has died a slow horrible death from neglect and/or lack of veterinary care because of this. Budgies produce a lot of powder down known as "bird dust," so people with respiratory issues may be susceptible to "Bird Fancier's Lung or Breeder's Lung Disease," a type of hypersensitivity pneumonitis caused by inhaling the dust.

QUICK FACTS 14.03 Budgerigar "Budgie" or Parakeet

Life Span	Height/Weight	Color	Genus Species	Other Information
18–21 years	5.9–7.9 inches 15–20 centimeters 0.05–0.07 pounds 25–36 grams Wild and "pet" Budgies are about half the size of "English" or show Budgies	Natural color: green body, yellow head, small dark purple patch on cheeks, and a series of black spots across throat Color mutations: divided into two color series, white-based and yellow-based, with 32 primaries enabling hundreds of possible secondary mutations and color varieties	*Melopsittacus undulatus*	Sexually Dimorphic: Males have blue cere Females are flesh color Young – pinkish-purple Monogamous until one partner dies

Genetic Disease and Conditions: Malnutrition; tumors; polyomavirus carriers; overgrowth of beak and toenails; egg binding; bumble foot; obesity; feather disorders; leg paralysis; scaly face mange; Chlamydiosis; bacterial, fungal, and viral infections; internal parasites; Psittacosis; Parrot fever; inhaled toxins; and fatty liver syndrome.

FIGURE 14.10
Budgerigar, Female and Male (Left to Right)

Cockatiel

Origins: Cockatiels are native to Australia with their preferred habitats being wetlands, scrublands, and bushlands. They tend to move to food and water sources in pairs or in large flocks. They are often described as pests in Australia as they will eat crops. They are of the Cockatoo family and were first described in 1793 by Robert Kerr.

Characteristics: Cockatiels are extremely social birds and require human interaction or another cockatiel to keep them company. If a pair is purchased and breeding is not in the plan, two males will get along better than two females as they tend to squabble. Also note that if a pair is desired,

FIGURE 14.11 Cockatiel, Natural Male

they should be youngsters as adults have a harder time adjusting to each other. If a pair of birds are purchased, neither one will welcome human interaction as much as a single bird would from its owner. There are a number of differences between male and female cockatiels in addition to their feathering. Males tend to have great vocal abilities being able to talk some (they are not known as great talkers) and whistle tunes, some learning a number of tunes. Females are fairly quiet; however, they tend to screech more and will hiss and bite more often. Males will often strut as they walk, lifting their wings and sticking their chests out; this is especially apparent when doing so in front of a mirror. Females will quickly loose interest in a mirror. Cockatiels tend to be more vocal in the morning and right before bedtime.

They will also call out to owners as they arrive home. Responding to cheerful chirps is fine, but owners should remember that responding to screeches will encourage the bird to screech for attention. Cockatiels have a neat little crest that can be used to read how they are feeling. If it is raised and fully extended it may be the bird is interested, alarmed, or excited. When it is flattened, that can indicate that it is angry or defensive. When it is relaxed and the entire face seems to flow into the crest, this is usually a sign that the bird is flirting with you. Cockatiels have long tail feathers that make up about half of its total body length. Cockatiels have been known to suffer from night terrors. These are not nightmares but something that has frightened them in the night so that they flap their wings furiously inside their cages. They have been known to actually fracture a wing during one of these terrors or at the very least break a blood feather and bleed. It is important to have some blood stop power or styptic sticks to try and stop the bleeding. The feather can be pulled and pressure applied over the hole if the bleeding cannot be stopped. This can be avoided by covering the cage at night or reducing the ambient light that may cause shadows to move at night. Man has created a number of color mutations that are quite stunning.

FIGURE 14.12 Cockatiel, Natural Female

QUICK FACTS 14.04 Cockatiel

Life Span	Height/Weight	Color	Genus Species	Other Information
10–15 years, some up to 30	12–13 inches 30–33 centimeters 0.16–0.38 pounds 75–178 grams	Natural Colors: **Males:** gray body, yellow head, and orange cheek patches **Females:** mostly gray, duller orange patches, and yellow horizontal stripes or bars on ventral surface of tail Color mutations: albinos, cinnamons, lutinos, pearls, pieds, white faced, yellow cheeked	*Nymphicus hollandicus*	Sexually dimorphic Both have crests, but the males' tends to be longer

Genetic Disease and Conditions: Malnutrition, obesity, egg-binding, upper respiratory issues, eye problems, Chlamydiosis, feather picking, lead or zinc poisoning, and injuries in coordination.

FIGURE 14.13 Cockatiel, Lutino

Cockatoo

Origins: Twenty-one species of Cockatoos are found in the Australia-Asia region. Eleven species are found in Australia; three in New Guinea and Australia; and seven in the Philippines, Indonesia, Papua New Guinea, and the Solomon Islands. Some species have been accidentally introduced in New Zealand, Singapore, Hong Kong, and Palau. The Palm Cockatoo is the largest, and the smallest is the Cockatiel. They have a wide range of habitats, and no species is found in all types of habitat. Several species have adapted to city life. Most of the species are considered threatened or endangered because of habitat loss and the wildlife trade. Several species have become pests and are often controlled by poisoning, gassing, shooting, or by non-lethal methods that include scaring,

FIGURE 14.14 Cockatoo, Umbrella or White

FIGURE 14.15 Cockatoo, Sulfur Crested

providing food dumps, sacrifice crops, or habitat manipulation. They cause damage by chewing on outdoor furniture, window and door frames, silicone sealant from plate-glass windows, external wirings, TV antennae, satellite dishes, tarpaulin, defoliate decorative trees, eating seed and fruit crops, or digging in the ground for edible roots. In their natural habitat, they typically feed on various seeds, nuts, and fruits. They also will feed on corn growing in fields, doing considerable damage and are, therefore, considered crop pests by farmers. They also eat large insects and skinks.

Characteristics: As pets, Cockatoos are beautiful, smart, and affectionate birds. However, they are not for the busy family or single person with little free time. They require a great deal of attention and interaction with their people, and if left in a cage by themselves, they can become very depressed and/or psychotic. This can manifest in feather picking, self-mutilization, biting, and screaming. It is a very sad situation for such an intelligent and sweet bird. They are also fairly picky about what they eat; if left to eat only what they want, they often develop health issues. Needless to say this is NOT the bird for first-time bird owners. Potential owners need to understand how to train these birds to prevent biting. The white Cockatoos such as the Umbrella, Sulfur-crested, Citron, and Moluccan are the Cockatoos most people recognize.

Of those the Moluccan tends to be the most laid back and reserved. The smaller Cockatoos such as the Goffins, Little Corellas, and the Galahs (pink and gray) are easier to keep as they don't require as large a cage and tend to be a bit more forgiving in their attention needs. Cockatoos seem to enjoy music and can "dance" to the beat. They require lots of toys to keep them entertained, and they can be taken outside with a harness to allow for wing exercise. Cockatoos will talk but are not as good at it as other birds, and they have a tendency to screech or scream, so it is also recommended that owners live in a single-dwelling home to reduce alienating the neighbors. Some Cockatoos will imprint on one person and be aggressive towards others. It is important that owners understand this trait and take steps to properly socialize and train their bird. Paying attention to the position of the crest can help a person understand what the bird is feeling. A fully raised crest can indicate curiosity, excitement, surprise, fear, or frustration. A lowered, relaxed crest can indicate the bird is calm and friendly. Cockatoos also seem to produce more powder down than other birds, so people with respiratory issues should steer clear.

QUICK FACTS 14.05 Cockatoo

Life Span	Height/Weight	Color	Genus Species	Other Information
30–70+ years	12–24 inches 30–60 centimeters 0.66–2.65 pounds 300–1200 grams	Body colors of white, gray, and black with various colored erectile crests, cheeks, and tail, large beak in various colors	*Cacatua sp.* Sulphur Crested *C. galerita* Galah *C. roseicapilla* White (Umbrella) *C. alba* Moluccan *C. moluccensis* Little Corella *C. sanguinea*	Monogamous Males eye color is black and females reddish brown, however, not always true Minimum Cage: 4'H x 6'L x 2'W

FIGURE 14.16 Cockatoo, Galah

Genetic Disease and Conditions: Behavioral (screaming, biting, feather picking, self-mutilation, mate aggression), cloacal prolapse and papillomas, pododermatitis, localized herpes infections (on their feet), obesity, nutritional deficiencies, lipomas, upper respiratory disease, oral abscesses, idiopathic liver cirrhosis, PBFD, protozoan infections (haemoproteus, plasmodium), fungal infections (Aspergillosis), pneumonia, inhaled toxins, sarcocystosis, tape worms, kidney disease, cancer, and fibromas.

Conure

Origins: In the wild, Conures live in Central and South America. They are a diverse group, and there is some debate as to whether they are small parrots or large parakeets. The term Conure is used in aviculture to describe these smaller parrots. Most sources combine the genera *Aratinga, Pyrrhura, Eupsittula* and *Thectocercus* as Conures, and there is currently discussion about the division of birds between the genera. In an article entitled "DNA-sequence Data Require Revision of the Parrot Genus *Aratinga*" this division is described as follows:

- Aratinga as having violet-blue greater primary coverts, tail olive-green tipped with violet.
- Psittacara as having green greater wing coverts and tail.
- Thectocercus as having green great wing coverts and a tail with red in the inner webs.

FIGURE 14.17 Conure, Sun

• Eupsittula as having greater primary coverts mostly green with some blue on inner primary coverts and green tail.

For this book the genus ***Aratinga*** will be covered as these are the most commonly seen pet Conures.

FIGURE 14.18 Conure, Blue Crowned

Characteristics: If Budgerigars and Cockatiels are "beginner birds," the Conure is the next step up. Size wise, they are a bit bigger than a Cockateil. They possess all of the humor, fun-loving, clownish behaviors that the larger parrots are known for, and they are easily trained to do tricks, eliminate on command, and talk. Conures tend to be naturally loud and will need lots of training to redirect their vocalizations. They enjoy snuggling with their favorite person and will often snuggle into a pocket or hood as they like being in a "nest." They will also enjoy climbing under blankets or pillows, which should be kept in mind when they are out of their cage. Equipping their nest with a soft piece of fabric, fuzzy toy, or "bird cave" will be much appreciated by the Conure. Though easily trained, they can become a bit more nippy and more prone to behavioral issues than the Cockatiel or Budgie.

Conures are also more active and will play, explore, and chew lots of things. They can become bored and lonely if left alone for long periods of time during the day, so this is not the bird for active families that do not spend much time at home. However, if daily life is centered around the home, a Conure will fit right into the hubbub. Eventually, a Conure may pick a favorite person in the family. It is important that everyone continue to touch and handle the bird to maintain its friendliness to everyone. As a group these are small parrots with long tails, a wide range of colors, and a light build. The *Aratinga* are the ones found most commonly as pets. They are the largest birds in the group, and the species most commonly seen include the Sun, Jandaya, Blue Crowned, Cherry- or Red-headed and the Nanday. The Cherry-headed or Red-headed Conure, Blue Crowned Conure, and the Nanday have a green body with red, blue, and black colors respectively on their heads. The Sun and Jandaya Conures are mostly a yellowish-orange.

QUICK FACTS 14.06 Conure

Life Span	Height/Weight	Color	Genus Species	Other Information
23–30 years	8.5–12 inches 22–30 centimeters	Green body color with colored patches on head, lore, wings, or tail. Small gray or black beak.	*Aratinga sp.* Sun Conure Jenday Conure Nanday Conure Blue Crowned Red Masked	Sexually monomorphic Minimum Cage: 2' x 2' x 2'

Genetic Disease and Conditions: Fungal (Aspergillosis); bacterial, viral (Polyoma, Erythremic myelosis) and herpes (Pacheco's Disease) infections; Proventricular Dilation Disease; nutritional deficiencies; and inhaled toxicosis. Conure bleeding syndrome may occur, usually in chicks and in adults that are severely stressed.

FIGURE 14.19 Conure, Red Masked

Eclectus Parrot

Origins: The Electus Parrot hails from Northeastern Australia, New Guinea and surrounding islands, the Solomon Islands, Sumba, and the Maluku Islands (Moluccas). There are ten subspecies of Eclectus Parrots with three of these being most commonly kept as pets: *Eclectus roratus polychloros* (New Guinea Red-sided), *Eclectus roratus vosmaeri* (Vos or Vosmaer's), and the *Eclectus roratus solomonensis* (Solomon Island). These parrots are unique in that the males have green body feathers, and the females have red and purple/blue body feathers. The males forage in the canopy, so their feather coloration is protective against predators, and the females stay close to or in the nest tree hollow, so their red feathers

FIGURE 14.20 Eclectus Parrot – Male (Left) and Female (Right)

FIGURE 14.21 Eclectus Parrot, Male

blend with the bark. At one time it was thought they were two different species.

Characteristics: These parrots tend to be a bit calmer than others. They are often tentative, almost pensive, with new items or situations, but in a secure and well-adjusted household they will quickly investigate and come to terms with new items. When they feel comfortable and secure, they are just as playful and energetic as the other parrot species. However, if they never feel secure, they tend to be more fearful and resort to feather picking. Once started, it is very difficult to stop until the cause is addressed. This can be anything from diet imbalances to the wrong perch, health issues, hormonal changes, and even the edges of clipped wings irritating the sides of the bird. A trip to the veterinarian to rule out health issues is a start, and then ruling out other issues must begin.

Males can be more timid than females and should be socialized to a variety of situations at an early age and continued throughout their lifetime. Females tend to be a bit more aggressive than males, and although both genders can bite for no apparent reason, females are a bit more unpredictable, especially around strangers. Males tend to be better "talkers" than females, and they have a wide range of natural sounds that include bell or gong tones, coos, whistles, and squeals. If upset or frightened, they will screech and/or scream. Both sexes have been reported to develop a nervous habit referred to as "toe-tapping" or "wing flapping." It is not known what causes this, although some believe it is the start of a viral infection or a vitamin and mineral imbalance that does pass with time. Others believe it can be brought on by a buildup of chemicals from pelleted bird food, especially if it has artificial colors. It is recommended that a veterinary consult occur at the onset of this behavior. These beautiful birds have a smooth appearance to their feathers. They have oil glands rather than dust glands and don't produce any powder down. The Eclectus parrot has an unusually long digestive tract thus requiring a higher-fiber diet.

QUICK FACTS 14.07 Eclectus

Life Span	Height/Weight	Color	Genus Species	Other Information
20–30 years	12–14 inches 30–35 centimeters **Males:** 0.83–1.14 pounds 380–524 grams **Females:** 0.82–1.2 pounds 375–550 grams.	**Males:** bright green body, bright orange fading to yellow beak **Females:** red body with brighter red head, dark brown/black beak **Adults:** orange iris **Juveniles:** dark brown or black iris	*Eclectus roratus*	Sexually Dimorphic Minimum Cage 3'H x 11'L x 3'W

FIGURE 14.22 Eclectus Parrot, Juvenile Male

Genetic Disease and Conditions: Feather picking, hypovitaminosis A, proventricular dilatation disease, Chlamydiosis, bacterial infections, annular toe deformities (neonates), female aggression, Aspergillosis, inhaled toxins, sarcocystosis, pneumonia, wing flapping and toe tapping, nutritional deficiencies, PBFD, liver disease, and polyoma.

Lory and Lorikeet

Origins: Lories and Lorikeets are widely distributed throughout southeastern Asia, Polynesia, Papua New Guinea, Timor Leste, and Australia. Traditionally those with short, blunt tails are referred to as Lories, and the species with long tapered tails are Lorikeets. Both brightly colored, medium-to-small parrots are unique in that they have a "brush-tipped" tongue that is specialized for eating nectar, soft fruits, and berries. There are over 50 species and more sub-species of Lories and Lorikeets. The Ultramarine Lorikeet is endangered, and the Blue Lorikeet is vulnerable. Some make good pets, but all make great aviary subjects.

FIGURE 14.23 Lory, Rainbow, Showing Brush-Tipped Tongue

FIGURE 14.24 Lory or Lorikeet, Red

Characteristics: As pets Lories are smart and can be entertaining with their natural acrobatics and learned tricks, which include talking. They also have their natural whistles that can be loud and piercing, metallic pings, soft but high-pitched warbles, and chattering. Both Lories and Lorikeets can become quite nippy over territory and favored human. Training and guidance before they become adults is a must; otherwise, behavior modification will become necessary to avoid painful bites. Many species of Lories and Lorikeets are kept in aviaries. They make a colorful and entertaining addition to these "mini" environments. Some can even survive over winter in northern areas of the U.S. and Canada if provided proper housing. Research choices for aviaries if the plan is to mix species since some are very territorial and will kill other birds.

Selection of indoor caging is important to protect walls and floors. Their droppings are of a liquid consistency that is shot out with a bit of force. Placing the cage in areas that can be hosed down will facilitate cleaning. It is not recommended that plexiglass be used as twice daily scrubbing will be necessary to keep the cage clean enough to avoid causing bacterial infections. Also this type of caging does not allow the birds to climb, which is good exercise for their mental health. In outdoor aviaries this is usually not an issue as the waste absorbs into the ground cover. Commercially prepared nectars meet many of the dietary requirements for these birds. The liquid nectar needs to be changed at least twice daily and more often in warm weather. Otherwise, bacteria will grow in the nectar which can make the bird ill. Honey and fruits, such as apples, papaya, grapes, cantaloupe, pineapple, figs, kiwi, and corn-on-the-cob, and flowers such as pansies, nasturtiums, roses, hibiscus, marigolds, and dandelions provide pollen and other nutrients. Other food items that can be fed in small portions include brewer's yeast, oat or multi-grain flakes, and small quantities of millet spray, plus oats, canary grass seed, and some sunflower (sprouted). Careful monitoring of the amount of protein given is important to avoid issues like gout, kidney and liver disease. Dr. Rob Marshall, Avian Vet–(http://www.birdhealth.com.au) has developed an "Ongoing Health Program" that provides Lories and Lorikeets with the necessary vitamins and mineral for optimal health.

QUICK FACTS 14.08 Lory and Lorikeet

Life Span	Height/Weight	Color	Genus Species	Other Information
20 years	9.8–11.8 inches 25–30 centimeters 0.16–0.34 pounds 75–157 grams	Depends on the species	*Best Pet Species:* *Lorius sp.* Chattering Yellow-backed Chattering Black-capped *Eos* Red Lory Blue-streaked	Sexually Monomorphic Cage, single Lory: 36"H x 48"L x 24"W Cage dimensions for pair: 36"H x 60"L x 36"W

FIGURE 14.25 Lory or Lorikeet, Chattering

Genetic Disease and Conditions: Aspergillosis, bacterial infections, pneumonia, nutritional deficiencies Hypovitaminosis A, inhaled toxin, Sarcocystosis, hemochromatosis, and PBFD.

Lovebird

Origins: Central and southern Africa is home to nine species of the genus *Agapornis*. Feral populations are established in Phoenix, AZ, and San Diego, CA. Lovebirds form monogamous bonds, live in small groups, and eat fruits, vegetables, and vegetation. There are sexually diamorphic and monomorphic species; see the Quick Facts table for a list.

FIGURE 14.26 Lovebird, Peach faced

Characteristics: Lovebirds are small parrots with a stocky build and a short, blunt tail. They come in a variety of natural and man-made colors. They are entertaining pets that are somewhat noisy, making pleasant noises to occasional loud screeches throughout the day. They enjoy hiding, and nest boxes or "caves" are favorite places do so in their cages. They will often crawl into sleeves or pockets to hide on their humans, too. Because of their sociable nature, they require a great deal of human interaction if kept as a single. When left alone, it is suggested that the TV or radio be left on to provide background sound. If paired, they will be quite content to preen and "chat" with each other. They do require toys to chew, enjoy ripping up paper, and like bathing and sunning themselves. The three most common species kept

FIGURE 14.27 Lovebirds, Fischer's

as pets include Peach-faced (thought the smartest), Fischer's, and Masked Lovebirds (thought to be the gentlest and more pleasant).

Lovebirds are territorial, and introduction to other birds and pets must occur slowly. They have been known to nip the toes off "rival" birds or to attack cats and dogs if they are too close to the bird's favorite person. If you happen to get a male and female pair, they may start mating. During this time the female will tuck strips of nesting material into her wing feathers. Once the nest is built, mating occurs with eggs following in three to five days. The eggs should hatch within 30 days.

QUICK FACTS 14.09 Lovebird

Life Span	Height/Weight	Color	Genus Species	Other Information
10–15 years	5–7 inches 12–18 centimeters 0.08–0.13 pounds 38–60 grams	Peach/Rosy faced: green body with peach-colored face Black Masked: green body with black head and yellow chest Fischer's: green body with an orange-yellow head and shoulders	*Agapornis sp.* Peach/Rosie Faced *A. roseicollis* Black Masked or Masked *A. personata* Fischer's *A. fischeri*	Sexually Dimorphic Species: Madagascar, Abyssinian, Red-Headed Sexually Monomorphic Species: Peach-faced, Masked, Fischer's, Lilian's, and Black-Cheeked Minimum cage space per bird 1m × 1m × 1m 3' x 3' x 3'

Genetic Disease and Conditions: Feather and skin abnormalities, self-mutilation, bacterial and fungal infections, Circovirus, Chlamydiosis, aggression, cannibalism, fungal infections, egg binding, obesity, lipomas, and iron overload. Man-made issues from inbreeding include immunosuppression, reduced longevity, and hatching defects. Boredom or lack of social interaction can result in nervousness, feather picking, depression, aggression, screaming, and immune-system suppression.

Macaw

FIGURE 14.28 Lovebird, Black Masked

Origins: Mexico and Central and South America is home to these large, colorful birds. There are 19 species of Macaws with most living in rainforests, while others prefer woodland or savannah-like habitats. Most of the Macaws in the wild are endangered, and a few are extinct. This is due to deforestation and illegal trapping for the pet trade. They naturally come in a wide variety of colors, and most have a bare facial patch around the eyes that ranges from quite large to just a small rim of yellow feathers. Each bird's patch will have unique feathering much like a finger print.

FIGURE 14.29 Macaw, Scarlet, Blue, and Gold, Military (Left to Right)

Characteristics: Macaws are extremely intelligent, curious and social birds. They are monogamous, and even when flying with the flock will often fly so close to their mate as to almost touch wings. This is important to remember as a single bird will need that kind of bond and socialization as a pet. This will require extensive amounts of time spent with the bird every day. If this isn't something that the owner can do, these birds become very sad and depressed, which often leads to feather picking. If a household has small children, it is recommended that you pass on a Macaw. Their natural defense mechanism kicks in when fast moving objects come at or past them, and they will bite. Their beaks are strong enough to mangle and break an adult finger, so little fingers don't stand a chance. Another issue to take into consideration when thinking of adopting this bird is their longevity. Some have lived for over 80 years. Arrangements must be in place when they outlive you.

FIGURE 14.30 Macaw, Hyacinth

These amazing birds have a slim body and a tail that is long or longer than its body. They are agile and fast fliers because of large, pointed wings. They have large, pointed beaks that are strong enough to easily crack open nuts and seed pods. Their tongue has a bone inside of it that allows them to manipulate seeds inside nut shells and other food items. They require a variety of food as well as unlimited toys and objects to chew upon, not only for entertainment but to keep their beaks in good shape. Macaws are LOUD and have a squawky voice when they talk. They also make all kinds of other noises like trills, blurbs, "raspberries," and screeches. Single-family dwellings are recommended as early morning and late afternoon calls are a natural part of this bird's daily routine.

These large birds require a lot of room and because their tails don't fit into any kind of cage. It is a sad, sad sight to see that magnificent tail broken off because it doesn't fit the cage in which it is housed. An aviary is more appropriate for them so that they can exercise their wings and move from roost to forage areas. Macaws eat a wide variety of items as discussed in the introduction to birds.

QUICK FACTS 14.10 Macaw

Life Span	Height/Weight	Color	Genus Species	Other Information
Up to 80 years	**Small** (Hahn's or Noble Macaws): 12 inches 31 centimeters 0.4 pounds 165 grams **Medium** (Yellow-Collared or Illiger's Macaw range): 15.8 inches 40 centimeters 0.55–0.58 pounds 250–265 grams **Large** (Scarlet, Green Winged, Blue & Gold, Caninde, or Military Macaws range): 28–34 inches 70–90 centimeters 1.7–2.65 pounds 750–1100 grams **Extra Large** (Hyacinth Macaw): 40 inches 1 meter 2.65–3.2 pounds 1,200–1,450 grams	Scarlet: red body, blue wings with yellow coverts Green wing: red body, blue wings with green coverts Blue & Gold: blue wings, golden yellow head and body with black throat latch Military: green body with bright red lores Hyacinth: bluish purple with yellow around eyes	Scarlet *Ara macao* Green wing Ara chloropterus Blue & Gold *Ara ararauna* Military *Ara militaris* Hyacinth *Andodorhynchus hyacinthinus*	Sexually monomorphic Monogamous Cage Minimum: 3'x 4'x 5' 1m x 1.2m x 1.5 m Aviary Size: 10'W x 15'L x 8'H 3m x 4.5m x 2.5m

Genetic Disease and Conditions: Proventricular dilation disease, PBFD, Psittacosis (Chlamydiosis), Aspergillosis, bacterial and viral infections, pneumonia, nutritional deficiencies, hypovitaminosis A, inhaled toxins, chronic sinus infections, sunken eye syndrome (sinus infections), herpes (proliferative lesions and depigmentation), feather picking, heavy metal poisoning, allergies, cloacal papillomas, hypervitaminosis A (gout), lipomas, acne (ingrown feathers), beak malformations, and annular toe lesions.

FIGURE 14.31 Macaw, Green winged

Quaker Parrot

Origins: Quaker or Monk Parrots are native to Argentina, Bolivia, and Brazil. Feral flocks have established themselves in Puerto Rico and cities in the U.S., Caribbean Islands, Easter Islands, South Korea, Japan, and almost all of Europe. In the wild, Monk Parrots have been introduced around the world with population explosions in South America because of eucalyptus tree farming which supports nesting sites. As they are the only parrot that actually build nests verses living in tree hollows, access to unlimited trees allows them to expand. They live in large flocks with many pairs living in multi-room nests. They are considered a pest as they feed on crops, and many U.S. states have banned their presence, and others have restrictions on owning a Monk or Quaker Parrot. They are illegal in California, Georgia, Kansas, Kentucky, Hawaii, New Jersey, Pennsylvania, Tennessee, and Wyoming. They can be owned but not bred in Connecticut, and they have to be registered and banded in New York. They can only be owned in Virginia if owners are breeders or sellers. They have established wild colonies as far north as Chicago, IL, and north of Albany, NY.

FIGURE 14.32 Quaker Parrot, Feral

Characteristics: Quaker or Monk Parrots are sociable birds that are great talkers, developing a large vocabulary starting before they are even weaned. However, they also make loud and high-pitched cries when bored or upset. They are very intelligent and are often seen manipulating their toys to move them or place them in different locations within their cage. It is reported that they

FIGURE 14.33 Monk Parrot, Blue mutation

get bored with toys, and so rotation of toys is a good idea. They can figure out many different locking mechanisms that come on cages and will let themselves out for a stroll. Quakers can be sweet and loving, especially toward one person, or they can be aggressive and very naughty towards everyone else. Even though in the wild they live in large nesting "condominiums," they are extremely territorial about their cage. They have often been likened to the "pit bull" of the bird world because of this behavior. However, like the pit bull, it is usually a matter of training and how they were raised. Well socialized, hand-fed birds will make the best pets, and owners need to understand how to continue the socialization and training to keep these great little birds from becoming spoiled brats. One easy technique is to move the cage to different locations and rearrange the perches and toys frequently to reduce the territoriality.

If there are other birds in the household, it is important to monitor the Quaker if out of its cage. They have been known to injure or kill other birds especially if the others are smaller than themselves. Monks love to take baths or have a shower, and using this as a reward is worth cleaning up the mess they make with their exuberant splashing.

QUICK FACTS 14.11 Quaker Parrot

Life Span	Height/Weight	Color	Genus Species	Other Information
25–30 years	11–12 inches 28–30 centimeters 0.19–0.3 pounds 90 -140 grams	Bright green body, gray head, breast and legs, green-ish yellow abdomen, blue edged flight feathers, long tapering tail with blue color down center, orange beak	*Myiopsitta monachus*	Sexually monomorphic

FIGURE 14.34 Quaker Parrot, Female

FIGURE 14.35 Canary, Yellow

FIGURE 14.36 Canary, Red Factor

FIGURE 14.37 Canary, Norwich

Genetic Disease and Conditions: Hypovitaminosis A, fatty liver disease, obesity, bacterial and fungal infections, parasites, Chlamydiosis, Polyomavirus, Circovirus, feather picking, and self-mutilation.

Canary

Origins: Spain's Canary Islands, the Azores, and Madeira islands (Macaronesian islands) are the native homes of the Atlantic or Common Canary. *Serinus canaria* is the genus species of wild canaries which are part of the finch family. In the wild, their natural coloration is a grayish-greenish yellow with brown streaks on the back and yellow underparts. There are feral domestic populations which were introduced in Bermuda. They have been domesticated since the 1700s, and man has developed a multitude of colors and feather patterns over the years. Their appeal is the beautiful song the males sing to attract females. Historically they were used in mines to warn the miners of dangerous gas build-up. They became restless and agitated when exposed to the gas and died when the levels became high. This was the miner's early warning system, and they were still used in mines until 1987. They are shown worldwide after their annual molt with the world show being in Europe where more than 20,000 birds are brought together. These birds helped researchers study neurogenesis and as a model for how the brain learns.

Characteristics: Canaries are great for people that don't have hours of time to devote to socializing with their bird. They tend to be timid birds, and only with extreme patience and perseverance will one sit on a finger or learn a trick. These birds enjoy toys as much as the parrots; however, they don't chew on them like the parrots, but they do appreciate them being swapped out to relieve boredom. Care must be taken to avoid toys that can catch on the bird's toenails or wrap around their little legs. Canaries will get along with each other or with other species in an aviary where there is room for everyone. They are quite territorial when caged, and even a male and female pairing can end badly if one is dominant over the other. Pairs should be split from midwinter to early spring to give them a break from each other. Males are never happy if caged together; however, if they each have their own cage, it will spur them into a singing duel. They do require a cover over their cages to block out light at night, so the light stops with the sunset. This is important for keeping the bird in tune with the seasons. Canaries love and require a cold water bath every week if possible and especially during the molt. Cold water improves feather condition and encourages preening. Warm water baths strip essential oils and can cause itchiness.

There are three groups of canaries: song canaries that are bred specifically for their songs, colorbred canaries that have multiple color mutations, and type canaries that are bred for shape and conformation with little or no regard for song or color.

Although all canaries will sing, a number of canaries are bred for their songs that are unique to each breed, including the Spanish Timbrado, German Roller (Harz Roller), Waterslager (Malinois), American Singer, Russian Singer, and Persian Singer. Canaries can actually improve their song quality and imitate sounds heard in the household, especially if young.

Colorbred: Bronze, Brown, Eumo, Ino, Ivory, Mosaic, Onyx, Red Factor, and Satinette are all man-made colors.

Type: Australian Plainhead, Berner, Border, Espanola, Fife, Gibber Italicus, Gloster, Lancashire, Norwich, Raza, and Yorkshire have different feather patterns.

FIGURE 14.38 Finch, Zebra Male

QUICK FACTS 14.12 Canary

Life Span	Height/Weight	Color	Genus Species	Other Information
10–15 years	4.9–5.3 inches 12.5–13.5 centimeters 0.02–0.06 pounds 12–29 grams	Yellow – most common Red Factor – reddish orange body Norwich – feathers are fluffy Many others	*Serinus canaria*	Sexually monomorphic–Males tend to be the greatest

FIGURE 14.39 Finch, Zebra Female

Genetic Disease and Conditions: Aspergillosis, mycoplasma, pneumonia, nutritional deficiencies, air sac mites, upper-respiratory infections, inhaled toxins (carbon monoxide, methane, or carbon dioxide, plus household fumes such as non-stick pans, cigarettes, aerosol sprays, plug-in air fresheners, and scented candles), lymphoproliferative disease, Canary Pox, eye infections, feather mites, Knemidocoptes mites (scaly legs or face), proliferative conjunctivitis, keratis, corneal ulcers, cataracts, eyelid distortion, feather plucking, egg binding, genetic issues in in-bred canaries, and feather cysts.

FIGURE 14.40 Cut-throat Finches, Female and Male (Left to Right)

Finch

Origins: The Finches kept for pets are Passerines native to tropical areas of Africa, Asia, and Australia. There are multiple species in the wild as well as those kept for pets.

Characteristics: Finches have been kept in captivity for generations with the most common species such as the Zebra, Cut-throat, and Gouldian Finches being relatively easy to keep. They do not require the "hands-on" attention that parrots require although they are social and do best in pairs or small flocks in an aviary. Pairs do not have to be male and female as same-sex pairs will get along well together. If you plan on putting more than two birds together, keep in mind that they will bicker, and some species will be more dominant than others. If this is the plan, make sure the cage is large enough that each bird has room enough to escape and have access to food and water. See the Quick Facts table for the minimum space requirement for each pair of finches.

FIGURE 14.41 Gouldian, Red Head

If you do get a male and female, you need to be ready for eggs. If you take the eggs away, the hen will continue to lay eggs and can become deficient in calcium or egg-bound or start laying soft-shelled eggs. If there is no desire to raise chicks, there are two things that can be done to keep the hen from laying more eggs. As the eggs are laid and before incubation starts, when the eggs will be cool to the touch, vigorously shake them to disturb the contents. Then date the eggs and return them to the nest. Please do not do this if the eggs are warm, as this will damage the developing chick, and if allowed to remain in the nest, they can hatch with terrible deformities. It is better to destroy the eggs at this time. You can also replace the eggs with fake eggs of the same size and shape.

Remove the eggs after 28 days or if you see that the hen is no longer covering them. Male finches will sing with beeps and meeps and move into more complex songs. Youngsters will learn from other males as well as experiment with their own sounds until they are satisfied with their song. Finches are a good choice for new bird owners as they are not terribly expensive and if they have a friend will be content with each other.

Zebra Finch – Found in nature over most of Australia, Indonesia, and East Timor. They have also been introduced to Brazil, Puerto Rico, Portugal, and the U.S. The male feather pattern is crown, lores, cheeks, and neck of light gray, set off by two vertical black lines on each side of the head with rust red cheek patches. Under the orange beak there is silver gray and black cross barring down the front of the breast with a black stripe across the chest. The sides of the body have round white dots on a brown background. The female is similar to the males but without the stripes.

Cut-throat – Found throughout Africa in nature where it is often called a ribbon, bearded, or weaver finch. The males have a sandy brown body with small stripes of black all over, a dark brown tail, a white thin strip under the chin, and a bright red band under the white strip from ear to ear. Females do not have the red patch. The legs are pink, and the beak is a light tan color.

Gouldian – Native to Australia, their numbers in the wild are endangered; however, large numbers are bred in captivity. They are also referred to as Lady Gouldian, Gould's, or Rainbow finches. There are several varieties of Gouldian finches; they have red, black, or orange heads with a patchwork of other colors on chest, abdomen, and lower back. The back and wings are green, and the legs are flesh colored. Females are muted in similar colors. Some man-made colors in the U.S. include a blue, pastel green and blue, lutino, and albinos.

QUICK FACTS 14.13 Finch

Life Span	Height/Weight	Color	Genus Species	Other Information
4–7 years 12 years maximum	5–5.5 inches 13–14 centimeters 0.035–0.059 pounds 16–27 grams	See individual species for color descriptions	Zebra Finch *Taeniopygia guttata* Cut-Throat Finch *Amadina fasciata* Gouldian Finch *Erythrura gouldiae*	[illegible]

FIGURE 14.42 Pigeon, Rock Pigeon

Genetic Disease and Conditions: Feather loss from parasites; mites (Quill, Red, Scaly Leg, and Air Sac), fleas, and lice. Stressed birds will lose feathers,

FIGURE 14.43 Dove, Collared Dove Profile

and if there are others in the cage, the most common culprit is a dominant bird harassing it. Malnutrition can also cause feather loss. Increasing the quality of food and sunlight exposure will help. Iodine deficiencies can also result in poor feather condition, difficulty molting, infertility issues and breathing issues. Adding grated cuttlebone or crushed oyster shells on top of greens, sprouted seeds, or egg food supplements is a good source of iodine. Hens can lose feathers from hormonal changes during breeding season. Other ailments include fungal (Aspergillosis), bacterial (pneumonia and mycoplasma), and viral (Polyoma and herpes) infections; vitamin A deficiencies; strangulated legs and feet from fibers; conjunctivitis; or blepharitis. Changes in the droppings can be cochlosoma infections, vitamin E or selenium deficiency, enteritis, lack of grit, campylobacteriosis, or pancreatic insufficiency.

Pigeons and Doves

FIGURE 14.44 Pigeon, Fantail

Origins: There are over 300 species of pigeons and doves found in all areas except for the driest or coldest places in the world. Though used interchangeably, doves are usually the smaller species, and the common or feral pigeon is the Rock Pigeon from which all "man-made" pigeons have evolved. Pigeons and doves have been kept for centuries and are used in competitive breeding, homing, racing, as pets, and for food. They can be very affectionate and loyal, and millions of enthusiasts have learned to love and appreciate them as companion birds. Several species of pigeons and doves are used as food; the powerful breast muscles make up 31–40% of the body weight and make excellent meat that is served a scaub, the same name used for their young.

Characteristics: They have stout bodies, long necks, slender short bills, fleshy cere, and, thanks to man, a variety of plumage patterns and colorations. Their calls are pleasant coos to blats or trumpets depending on the variety. Pigeons are among the most intelligent birds. Research has shown that they learn complex actions and responses which they can change when confronted with different situations. They have shown that they can recognize the English alphabet and different humans, and they can remember images for a long period of time. Pigeons see in color as well as in the ultra-violet spectrum which was put to use during rescue missions at sea during World War I and II. Their most unique and notable ability is homing. This is their ability to find their way back home from long distances. Humans have utilized this ability to carry messages across vast distances in a relatively short period of time. Of course, the birds would have to be wary of natural predators like eagles and hawks.

In nature in order to attract a mate, males will strut with swelled neck, fluffed out chest feathers, lowered

FIGURE 14.45 Pigeon, Ice Pigeon, Muffed, Silver Chequer

FIGURE 14.46 Pigeon, Italian Owl, Ice Silver

FIGURE 14.47 Pigeon, American Show Racer

FIGURE 14.48 Pigeon, Brunner Pouter

FIGURE 14.49 Pigeon, English Long Faced, Muffed

FIGURE 14.50 Pigeon, Ancient Tumbler

FIGURE 14.51 Pigeon, Carneau

wings, and fanned tails; they leap into the air in a noisy, climbing flight, then crack their wings like a whip, and gently glide down to hopefully an impressed female. This display has been capitalized upon by Pigeon fanciers who have enhanced these characteristics and developed over 200 different breeds, in infinite varieties of colors, markings, and feather patterns. Each country has its own grouping, and the following are the European's grouping:

Asian feathered and voice pigeons were developed for ornate feathering and trumpeting or laughing voices. They include the Fantail, Frillback, Jacobin, Lahore, and Trumpeter Pigeons.

Color pigeons, many of which originated in Germany, have a wide variety of colors and markings and include Archangel, Danish Suabian, Saxon Field, Starling, Swallow, Thuringen Field, and Ice Pigeons.

Frills and Owls are noted for their short beaks and having reversed feathering on the chest and include Aachen Lacquer Shield, African, Chinese, Italian and Old German Owls, Oriental Frill, and Turbit Pigeons.

Homer and Hen Pigeons were originally developed for their homing ability, but now this category includes show types and racing pigeons as well, such as American Show Racers, Dragoon, English Carrier, German Beauty Homer, and Homing Pigeons.

Pouters and Croppers are pigeons that are able to inflate their crops to immense proportions and include English, Brunner, Gaditano, Horseman, Pigmy Pouter and Pouter Pigeons and Holle, Norwich, Voorburgshield, and Old German Cropper Pigeons.

Exhibition Tumblers consist of tumbler breeds that are purely ornamental and/or exhibition breeds: Budapest Short Face, English Long Faced and English Short Faced Tumbler Pigeons and Helmet, English Magpie, and Nun Pigeons.

Flying Tumblers and Highfliers are dual-purpose pigeons in that they can be exhibited and/or used in acrobatic or sporting competitions and include Ancient Tumbler, Armenian Tumbler and Australian Performing Tumbler Pigeons, Danzig Highflyer, Donek, Roller, and Tippler Pigeons.

Utility pigeons were developed as a source of meat, which is called "squab," and include Carneau, French Mondain, King, Runt, and Strasser Pigeons.

QUICK FACTS 14.14 Pigeon and Dove

Life Span	Height/Weight	Color	Genus Species	Other Information
10–20 years	Pigeons 15–17.5 inches 38–44.5 centimeters Male 0.87–1 pounds 400–500 grams Female 0.87–0.98 pounds 400–450 grams Doves 13 inches 32 centimeters 0.27–0.52 pounds 124–240 grams	Multiple colors and varieties of feather patterns	*Columba livia* Rock Pigeon *Streptopelia sp.* Collared Dove	Sexually monomorphic Monogamous Caging minimum: 3' x 3'x 6' per 1 to 2 small, flat-raised platforms vs. perches Outdoors Loft: shelter they can fly in and out of, come home every night

Genetic Disease and Conditions: Viruses, bacteria, fungi and parasites, malnutrition, Trichomoniasis (pigeon canker), toxic or poisonous elements in the food. Signs of illness include weakness, sleeping more than usual, loss of appetite, rotting or bruised-looking skin, milky eyes, thinned and/or fluffed feathers.

Abyssinian Cat Association www.abyworld.com/acca/
Afghan Hound Club of America, Inc. http://afghanhoundclubofamerica.org/
African Grey Parrots http://www.african-grey-parrot.com/
Airedale Terrier Club of America, Inc. http://www.airedale.org/
Akita Club of America, Inc. http://www.akitaclub.org/
Alaskan Malamute Club of America http://alaskanmalamute.org/
Amazon Parrots http://beautyofbirds.com/amazons.htm
American Belgian Malinois Club http://www.malinoisclub.com/abmc/
American Belgian Tervuren Club http://www.abtc.org/
American Bloodhound Club http://www.americanbloodhoundclub.org/
American Bouvier des Flandres Club http://www.bouvier.org/
American Boxer Club http://americanboxerclub.org/
American Brittany Club http://clubs.akc.org/brit/
American Brussels Griffon Association http://www.brussels-griffon.info/
American Bullmastiff Association http://www.bullmastiff.us/
American Cavalier King Charles Spaniel Club, Inc. http://www.ackcsc.org/
American Chesapeake Club http://www.amchessieclub.org/health/health.html
American Chinese Crested Club http://www.chinesecrestedclub.info/
American Eskimo Dog Club of America http://www.aedca.org/home.html
American Fancy Rat & Mouse Association http://www.afrma.org/
American Fox Terrier Club http://www.aftc.org/
American Kennel Club http://www.akc.org/breeds
American Lhasa Apso Club http://www.lhasaapso.org/
American Maltese Association http://www.americanmaltese.org/
American Miniature Schnauzer Club http://www.amsc.us/
American Pointer Club, Inc. http://www.americanpointerclub.org/
American Pomeranian Club http://www.americanpomeranianclub.org/
American Rottweiler Club http://www.amrottclub.org/
American Shih Tzu Club http://www.americanshihtzuclub.org/
American Spaniel Club, Inc. http://www.asc-cockerspaniel.org/

American Shetland Sheepdog Association, Health Issues/Research http://www.assa.org/health.html#notice

American Water Spaniel Club http://www.americanwaterspanielclub.org/

American Whippet Club http://www.americanwhippetclub.net/

Animals, General: Animal-World, Pet and Animal Information http://animal-world.com/

Association of Wirehaired Pointing Griffon http://www.awpga.com/

Australian Cattle Dog Club of America, Inc. http://www.acdca.org/

Australian Shepherd Club of American, Inc. http://www.asca.org/education/health

Basenji Club of America https://www.basenji.org/joomla/

Basic Pet Care (ASPCA) http://www.aspca.org/pet-care

Basset Hound Club of America, Inc. http://192.220.72.203/

Bearded Collie Club of America http://www.bcca.us/

Beauty of Birds http://beautyofbirds.com

Bedlington Terrier Club of America, Inc. http://www.bedlingtonamerica.com/

Belgian Sheepdog Club of America http://bsca.info/

Bernese Mountain Dog Club of America, Inc. http://www.bmdca.org/

Bichon Frise Club of America, Inc. http://www.bichon.org/

Biomethodology for Laboratory Mice http://www.theodora.com/rodent_laboratory/identification.html

Border Collie Collapse http://www.vetmed.umn.edu/departments/veterinary-and-biomedical-sciences/research/canine-genetics-lab/genetic-research/border-collie-collapse

Border Terrier Club of America http://www.btcoa.org/

Borzoi Club of America Inc. http://www.borzoiclubofamerica.org/

Breed Longevity Survey Results http://users.pullman.com/lostriver/longhome.htm

Briard Club of America http://briardclubofamerica.org/bca/

Bulldog Club of America http://www.bulldogclubofamerica.org/

Bull Terrier Club of America, Inc. www.btca.com/

Cairn Terrier Club of America http://www.cairnterrier.org/

Canine Health Information Center – central information database http://www.caninehealthinfo.org/brdreqs.html?breed=ACD

Canine Terminology http://www.dogbreedinfo.com/terms.htm

Cardigan Welsh Corgi Club of America http://www.cardigancorgis.com/

Cat Body Types http://pictures-of-cats.org/cat-body-types.html

Cat Breed Info Center http://cattime.com/cat-breeds/abyssinian-cats

Cat Breed List http://www.catbreedslist.com/

Cat Colors FAQ http://fanciers.com/other-faqs/colors.html

Cat Fancier's Association http://www.cfainc.org

Cat Time http://cattime.com/

Cats by Color http://commons.wikimedia.org/wiki/Category:Cats_by_color

Catster http://www.catster.com/lifestyle/cat-blood-type

Cattledog.com – History of Australian Cattle Dog http://www.cattledog.com/misc/history.html

Chihuahua Club of America http://www.chihuahuaclubofamerica.com/

Chinchilla Care Organization http://chinchillacare.org/

Chinchilla Cronicles http://www.chinchillachronicles.com/chinchilla_teeth.html

Chinese Shar-Pei Club of America, Inc. http://www.cspca.com/

Chow Chow Club, Inc. http://chowclub.org/ccci/

Collie Club of America http://www.collieclubofamerica.org/akc-meet-the-breed04.pdf

Collie Health Foundation http://www.colliehealth.org/tableoftestandstudies2.pdf

Dachshund Club of America, Inc. http://www.dachshundclubofamerica.org/

Dalmatian Club of America http://www.thedca.org/

Doberman Pinscher Club of America http://www.dpca.org/

Dogs, General Information:

- Doctor Barkman Speaks, Dog Tail Shapes and Sizes, 24 December 2012 http://doctorbarkman.blogspot.com/2012/12/dog-tail-shapes-and-sizes.html
- Dog Breed Info Center http://www.dogbreedinfo.com
- Dog Coat Types, Colors, Patterns http://www.dogbreedinfo.com/terms.htm
- Dog Ear Shapes, Ear Shapes, "What Shape is Your Dog's Ear?" http://www.psychologytoday.com/blog/canine-corner/201208/what-shape-is-your-dogs-ear
- Dog Head Shapes http://www.psychologytoday.com/blog/canine-corner/201401/is-dogs-head-shape-related-his-intelligence
- Dog Tail Types http://doctorbarkman.blogspot.com/2012/12/dog-tail-shapes-and-sizes.html
- Dogs in Depth. Com http://www.dogsindepth.com/index.html
- Dog's Best Life, Karen A. Soukiasian, 13 October 2014 http://www.dogsbestlife.com/home-page/dog-tail/.

English Cocker Spaniel Club of America http://www.ecsca.info/

English Setter Association of America http://www.esaa.com/BreedInfo/BreedStandard.html

English Springer Spaniel Field Trial Association http://www.essfta.org/

Eye Disease ("An Explanation of Eye Disease" by Gayle Kaye, edited by Nancy M Bromberg, VMD, MS, DACVO) http://www.colliehealth.org/explanationofeyedisease.pdf

Feline Blood typing http://www.rapidvet.com/feline.html

Feline Heat Cycles (Revival Animal Health) http://www.revivalanimal.com/store/pg/1560-The-Feline-Heat-Cycle.aspx

Ferrets:

- Ferret Dentistry http://www.dentalvet.com/vets/ferrets/ferret_dentistry.htm
- 14 things you should know about Ferrets http://www.seniorlink.co.nz/interests/pets/ferret.shtml
- What you should know about Ferrets http://www.drsfostersmith.com/pic/article.cfm?articleid=1850

Finch Information Center http://www.finchinfo.com/

Flat-Coated Retriever Society of America, Inc. http://www.fcrsainc.org/

French Bull Dog Club of America http://www.fbdca.org

General Pet Health:

- Vet Street http://www.vetstreet.com/
- WebVet.com http://www.webvet.com

Gerbils Care Organization http://petgerbils.org/

German Shepherd Dog Club of America https://gsdca.org/

German Wirehaired Pointer Club of America http://www.gwpca.com/

Giant Schnauzer Club of America http://www.giantschnauzerclubofamerica.com/

Golden Retriever Club of America http://www.grca.org/

Gordon Setter Club of America http://www.gsca.org/

Great Dane Club of America http://www.gdca.org/

Great Pyrenees Club of America http://www.gpcaonline.org/

Greyhound Club of America http://www.greyhoundclubofamericainc.org/

Guinea Pigs http://www.calicavycollective.com/

Guinea Pigs as Pets http://exoticpets.about.com/od/guineapigs/tp/aboutgp.htm

Guinea Pigs Today http://www.guineapigtoday.com/2011/09/14/10-things-you-might-not-know-about-guinea-pigs/

Hamster Care Organization http://pethamstercare.com/

Havanese Club of America http://www.havanese.org/

Hedgehog Care Organization http://hedgehogcare.org/

Irish Setter Club of America, Inc. www.irishsetterclub.org

Irish Wolfhound Club of America http://www.iwclubofamerica.org/index.html

Italian Greyhound Club of America http://www.igrescue.com/

Japanese Chin Club of America http://japanesechinclub.org/

Keeshond Club of America http://www.keeshond.org/keeshond-breed-standard.html

Labrador Retriever Club www.thelabradorclub.com

List of Animal Names http://blog.oxforddictionaries.com/2016/04/male-and-female-animals/

Mastiff Club of America http://www.mastiff.org/

Merck Veterinary Manual http://www.merckvetmanual.com/mvm/index.html

Mice as Pets http://exoticpets.about.com/cs/careofmice/a/mousecare.htm

Millermeade Farm's Hedgehog Headquarters http://hedgehogheadquarters.com/secure/characteristics.htm

Miniature Bull Terrier Club of America http://www.minibull.org/

Miniature Pinscher Club of America, Inc. http://www.minpin.org/

"Multidrug Sensitivity in Dogs" College of Veterinary Medicine, Veterinary Clinical Pathology Lab, Washington State University, Pullman, WA http://www.vetmed.wsu.edu/depts-VCPL/

National Beagle Club of America, Inc. http://clubs.akc.org/NBC/

National Shiba Club of America http://www.shibas.org/index.html

Newfoundland Club of America http://www.ncanewfs.org/

Norwegian Elkhound Association of America http://neaa.net/

Nova Scotia Duck Tolling Retriever Club http://nsdtrc-usa.org/

Old English Sheepdog Club of America http://www.oldenglishsheepdogclubofamerica.org/

Papillon Club of America http://www.papillonclub.org/

Pekingese Club of America http://www.thepekingeseclubofamerica.com/

Pembroke Welsh Corgi Club of America Inc. http://www.pwcca.org/

Pet Eduction http://www.peteducation.com/article.cfm?c=18+1798&aid=3078

Pet Rabbit Care Organization http://petrabbitcare.org/

Pet Rat Care Organization http://petratcare.org/

Petit Basset Griffon Vendéen Club of America http://www.pbgv.org/

Poodle Club of America http://www.poodleclubofamerica.org/

Portuguese Water Dog Club of America http://www.pwdca.org/

Pug Dog Club of America http://pugs.org/

Puli Club of America, Inc. http://puliclub.org/

Rhodesian Ridgeback Club of the United States http://www.rrcus.org/

"Rodent Dentistry" http://www.chincare.com/HealthLifestyle/DHdocs/rodentdentistry.pdf

Saint Bernard Club of America http://saintbernardclub.org/

Saluki Club of America http://www.salukiclub.org/

Samoyed club of America http://www.samoyedclubofamerica.org/

Schipperke Club of America http://www.schipperkeclub-usa.org/

Shetland Sheepdog ("A Brief History of the Shetland Sheepdog," ASSA Judge Charlotte McGowan) http://www.assa.org/history.pdf
Siberian Husky Club of America, Inc. http://www.shca.org/
Silky Terrier Club of America, Inc. http://silkyterrierclub.yolasite.com/
"Small Mammal Dentistry" Istace, CVT, VTS (Dentistry) Kathy, Vet Learn.Com https://s3.amazonaws.com/assets.prod.vetlearn.com/ff/ee66e0c8d811e087120050568d3693/file/VT0911_Istace_DC%20revised.pdf
Soft Coated Wheaten Terrier Club of America, Inc. http://www.scwtca.org/
Staffordshire Bull Terrier Club of America http://www.sbtca.com/
Staffordshire Terrier Club of America http://www.amstaff.org/legislative.html
Standard Schnauzer Club of America http://www.standardschnauzer.org/
Sugar Glider Care Organization http://sugarglidercare.org/
Tibetan Spaniel Club of America http://www.tsca.ws/
Tibetan Terrier Club of America http://www.ttca-online.org/
United States Australian Shepherd Association http://australianshepherds.org/
United States Border Collie Club http://www.bordercollie.org/
Vizsla Club of America, Inc. http://www.vcaweb.org/
Weimaraner Club of America http://weimaranerclubofamerica.org/main/index.php?lang=en
Welsh Terrier Club of America, Inc. http://clubs.akc.org/wtca/
West Highland White Terrier Club of America http://www.westieclubamerica.com/
Yorkshire Terrier Club of America http://www.theyorkshireterrierclubofamerica.org/

Abscess – a swollen pocket of pus in the body

Achondroplasia – a genetic disorder caused by a mutation of the fibroblast growth factor receptor gene, resulting in dwarfism, shortened limbs with a normal torso

Acne – inflammation of the skin glands and hair follicles resulting in pimples or blackheads seen with birds as ingrown feathers and on cat chins as blackheads often associated with dirty food dishes

Addison's disease – the production of too little cortisol and insufficient levels of aldosterone, also called adrenal insufficiency

Adenocarcinoma – cancer originating in glandular tissue

Agouti – two or three bands of contrasting color on each hair shaft, as in the Abyssinian.

Albinos – lacking normal pigmentation resulting in pale or white skin and hair, pink, or blue eyes with dark red pupils

Aleutian disease – progressive parvovirus infections that causes weight loss, reproductive failures, and death from renal failure in the Aleutian mink

Alopecia – hair loss

Alpha-mannosidosis – lysosomal storage disease caused by a defect in the metabolism of mannose containing glycoproteins resulting in coarse faces, mental retardation, upper-respiratory congestion, infections, hepatosplenomegaly, cataracts, and skeletal abnormalities

Alter – a neutered or spayed cat

Altricial – offspring of rodents and birds that are born with their eyes closed and no hair or feathers

American Kennel Club (AKC) – "is dedicated to upholding the integrity of its Registry, promoting the sport of purebred dogs and breeding for type and function."

Amyloidosis – abnormal deposits of protein in the body systems or organs, incurable and progressive

Anaphylactic reactions – allergic reaction that progresses rapidly and is life-threatening if not treated immediately

Annular toe deformities – constricting bands of tissue around toes causing swelling and necrosis if not treated, seen on young birds

Anogenital distance (AGD) – the distance between the anus and genitalia on male and female animals, usually used to determine sex of rodents and kittens

Anterior Cruciate Ligament – a ligament that crosses from the anterior intercondylar area of the tibia to the posterior part of the lateral condyle of the femur, injuries occur from sudden stops or changes in directions

Aortic stenosis – pathological narrowing of the aortic valve orifice

Aplastic anemia – disorder of the bone marrow that drastically reduces or stops the production of all types of blood cells

Arboreal – living in trees

Arthritis – inflammation of a joint, evidenced by swelling, pain, stiffness caused by injury, infection, degenerative changes, metabolic disorders

Arthropathy – disease or abnormality of a joint

ASCOB – Any Solid Color Other than Black (used in reference to cocker spaniels)

Aseptic meningitis – a viral inflammation of the meninges

Aspergillosis – fungus that can invade lungs, ear canals, eyes, nose, sinus cavities, brain, and spinal cord

Asthma – chronic inflammatory disease of the airways, causes spasms and swelling that narrows airways causing the animal to wheeze or gasp for breath

Ataxia – muscle incoordination, inability to control movement

Atopic dermatitis – a hereditary hypersensitivity or allergy that results in pruritus of feet, face, and ventral body with self-trauma from the intense itching, often seasonal with secondary infections common

Aural hematoma – a pool of blood that collects between the skin and the cartilage of an ear flap, typically caused by aggressive ear scratching or head shaking

Autoimmune disease – the body's immune defenses turns against the body's own tissues

Autoimmune hemolytic anemia – "self" destruction of red blood cells by the body's immune system

Autoimmune thyroiditis – also known as lymphocytic thyroiditis, the body's lymphocytes invade the thyroid gland and the damaged tissue is replaced by fibrous tissue which damages the thyroid causing a hypothyroid disease

Autosomal recessive disorder – two copies of an abnormal gene are present causing a disease or trait to develop

Avian – of or relating to birds

Avian Proventricular Dilatation Disease ("Macaw Wasting Disease") – Digestive disorders in birds are due to various reasons, including viral infection, low immunity, and injury, affects the nerves of the stomach which stretches and loses the ability to contract

Bacteruria – presence of bacteria in the urine

Basal cell carcinoma – skin tumors thought to be caused by long-term exposure to ultraviolet radiation from the sun

Basophilic stippling – presence of blue nucleic acid remnants in red blood cells often seen with lead poisoning

Beige – used to describe a coat color that is a mixture of black and reddish brown fur

Belton – an intermingling of colored and white hairs, as blue, lemon orange, or liver belton often associated with English Setters; if there is more color than white, then the animal is called a roan

Bitch – a female dog

Blaze – a white stripe running from the nose up the center of the face between the eyes

Blenheim – pearly white body colored hair, with chestnut colored hair surrounding both eyes, ears, and on the body, associated with Cavalier King Charles Spaniels

Blepharitis – inflammation of the eyelids

Bloat – distension of the stomach with gas and fluid, can progress to gastric dilation and volvulus if not treated

"Blue Doberman syndrome" – color mutant alopecia, a clinical condition of blue or fawn coat colored dogs caused by a dilution gene at the D locus; bacterial folliculitis, scaling, and hair loss is commonly seen in Blue and Fawn Doberman Pinschers, Fawn Irish Setters, and a number of cattle breeds

Blue Merle – mottled or marble coloring, blue-gray/black, illusion of blue coloring on a solid or piebald coat

"Border Collie Collapse Syndrome" – episodic nervous-system disorder triggered by strenuous exercise, including disorientation, dullness or loss of focus, swaying, staggering, exaggerated movements; episodes last from 5–30 minutes with complete recovery; Australian Cattle Dogs, Kelpie, and Shepherds are affected as well as Bearded Collie, Collies, and Shetland Sheepdogs

Bordetella – a extremely contagious respiratory disease in dogs and cats caused by the Bordetella bronchiseptica bacteria spread by direct contact, airborne or fomites; symptoms include dry hacking cough, retching, sneezing, nasal discharge with progression into pneumonia; vaccines are available

Brace – two dogs of the same breed

Brachycephalic – a broad, short skull that is wider than it is long, as shown, for example, in a Bulldog or Persian cat

Brachygnathism – refers to shortening of the upper jaw, a congenital condition in non-brachygnathism breeds (see Undershot bite)

Breed – animals of like color, type, and other characteristics which are similar to those of parents or past generations

Brindle – a fine even mixture of black hairs with hairs of a lighter color, usually tan, brown, or gray

Bronchial dysplasia – abnormal development of the bronchus resulting in an alteration of size and shape

Bronchitis – inflammation of one or more bronchi with irritating chronic or acute cough and fever caused by air pollutants, infectious agents; in birds it is often caused by a coronavirus

Brood Bitch – a female used for breeding

Brush – A bushy or plume-like tail of a longhaired cat often likened to a "bottle-brush"

Butterfly – The pattern of striped markings on the shoulders of tabbies in the shape of butterfly wings

Butterfly nose (or parti-colored nose) – a dark colored nose spotted with flesh colors, seen in often Australian Shepherds, Papillons, and Yellow Labradors but can occur in any breed or mixed breed dog

Button – A single white patch anywhere on the body on an otherwise solid colored cat; if on the front of the chest, it is often referred to as a "locket"

Calcium oxalate stones – a type of stone found in the kidney or urinary bladder

Calling – sound made by a female cat when in season (heat) to attract a male

Campylobacteriosis – a gram-negative, microaerophilic, oxidase-positive, non-fermentative bacteria that is spiral shaped with uni- or bi-polar flagella, often causing diarrhea, abdominal pain, cramping, and fever

Canary Pox – contagious viral infection transmitted by mosquitoes and mites; the most common condition is dry nodules or crusty areas about the face and legs and a "wet" form that causes thick plaques inside the mouth, throat, or sinus cavity; may also see weight loss, rapid breathing, listlessness, fluffed feathers, and sudden death

Canine – of or related to dogs

Canine Cyclic Neutropenia – "Gray Collie Syndrome" (GCS) an autosomal recessive disease, where the neutrophil (white blood cells that are the first line of defense against pathogens) numbers drop in a cyclic manner about every 10–12 days; if exposed to pathogens during this drop, the animal has a higher incidence of succumbing to diseases

Canine discoid lupus erythematosus ("Collie Nose") – an immune-mediated disease that is confined to the skin and almost always to the nose leather; lesions start with the loss of pigment on the nose, progressing to scaling and cracking with ulcerations occurring if left untreated; usually responds to immuno-suppressive drugs

Canine Epileptiod Cramping Syndrome (CECS) – linked to high levels of exertion after which abrupt, non-sustained contraction of muscle groups (chorea) and sustained involuntary contractions of muscle groups (dystonia) often seen with trembling, staggering, exaggerated stretching, lip smacking or licking, borborygmi, and an inability to stand

Canine neuronal ceroid lipofuscinosis ("Batten Disease") – see lysosomal storage disorders

Canine pemphigus – autoimmune disease that presents as ulcerations and crusting of the skin with or without cysts and pustules

Canine Renal Dysplasia – abnormally developed kidney(s) that seems to be an inherited disorder with non-specific symptoms such as poor hair coat, stunted growth, urinary abnormalities, weight loss, weakness, lethargy, depression, vomiting, diarrhea, oral ulcers, and pain

Cannibalism – eating one's own kind

Cardiomyopathy – hereditary or acquired disease of the heart muscle

Carnivore – one who eats flesh

Cataract – the lens of the eye become opaque or cloudy

Cavy – a term used for a family of rodents native to South America including Guinea pigs, cavies, and capybara

Celiac disease – an immune reaction to gluten, a protein found in wheat, barley, and rye

Cesarean – the surgical removal of a fetus through the abdominal and uterine walls

Cerebellar ataxia – hereditary defects, tumors, or infections of the brain that are progressive and fatal, starting with clumsiness and swaying and progressing to losing balance, falling, and head and eyes moving rapidly and randomly

Cerebellar cortical abiotrophy – rare degenerative disease where the cerebellum degenerates over time

Cerebrovascular disease – rupture or obstruction of blood vessels in the brain with acute neurological signs including seizures, depression, and circling, often regressing after 24–72 hours depending on if the bleeding stops

Cervical disc syndrome – occurs when an intervertebral disk presses on the spinal cord causing pain and pressure, can range from mild (pain) to severe (complete paralysis)

Cervical lymphadenitis (CL or Lumps) – swelling of the lymph nodes in the neck of a Guinea pig, linked to a gram-positive streptococcus and other bacterium transmitted through bite wounds or aerosol

Chagas disease – an inflammatory, infectious disease caused by the reduviid (kissing) bug, also known as trypanosomiasis; it can be acute or chronic and manifests as swelling at the infection site, fever, fatigue, rash, loss of appetite, diarrhea or vomiting, swollen lymph nodes, and if chronic can lead to heart damage and/or abdominal pain

Chédiak-Higashi syndrome – in Persian cats with a dilute smoke-blue coat and yellow-green iris causes clotting issues and photophobia

Cherry eye – prolapse or eversion of the nictitating membrane or third eyelid in dogs

Chiari-like malformations – congenital birth defect where the back of the skull is too small for the cerebellum and brainstem, causing it to push against the foramen magnum causing an obstruction of the cerebrospinal fluid; Cavalier King Charles spaniels are most consistently affected

Chlamydiosis – intracellular bacteria that causes upper respiratory disease, conjunctivitis. In birds it is a zoonotic disease known as parrot fever, psittacosis or ornithosis, and it is transmitted by inhalation of the Chlamydia psittaci bacteria; signs range from sudden death to respiratory disease, conjunctivitis, liver, or intestinal disease

Choanal atresia – congenital blockage of the nasal passage (choanae) by bony or soft tissue

Choke collar – leather or chain collar fitted to the dog's neck in such a manner that the degree of tension exerted by the hand tightens or loosens it; when used correctly it is an effective training tool

Chondrodyplasia – abnormal development of bones and cartilage, form of dwarfism

Chorioretinal change – an area of pigmentary change or fibrosis located on the inside surface of the eye, can be from an infection or injury

Choroidal Hypoplasia – "Collie eye" – incomplete development of the eye

Chylothorax – pleural effusion of lymph formed in the digestive system called chyle accumulating in the pleural cavity from a disruption or obstruction of the thoracic duct

Circovirus – infects dogs, pigs, and birds, causing vomiting and diarrhea in dogs and beak and feather disease in Psittacine birds and deadly infections in pigeons, canaries, and finches.

Cleft palate or lip – abnormal opening in the roof of the mouth connecting to the nasal passages, can result in runny nose, coughing, aspiration pneumonia, difficulty sucking and nursing; seen in brachycephalic breeds but also in beagles, cocker spaniels, dachshunds, German shepherds, Labradors, schnauzer, and Shetland sheepdogs

Cloacal papilloma – a benign tumor or wart derived from epithelium, caused by a species specific papillomavirus

Cloacal prolapse – cloacal tissues that protrude from the vent and can include intestines or uterus, caused by long-term straining to lay an egg or holding fecal material

Coat – the dog's hairy covering, usually a single coat that is often referred to as hair or double coat that has a thick under layer with guard hairs growing through it and is often called fur

Cobby – in reference to cat and rabbit breeds these are short-legged, thick-bodied animals, often with round heads

Coccidiosis – a protozoan parasite that causes enteritis in all species of animals

Colitis – inflammation of the colon, caused by many different etiologies

Collie Eye Anomaly, (CEA) – inherited autosomal recessive trait resulting in incomplete closure of the embryonic fissure; other defects that occur include scleral ectasia, scleral folds and detachment, coloboma of the optic disk, and microphthalmia

Coloboma – inherited absent or defective ocular tissue from an incomplete closure of the embryonic fissure

Congenital cataract – inherited opacity of the lens of the eye, capsule, or both

Congenital ectodermal dysplasia (skin fragility syndrome) – extremely fragile skin that is thin and tears easily, can also occur in older cats with hyperadrenocorticism, no breed, or gender predilection

Congenital hypotrichosis – animals born without the normal amount of hair or one that progressively loses hair starting at four weeks of age, usually associated with females with other defects of the reproductive organs, nails, teeth, eyes, and skeletal system

Congenital peripheral vestibular disease – conglomerations of neurological signs caused by a dysfunction in either the peripheral or central vestibular system

Congenital portosystemic shunts – failure of the ductus venosus in the fetus to close down shortly before or after birth, also called patent ductus venosus

Congestive heart failure – for a variety of reasons the heart can't deliver enough blood to the body, so fluid backs up into the animal's lungs, most often caused by hypertrophic cardiomyopathy, thyroid disease, *high blood pressure,* or birth defects among the possibilities

Conjunctivitis – inflammation of the conjunctiva or membrane covering the back of the eyelids and surface of the eyeball up to the cornea, symptoms are red and swollen tissue

Contact dermatitis – a reaction to something in the animal's environment, either allergy or irritant related; small red, itchy bumps often on the paws, abdomen, muzzle, and lip

Coprophagia – ingestion of feces, for some animals like rabbits it is a necessity in order to receive the nutrients from their food

Copper toxicosis – autosomal recessive inherited disorder of copper accumulation in the liver leading to chronic hepatitis and cirrhosis

Corneal dermoid – a congenital eye disease where normal tissue is in an abnormal location (choristoma), often found on the eyelid, conjunctive, or cornea

Corneal dystrophy – fairly common inherited progressive disease of the cornea, three types of disease categorized by location: epithelial (cell formation is affected), stromal (cornea become cloudy), and endothelial (lining of the cornea is affected)

Corneal lacerations – a penetrating injury or foreign object to the cornea

Corneal sequestration – development of an opaque, dark brown, or black plaque on the cornea occurring in cats of all ages and breeds

Corneal ulcers – deep layers of the cornea are ulcerated, caused by trauma, disease, tear deficiency, foreign body or chemical; short nosed dogs with prominent eyes are more likely to develop this problem

Coronavirus – an enveloped, single-stranded, plus-sense RNA virus that causes avian infectious bronchitis, hepatitis in mice, canine coronavirus gastroenteritis, and feline infectious peritonitis

Craniofacial abnormality – birth defects of the head and face

Craniomadibular osteopathy – bone disease affecting the mandible, tympanic bullae, and the temporal region consisting of irregular enlargements of the affected bones

Crepuscular – appearing or active in twilight

Cropped – surgically shaping the ear on certain breeds of dogs to make them stand up

Cruciate ligament rupture – a rip, or tear, of one of the cruciate ligaments, most common cause is by any movement that puts excessive pressure on the joint such as stepping in a hole while running, twisting while the paw remains planted, landing wrong when jumping, or general rough play

Cryptorchidism – the failure of one or both testicles to descend into the scrotum

Cryptosporidium – a microscopic protozoan parasite that causes diarrhea

Cur – a mongrel or a dog of mixed parentage

Cushing's disease – an increase in cortisone due to a benign tumor of the pituitary gland

Cystinuria – stones made from an amino acid called cystine form in the kidney, ureter, and bladder, inherited condition

Dam – a female parent

Decubital ulcers – Abscess or pressure wound occurs when a bony part of the body rests for long periods against a hard surface, restricting blood flow to the area

Degenerative myelopathy – neurological condition that eventually paralyzes Boxers, cause is unknown

Dehydration – the loss of more fluid than what is taken in, so the body doesn't have enough water to carry out normal functions

Demodectic mange – a skin disease caused by the Demodex canis mite; various species live on adult dogs and most human beings without causing any harm or irritation unless the animal or person become immunocompromised, also known as red mange, follicular mange, or puppy mange

Dermatitis, contact – see contact dermatitis

Dermatitis, moist – Also known as "hot spots" these circular lesions are found on the head, hip, and chest area and are moist, inflamed, often hairless, and quite painful; they are most often seen on long haired or double coated dogs, caused by a local allergic reaction to insect bites (fleas, mites, spiders), poor grooming, plant awns or burs; dogs will lick the area constantly making the inflammation worse, also known as acute moist dermatitis

Dermatomyositis – chronic inflammatory disease of skin and muscle, evidenced by scaly rash;

muscle inflammation occurs without rash and is referred to as polymyositis

Dermatophytosis – clinical condition caused by fungal infection of the skin commonly referred to as "ringworm" because of the typical circular pattern that develops as the fungi feed on keratin, the material found in the outer layer of skin, hair, and nails

Dermoid cysts – a *cystic teratoma* that contains an array of developmentally mature, solid tissues such as *skin*, clumps of long *hair*, pockets of sebum, blood, fat, bone, nails, teeth, eyes, cartilage, and thyroid tissue

Dermoid sinus – a tube-like sinus that is skin-like so it is lined with hair, and as the hair is shed, it builds up, and the body tries to flush it with a build-up of serum, but not all sinuses have an opening, so it cannot drain and an abscess forms; these sinuses are found along the midline of the neck, back and tail along the spinal column; this is an inherited condition

"Devon Rex hereditary myopathy" – a degenerative autosomal-recessive *congenital myopathy* evidenced by a passive ventroflexion of the head and neck, high-stepping forelimb gait, head bobbing, exercise intolerance, difficulty swallowing food

Dew claw – a vestigial claw or functionless digit on the inside of the leg; a rudimentary fifth toe that does not touch the ground

Dewlap – loose, pendulous skin under the throat

Diabetes mellitus – a group of metabolic diseases in which an animal has high blood glucose (blood sugar), either because insulin production is inadequate, or because the body's cells do not respond properly to insulin, or both

Diarrhea – rapid evacuation of water and feces, reducing nutritional absorption and loss of electrolytes, major cause is inflammation of the intestinal mucosa by infections or chemicals

Dilated cardiomyopathy – inflamed heart muscle where the heart chambers increase in size and the muscles stretch thin, one of the causes of Congestive Heart Failure (CHF); usually affects the left side of the heart but can affect both sides of the heart, large and giant breed dogs are at greater risk of developing DCM

Dingos – Canis lupus dingo is a free-ranging dog found mainly in Australia

Diphyodont – any animal with two successive sets of teeth, the "deciduous" set (baby teeth), and consecutively the "permanent" set (adult teeth)

Distal neuropathy – Uncommon, generally progressive peripheral nerve disease in which the primary changes appear in the nerve cell

Diurnal – activity during the day and sleeping at night

Dock – to shorten the tail by cutting it off between vertebras

Dog – a male dog; also used collectively to designate both male and female canines

Dolichocephalic – an extremely long nose as compared to skull length

Double-coated – an outer coat resistant to weather and protective against brush and brambles, together with an undercoat of softer hair for warmth and waterproofing

Dudley nose – flesh colored nose

Dwarfism – animal is significantly smaller than standard members of their species, pathogenic changes in the structure of an organism

Dystocia – slow or difficult labor or delivery

Eclampsia – life threatening drop in blood calcium levels that occurs in nursing mothers; signs include tremors, weakness, and a form of paralysis called puerperal tetany characterized by still limbs and an inability to stand or walk, emergency treatment is necessary

Ectodermal cysts – a benign cyst usually found on the skin

Ectopia lentis – displacement or malposition of the eye's lens from its normal location

Ectopic ureters – the *ureter* terminates at a different site other than the bladder

Ectropion – the lower eyelid turns outwards, can be surgically repaired, it is a genetic disorder

Eczema – inflammation of the skin characterized by redness, skin *swelling*, itching and dryness, crusting, flaking, blistering, cracking, oozing, or bleeding

Egg-binding – reasonably common and potentially serious condition where the female is unable to pass an egg, often becoming stuck near the *cloaca* or further inside leading to infection or damage to internal tissues

Elbow dysplasia – condition involving multiple developmental abnormalities, specifically the growth of cartilage or the structures surrounding the elbow joint

Elongated soft palate – the soft palate on the roof of the mouth is so long it partially blocks the entrance to the trachea

Encephalitis – an acute inflammation of the brain, symptoms include fever, confusion, drowsiness, fatigue, seizures, hallucinations, stroke, and hemorrhaging

Endocardial fibroelastosis (EFE) – a thickening of the innermost lining of the heart chambers (endocardium) from an increase in the amount of supporting connective tissue and elastic fibers

Enterotoxemia – an infection by Clostridium perfringens, also known as overeating disease killing an animal within two hours

Entire – A male cat with both testicles

Entropion – the eyelid folds inward causing the eyelashes to rub against the cornea causing irritation, this is an inherited condition

Epilepsy – a group of neurological disorders characterized by epileptic seizures, which can vary from brief and nearly undetectable to long periods of vigorous shaking, cause of most cases of epilepsy is unknown

Epileptiform convulsions – involuntary contraction or series of contractions of the voluntary muscles

Epiphora – overflow of tears onto the face, a clinical sign of insufficient tear drainage from the eyes through the nasolacrimal system

Episodic falling – increased muscle tone where the muscles cannot relax, previously thought to be a muscular disorder, now known to be neurological, "exercise-induced paroxysmal hypertonicity"

Epizootic catarrhal enteritis (ECE) – viral inflammation of the mucous membranes in the intestine, symptoms are severe diarrhea (often of a bright green color), loss of appetite, and severe weight loss

Esophageal chalasia – a rare degeneration disease of the esophageal muscles and nerves and the lower esophageal sphincter, swallowing difficulty, chest pains, and regurgitation of food and liquids are common signs

Escherichia coli – commonly abbreviated E. coli – a gram-negative, facultatively anaerobic, rod-shaped bacterium commonly found in the lower intestine of warm-blooded animals, most strains are harmless, but some can cause serious food poisoning

Estrous cycle – recurring physiologic changes induced by reproductive hormones in most mammalian therian females; cycles start after sexual maturity and are broken into four phases: anestrous (resting) phase, proestrus (preparation) phase, estrus (receptive) phase, metestrus (recovery) phase or if bred pregnancy

Euthanasia – the practice of intentionally ending a life in order to relieve pain and suffering

Eviscerate – the removal of internal organs (viscera), usually those in the abdominal cavity

Facial fold pyoderma – see pyoderma

Factor VII Deficiency – a lack of the blood-coagulation factor VII, one of the proteins that causes blood to clot in the coagulation cascade

Familial Shar-Pei Fever – an inherited condition seen in Shar Pei characterized by recurring fever and swelling of the hocks; can result in renal and liver failure through accumulation of amyloid in those organs (amyloidosis)

Fanconi syndrome – a disease of the proximal renal tubules of the kidney in which glucose, amino acids, uric acid, phosphate and bicarbonate are passed into the urine instead of being reabsorbed; may be inherited or caused by drugs or heavy metals

Fatty liver syndrome – in birds it is a slow progressive disease where the liver tissue is replaced with adipose (fat) tissue

Fawn – light yellowish tan color

Feathering – long strands of hair found on the back of legs, underside of the abdomen, chest, and tail, this fringe or feathery tufts of hair can occur on one area or all areas depending on breed

Feline – of a cat or the cat family

Feline acne – blackheads (comedones) developing on the chin of a cat; cause is unknown, but several factors appear to be associated with development including stress, a suppressed immune system, poor grooming habits, the presence of other diseases, contact or atopic dermatitis, and skin conditions in which abnormal amounts of oils are produced and the hair follicles do not function properly

Feline acromelanism – a pigmentation pattern affecting the legs, ears, tail, and face where cooler temperatures influence the color of the coat on those areas; this is why Balinese and Siamese have the pointed coat colors

Feline infectious peritonitis (FIP) – a fatal, incurable disease that affects *cats*, caused by Feline Infectious Peritonitis Virus, a mutation of Feline Enteric Coronavirus

Fibromas – benign tumors composed of fibrous or connective tissue, can be found in all organs

Flat-chested kitty syndrome – kittens develop a compression of the thorax caused by vertebral malformation, most frequently caused by collapsed lungs

Flatulence – when gases generated in the intestine or stomach are expelled through the anus

Flews – overhanging upper lips

Fontanel (soft spot) – an anatomical feature of the skull comprising any of the soft membranous gaps (sutures) between the cranial bones that make up the skull of a fetus or newborn

Fragmented coronoid process – developmental defect of the one of the coronoid processes of the ulna; a fissure or crack develops and separates from the rest of the bone, causing pain and joint instability

Frill or Ruff – long hair around the neck and chest, often brushed up to frame the face in show dogs and cats

Fur (hair) ball – Fur that is swallowed by a cat or dog when cleaning itself and can form a felt-like mass in the stomach, usually passed through the digestive system without trouble, but on occasion it can get lodged in the stomach forming a hairball too large to pass normally, which is then vomited up

Gaily carried tail – a short tail that is held straight up while in motion or if dog is excited

Gangliosidosis – a lipid storage disorder caused by the accumulation of lipids known as gangliosides; two distinct genetic causes, both autosomal recessive, affect males and females equally

Gastric dilation volvulus (GDV) – also known as twisted stomach, gastric torsion, a condition in which the stomach becomes overstretched and rotated by excessive gas content

Gastritis – inflammation of the lining of the stomach

Gastroenteritis – a medical condition from inflammation of the gastrointestinal tract involving both the stomach and small intestine, causes some combination of diarrhea, vomiting, and abdominal pain and cramping

Gay Tail – a tail that is held horizontally level with the back

Geriatric onset laryngeal paralysis polyneuropathy – a common condition affecting large-breed geriatric dogs where the innervation of the larynx degenerates, signs include respiratory stridor, exercise intolerance, dyspnea, and throat clearing

Gestation – time during which an animal is pregnant

Gingivitis – a non-destructive periodontal disease in response to bacterial plaque that adheres to tooth surfaces

Glanzman's thrombasthenia – an abnormality of the platelets

Glaucoma – a group of ocular disorders that results in optic nerve damage, often associated with increased intraocular pressure

Globoid cell leukodystrophy (Krabbe disease) – a rare, inherited autosomal recessive degenerative disorder that affects the myelin sheath of the nervous system

Glycogen storage disease type IV – a rare inherited disorder with various types, all of which are characterized by deficient or defective activity of the enzymes responsible for metabolizing glycogen in the body leading to the accumulation of glycogen

GM-1 Storage Disease – a recessive genetic disorder, one of a family of conditions called GM1 gangliosidosis, resulting in abnormal storage of acidic lipid materials in the nerve cells, is inevitably fatal

Goiter – an enlargement of the thyroid gland that is not functioning properly, signs include swelling of the neck or larynx

Grizzle – bluish-gray colored hair coat color

Groups – dog breeds as grouped in seven divisions to facilitate judging

Guard Hairs – the longer, smoother, stiffer hairs which grow through, cover and protect the soft undercoat

Hackles – erectile hair on neck and back raised involuntarily in fright or anger

Hairballs – See fur balls

Harderian gland – found within the eye's *orbit* which occurs in animals that possess a nictitating membrane; the fluid it secretes varies between different groups of animals; in rodents it is a red color

Harlequin – patched or pied coloration, usually black on white

Haw – a transparent third eyelid which keeps the eye clean and moist; it is located at the inner side or lower lid of the eye, also called the nictitating membrane

Heat stroke – see hyperthermia

Heart murmurs – abnormal heart sounds produced when blood flows across one of the heart valves; two types of murmurs: functional murmur or "physiologic murmur" is a murmur primarily due to physiologic conditions outside the heart, while other types of murmurs are due to structural defects in the heart itself

Hematuria – the presence of erythrocytes in the urine

Hemivertebra – wedge-shaped vertebrae that cause an angle in the spine; congenital types cause neurologic problems

Hemochromatosis (Hereditary hemochromatosis) – causes a body to absorb too much iron from foods; it is stored in the organs, especially liver, heart and pancreas which in turn poisons these organs, leading to life-threatening diseases

Hemolytic anemia – a form of anemia due to hemolysis, either in the blood vessels or elsewhere in the body, either inherited or acquired

Hemophilia – a group of inherited disorders that impair the body's ability to control blood clotting; Hemophilia A is the most common form of the disorder

Hepatic lipidosis – commonly known as fatty liver causing severe feline liver diseases in anorexic cats

Herbivore – an animal anatomically and physiologically adapted to eating plant material

Hereditary cataract – a breakdown of tissue in the eye causing clouding of the lens, often familial in dogs

Hereditary non-inflammatory myopathy – a muscular disease in which the muscle fibers do not function properly leading to overall muscle weakness; starts to develop between ages of 3–4 months and worsens with cold weather, exercise, and excitement

Hermaphroditism – a form of sexual reproduction in which both partners can act as the male or female

Hibernation – a state of inactivity and metabolic depression, characterized by low body temperature, slow breathing and heart rate, and low metabolic rate

Hierarchy – a system in which one animal is ranked above another according to status

Hip dysplasia – a genetic trait that causes abnormal formation of the hip socket that can cause crippling lameness and painful arthritis

Hot spots – see dermatitis, moist

Hound-marked – a dog coat coloration composed of white as the base color and may be marked with tan and/or black patches on the head, back, legs, and tail

Hydrocephalus – a condition in which fluid accumulates in the brain

Hygroma – is a subdural body of cerebrospinal fluid (CSF) without blood

Hyperesthesia syndrome – a rare malady in domestic cats that causes episodes of agitation, self-mutilation, and a rippling of the skin when touched, also known as rolling skin disease

Hyperkeratosis – abnormal thickening of the outer layer of the skin; examples include corns and calluses

Hyperlipidemia – elevated levels of cholesterol in the blood

Hyperphagia – abnormally increased appetite frequently associated with hypothalamus injuries

Hyperthermia – elevated body temperature due to failed thermoregulation that occurs when a body produces or absorbs more heat than it dissipates

Hypertrophic cardiomyopathy (HCM) – the myocardium becomes abnormally thick making it harder for the heart to pump blood

Hypertropic osteodystrophy – a bone disease of fast-growing large and giant breed dogs, controlled by strict exercise and nutritional requirements

Hyperuricosuria – elevated levels of uric acid in the urine, predisposing dogs to form uroliths in bladders or kidneys, is an inherited simple autosomal recessive trait

Hypocalcaemia – deficiency of calcium in the bloodstream

Hypoglycemia – deficiency of glucose in the bloodstream

Hypokalemia – deficiency of potassium in the bloodstream

Hypokalemic polymyopathy – a metabolic disorder of cats characterized by muscle weakness associated with intermittent hypokalemia

Hypothermia – an abnormally low body temperature, induced by exposure or during anesthesia when thermoregulation is affected

Hypothyroidism – abnormally low activity of the thyroid gland

Hypo-vitaminosis – Insufficient absorption or availability of one or more essential vitamins

Idiopathic facial dermatitis – facial pruritus, Persian and Himalayan cats are predisposed

Idiopathic liver cirrhosis (Fibrosis) – accumulation of tough scar tissue that replaces normal liver tissue, signs include ascites, hepatic encephalopathy, weight loss, vomiting, and diarrhea

Idiopathic pulmonary fibrosis – a chronic and ultimately fatal disease characterized by a progressive decline in lung function from the scarring of lung tissue by an unknown agent or disease process

Ileitis – inflammation of the ileum resulting from infection, obstruction, severe irritation, or faulty absorption

Immune mediated hemolytic anemia – immune system destroying its own red blood cells

Immunosuppression – an act that reduces the activation or efficacy of the immune system

Impacted anal glands – impactions, infections, and abscesses can occur for a variety of reasons, such as the conformation of the animals, the thickness of the gland's secretions, or the softness of the stool, may also become infected and abscess

Inbreeding – mating of closely related individuals, as cousins, sire-daughter, brother-sister, which tends to increase the number of individuals that are homozygous for a trait and therefore increases the appearance of recessive traits, utilized often with mice in a research setting

Inguinal hernias – a protrusion of abdominal-cavity contents through the inguinal canal

Insulinoma – a tumor of the pancreas that is derived from beta cells and secretes insulin

Intermingled – Fur of two colors mingled together, as required in the Blue-Creams

Intussusception – the sliding or telescoping of the intestine within itself, an uncommon but potentially life-threatening condition that can occur in dogs and cats of all ages

Intertrigo – a rash between adjacent areas of skin or body folds

Intervertebral disc disease – a serious condition where cushioning pads of fibrocartilage that sit between most of the vertebra of the spinal column are damaged or slipped, seen more often in dogs than cats

Iris coloboma – occurs when part of the iris fail to develop, seen as small notches out of the inner edge of the iris, pie slices, or massive holes, occasionally so large it appears the dog has no iris at all; present at birth and in most cases the effect on vision is minimal

Junctional epidermolysis bullosa – a group of genetic conditions that cause the skin to be very fragile and form blisters in response to minor injury or friction, such as rubbing or scratching

Keratitis – inflammation of the cornea

Keratoconjunctivitis sicca – results from lack of tearing

Kink – a malformation or thickening of a joint in the tail.

Kiss marks – tan spots on the cheeks and over the eyes.

Kitten – a cat up to the age of nine months in Britain or eight months in the U.S.

Krabbe-type disease – a demyelinating disease of the nerves (see globoid cell leukodystrophy)

Lacrimal punctal aplasia – rare autosomal dominant disorder characterized by aplasia, atresia, or hypoplasia of the lacrimal and salivary glands

Lead – a strap, cord, or chain attached to the collar or harness for the purpose of restraining or leading the dog

Legg-Calve-Perthes disease – blood supply to the growth plate of the femur becomes inadequate, as a result the bone becomes necrotic

Lens luxation – occurs when one or more zonules (small fibers attached to the lens) become loosened or completely detach allowing the lens to fall forward

Leptospirosis – a bacterial disease of the genus Leptospira that affects humans and animals

Leukocyte adhesion deficiency – a rare autosomal recessive disorder characterized by immunodeficiency resulting in recurrent infections

Level bite – incisors meet edge to edge

Lip fold pyoderma – a condition affecting dogs with pendulous lower lips where they develop skin infections that produce pus

Listeria – foodborne bacterial illness that can be very serious for animals with impaired immune systems

Liver – deep, reddish brown color

Liver amyloidosis – See Amyloidosis

Liver shunt – See Portosystemic liver shunts

Locket – A white or contrasting color patch under the neck.

Longhairs – A term used in Britain and other countries for all cats with long fur.

Lymphedema – condition of localized fluid retention and tissue swelling caused by a compromised lymphatic system, which normally returns interstitial fluid to the thoracic duct and then the bloodstream; can be inherited

Lymphocytic choriomeningitis (LCM) – causative agent is the Lymphocytic Choriomeningitis Virus (LCMV), a rodent-borne infectious viral disease that presents as aseptic meningitis, encephalitis, or meningoencephalitis

Lymphomas – a group of blood cell tumors that develop from lymphatic tissues

Lymphoproliferative disorders – several conditions in which lymphocytes are produced in excessive quantities; occur in animals with compromised immune systems

Lysosomal storage disease – lysosomes inside cells break down waste products and other byproducts; if one or more of the enzymes necessary for this breakdown is missing, the cells store this debris, and if it occurs in neuronal (nerve) cells, the animal's mental and motor functions deteriorate; this inherited disorder can affect animals as early as four months of age to 4–6 years depending on which type of enzyme is missing

Macaw Wasting Disease – See Avian Proventricular Dilation Disease

Macrothrombocytosis – congenital abnormality of exceptionally large platelets, found in at least a third of Cavilier King Charles Spaniels

Malassezia Pachydermatis – normal yeast found on the skin and ears of dogs; however, an abnormal overgrowth can cause dermatitis, or inflammation

Malignant Fibrous Histiocytoma – subtype of cancer, originating from a primitive cell, which can develop into either fibrous tissue or histocytes

Malignant neoplasms – composed of cells that look less like the normal cell of origin, a high rate of proliferation, potentially invade and metastasize; malignant neoplasms derived from epithelial cells are called carcinomas, derived from connective tissue cells are called sarcomas

Malocclusions – imperfect positioning of the teeth when the jaws are closed.

Marsupial – mammal whose offspring are born incompletely developed and are typically carried and suckled in a pouch

Mask – dark shading on the foreface (Mastiff, Boxer, Pekingese, and some cats, such as Siamese and Himalayans

Mast cell tumors – the most common skin tumor in dogs, it is a type of blood cell cancer normally involved in the body's response to allergens and inflammation, and can also affect the spleen, liver, gastrointestinal tract, and bone marrow

Megaesophagus – enlarged esophagus; signs include regurgitation, loss of weight, and recurrent episodes of aspiration pneumonia; two causes: a failure of the esophagus to contract and propel food into the stomach, which is an inherited disorder; second cause is a blockage of the esophagus by a foreign body or a developmental problem with abnormal blood vessels encircling the esophagus

Melanoma – tumor of melanin-forming cells, typically malignant tumor

Meningitis – inflamed meninges which are membranes that envelops the central nervous system

Meningoencephalitis – inflammation of the meninges and brain

Meningomyelitis – inflammation of the meninges and spinal cord

Mesocephalic – head is of medium proportions, skull and nose of equal lengths

Microphthalmia – abnormally small eyeball(s) congenital abnormality that arises before birth

Mink enteritis virus – species of parvovirus that infects mink and causes enteritis

Mitral valve dysplasia – congenital malformation or degeneration of the mitral valve and its supporting structures, resulting in valvular regurgitation

Mongrel – a dog whose parents are of mixed-breed origin.

Monorchidism – rare developmental condition in only one testicle develops

Mutant MDR1 gene – 70% of Collies have this gene which predisposes them to multiple drug sensitivities; adverse drug reactions to over a dozen different drugs have been documented; tests are available for the mutant MDR1 gene

Mutt – a crossbred dog of unknown parentage

Mycoplasma – simple one-celled bacteria without outer membranes that penetrates and infects individual cells

Necrotic myelopathy ("Afghan hound hereditary myelopathy") – an autosomal-recessive genetic disease of the Afghan Hound breed characterized by a rapidly progressive symmetrical paralysis of the limbs

Nephritis – inflammation of the kidneys

Nephrosis – non-inflammatory nephropathy or degenerative disease of the renal tubules that can be caused by kidney disease or can be secondary to another disorder

Neuromuscular degenerative disease – a very broad term that encompasses many diseases and ailments impairing muscle function, either directly being pathologies or indirectly being pathologies

Nocturnal – active at night

Nodular Fasciitis – tumor-like proliferation of fibroblasts with inflammatory exudation occurring in fascia

Nystagmus – rapid involuntary eye movements

Obesity – condition of being grossly overweight

Oligodontia (hypodontia) – congenital absence of six or more teeth in permanent dentition; relatively rare, occurring either as an isolated finding or as part of a syndrome

Omnivore – an animal that eats food of both plant and animal origin

Ophthalmorrhexis – ruptured eyeball

Oriental – grouping of some specific shorthaired varieties such as the Siamese, also the shape of some cats' eyes

Orofacial pain syndrome – syndrome characterized by acute and severe oral and facial pain, signs of affected cats present with face and tongue mutilation and often with exaggerated licking and chewing movements, this disease is most likely a neuropathic pain disorder similar to trigeminal neuralgia

Osteoarthritis (also known as degenerative arthritis, degenerative joint disease and osteoarthrosis) – a group of mechanical abnormalities involving degradation of joints, including articular cartilage and subchondral bone; signs include joint pain, tenderness, stiffness, locking, and effusion

Osteochondrodysplasia – a growth and developmental abnormality of the bone and cartilage which results in lack of normal bone growth and bone deformities

Osteochondrosis – disease that affects cartilage formation; the cartilage thickens rendering it unable to receive nutrients from the joint fluid, causing it to become weaker and more susceptible to damage; signs include pain, lameness, and arthritis, also known as *OCD, osteochondrosis, osteochondritis dissecans, shoulder/elbow/stifle dysplasia*

Osteoporosis – bones become brittle and fragile from loss of tissue often as a result of hormonal changes, or deficiency of calcium or vitamin D

Osteosarcoma – malignant tumor of bone

Otitis externa – inflammation of the outer ear and ear canal

Otitis media – a group of inflammatory diseases of the middle ear

Out-crossing or outbreeding – introducing unrelated animals into a breeding line to increase the genetic diversity and hopefully reducing the probability of congenital disease or genetic abnormalities

Ovarian cysts – fluid-filled sacs within or on the surface of an ovary

Overshot jaw – maxilla is longer than mandible, see Prognathic

Pack – a group of dogs, usually associated with wild dogs; if the group is made up of hounds it is called a cry, mute, or pack

Palsy – involuntary tremors

Pancreatic insufficiency (Exocrine Pancreatic Insufficiency) – failure of the pancreas to produce digestive enzymes and insulin, causing weight loss and chronic diarrhea, thought to be a hereditary disease in German Shepherds

Pancreatitis – inflammation of the pancreas

Pannūs (chronic superficial keratitis) – progressive inflammatory disease of the cornea; signs include brown discoloration, vascularization, and haziness of the cornea; may lead to scarring and progress to severe visual impairment or blindness

Panosteitis (hypertrophic panosteitis) – self-limiting and painful condition characterized by limping and lameness, affecting the long bones in young dogs between the ages of 5 and 18 months

Papillomas – oral warts are small benign tumors of the mouth caused by the species-specific papilloma virus found on lips, gums, mouth, and sometimes other mucous membranes; young dogs are more susceptible because their immune system is not fully developed; transmitted through direct contact

Paraplegia – paralysis of the legs and lower body typically caused by spinal injury or disease

Parrot fever (Psittacosis) – a zoonotic infectious disease caused by *Chlamydia psittaci* and contracted from parrots

Patagium – membranous structure that assists an animal in gliding or flight

Parti-colored – variegated in patches of white and another color

Parturition – the act of giving birth

Passerine – any bird of the order Passeriformes; includes more than half of all bird species, often referred to as seed eating birds

Pasteurella multocida – a gram-negative, non-motile, penicillin-sensitive coccobacillus, the cause of a range of diseases in mammals and birds

Patched – Fur that has patches of different coloring, as in the Tortoiseshell and Whites

Patellar luxation – when a dog's patella (kneecap) is dislocated from its normal anatomical position in the groove of the femur (thigh bone), most common in toy and miniature dog breeds

Patent Ductus Arteriosus (PDA) – congenital disorder in the heart in which a neonate's ductus arteriosus fails to close after birth

Pattern baldness – gradual, symmetric thinning and loss of hair, starting at 6–9 months and progressing until affected areas are bald, common in Dachshunds, American Water Spaniels, and Portuguese Water Dogs

Pemphigus *vulgaris* – chronic blistering skin disease with lesions that are rarely pruritic but are often painful

Penciling – The delicate, pencil-like markings on the faces of Tabbies.

Precocial – young bird or other animal that hatches or is born in an advanced state, able to feed itself almost immediately

Periodontal disease – common clinical condition occurring in adult dogs and cats in which bacteria in the mouth forms a substance called plaque that sticks to the surface of the teeth, minerals in the saliva harden the plaque into dental calculus which is firmly attached to the teeth; entirely preventable with brushing and chewing sticks

Peritoneopericardial diaphragmatic hernia – common congenital pericardial disease in dogs and cats, results from abnormal development of the dorsolateral septum transversum or from failure of the lateral pleuroperitoneal folds and the ventromedial pars sternalis to unite which results in the herniation of abdominal viscera into the pericardial sac

Persistent hyperplastic primary vitreous – rare congenital developmental anomaly of the eye that results following failure of the embryological, primary vitreous, and hyaloid vasculature to regress

Phosphofructokinase deficiency (Glycogen storage disease type VII or Tarui's disease) – autosomal recessive inheritance metabolic disorder

Piebald – comparatively large patches of two or more colors.

Pneumonia – lung inflammation caused by bacterial or viral infection in which the air sacs fill with pus and may become solid

Pododermatitis – inflammation of the skin on the foot, many potential causes

Points – dark or darker coloring on face, ears, legs, and tail, as seen in Siamese and other cats

Polycystic kidney disease – cystic genetic disorder of the kidneys

Polydipsia – abnormally great thirst

Polydontia – more than the normal number of teeth

Polyomavirus – small DNA-containing viruses of the family Papovaviridae capable of producing a variety of tumors in mice, hamsters, rabbits, and rats

Porphyria – the blood pigment hemoglobin is abnormally metabolized and so porphyrins are excreted in the urine; other symptoms include mental disturbances and extreme sensitivity of the skin to light; rare hereditary disease

Porphyrin – in rat tears is mostly protoporphyrin IX, an organic compound found all over the living world

Portosystemic shunts – a bypass of the liver by the body's circulatory system; can be either a congenital or acquired condition, also known as a liver shunt

Primary hyperparathyroidism – disorder of the parathyroid glands in which one or more of the glands are enlarged, overactive, and secrete too much parathyroid hormone

Primary seborrhea – excessive discharge of sebum (an oily secretion) from the sebaceous glands

Prognathic – a lengthening of the jaw relative to the opposite jaw, i.e., "overshot jaw," in which the upper maxilla is longer than the mandible

Prognathism – congenital defect in dogs associated with delayed closure of the parietal fontanelles

Progressive Retinal Atrophy – group of genetic diseases seen in certain breeds of dogs and cats, similar to retinitis pigmentosa in humans, characterized by bilateral degeneration of the retina, causing progressive vision loss culminating in blindness

Prolapse – slipping forward or down of an organ or body part

Protein-losing enteropathy – the severe loss of serum proteins into the intestine

Protein-losing nephropathy – disease where the filtering mechanism of the kidneys is defective, and the protein is lost in the urine

Proteus mirabilis – gram-negative, facultative anaerobic, rod-shaped bacterium, causing 90% of all Proteus infections, widely distributed in soil and water

Pruritus – Itching from drug reactions, food allergies, kidney or liver diseases, cancers, parasites, aging, or dry skin

Pseudohyperkalemia – unexplained elevation of potassium serum concentrations which occurs when potassium is released from the blood cells in a collected sample

Pseudomonas – gram-negative, aerobic gammaproteobacteria

Pseudotuberculosis – acute, sometimes fatal disease of rodents, birds and other animals caused by Yersinia bacterium, characterized by formation of nodules in the lungs resembling those that result from tuberculosis

Psittacine beak and feather disease – an infectious disease primarily affecting young birds just prior to fledgling age will present with acute lethargy, crop stasis, and death within 24–48 hours; adults may be asymptomatic carriers

Psittacines – the parrots, roughly 372 species that make up the order Psittaciformes, found in most tropical and subtropical regions

Pulmonary stenosis – congenital defect from abnormal development of the fetal heart

Puppy – any dog less than twelve months of age

Purebred – animal of recognized breed which is eligible for registry in the official breed association

Pyloric dysfunction – narrowing of the pyloric canal from an overgrowth of muscles in the region of the stomach that connects with the duodenum; cause is unknown, rarely found in cats, also called pyloric stenosis, or chronic hypertrophic pyloric gastrophy

Pyoderma – any skin disease that is pus forming (pyogenic); pus is an exudate that is yellowish-white to yellowish-brown of a thick liquid or clotted cream consistency

Pyometra – uterine infection, acute onset and life threatening if left untreated

Pyruvate kinase deficiency – enzyme deficiency that impairs the red blood cells' ability to metabolize; can cause anemia and other blood-related issues

Queen – a female cat used for breeding

Renal cortical hypoplasia – cortex of the kidneys is improperly developed, resulting in improper filtration of toxins from the bloodstream, renal failure is inevitable

Retinal detachment – part of the retina pulls away from supportive tissue; this is an emergency situation

Retinal dysplasia – non-progressive eye disease affecting the retina caused by viral infections, drugs, vitamin A deficiency, or genetic defects, characterized by folds or rosettes of retinal tissue

Retinal folds – folds in the inner retinal layer

Reverse sneezing – rapid, long inspirations where the dog will stand still and extend its head making a loud snorting sound; this is a pharyngeal gag reflex which is a reflex contraction at the back of the throat

Rhinitis syndrome – irritation and inflammation of the mucous membrane inside the nose

Ringworm – see dermatophytosis

Roan – a fine mixture of colored hairs with white hairs; blue, orange, lemon (English Cocker Spaniel)

Roman Nose – a hump or convex shaped nose

Ruff or Frill – long hair around the neck and chest, often brushed up to frame the face in show dogs and cats

Sable – a lacing of black hairs over a lighter ground color; in Collies and Shetland Sheepdogs it is a golden to mahogany brown color

Saddle – black marking over the back, like a saddle

Salmonella – gram-negative, rod-shaped bacteria of the Enterobacteriaceae family

Salt & Pepper – a mixture of black and white hairs with slightly more black hair than white

"Samoyed Hereditary Glomerulopathy" – hereditary non-inflammatory disease of the renal glomeruli, inherited as a sex-linked, genetically dominant disease that affects male dogs to a greater extent than females

Sarcocystis – protozoan parasites that infect mammals, reptiles, and birds

Sarcocystosis – to be infected with Sarcocystis

Scissor bite – upper incisors overlap the lower incisors, normal bite in dogs and cats

"Scottie Cramp" – hereditary neuromuscular disorder characterized by periodic cramps lasting up to 30 minutes seen after exercise or if the dog becomes overly excited

Scurvy – vitamin C is quickly cleared by the body and if not replenished it affects the body's ability to manufacture collagen, which is important for the formation of bone and tissue, in turn causing blood clotting issues, joint stiffness, lethargy, loss of appetite, diarrhea, and skin problems

Sebaceous adenitis – a rare idiopathic autoimmune skin disease against the sebaceous glands which can lead to the destruction of the gland

Sebaceous cysts – blocked hair follicle with dirt, infection, scar tissue, or even thick sebum

Sebaceous gland tumor – usually benign overgrowth of sebaceous gland cells

Seborrhea – sebaceous glands produce excessive sebum

Seborrhea oleosa – oily seborrhea that forms a yellow-brown crust, also known as exudative epidermitis

Semi-precocial – born or hatched with eyes open, covered with fur or down and capable of standing or leaving the nest soon after birth or hatching

Sendai virus – negative sense, single-stranded RNA virus of the family Paramyxoviridae, also known as murine parainfluenza virus type 1

Septicemia – a serious, life-threatening infection where bacteria is in the blood (bacteremia), also called sepsis

Skin fold dermatitis – abnormal rubbing combined with moisture, warmth, and lack of aeration are ideal conditions for the overgrowth of normal skin inhabitants like yeast and bacteria, producing a dermatitis which typically leads to pyoderma, common in breeds with pronounced facial, tail, and vulvar folds, though any deep skin fold can be susceptible

Slipped stifles – see Patellar Luxation

Snub – short noses seen in some of the longhaired cat breeds

Soft tissue sarcoma – sarcoma in connective tissue

Spay – removal of reproductive organs on female animals

Spectacles – shadings or dark markings around the eyes and then to the ears

Spiculosis – hyperkeratosis of the hair follicles resulting in abnormally thickened hair; the thickened hair bulb leads to formation of a keratin mass that grows out along the root sheath forming the thickened hair-like spicule, rare congenital skin disease of male Kerry Blue Terriers

Spinal muscular atrophy – genetic disease affecting the nervous system that controls voluntary muscle movement, atrophy of the muscles results from decreased movement because the muscle fibers are not receiving the signal to move

Spondylosis – defect of the lumbar or thoracic vertebra

Squamous cell carcinoma – uncontrolled growth of abnormal squamous cells

Squint – cross-eyed, the eyes are placed so that they seem to look permanently at the nose

Spraying – a male cat's habit of urinating to establish territory

Standard – the breed or variety characteristics required to be considered a purebred animal

Staphyloma – abnormal protrusion of the uveal tissue through a weak point in the eyeball, weakening caused by an inflammatory or degenerative condition

Stationary Night Blindness – autosomal dominant congenital disorder of the retina

Strains – accomplished through inbreeding; all members of a group are as nearly as possible genetically identical, in reference to rodents used in laboratory protocols

Subcutaneous cysts – see sebaceous cyst

Substrate – the substance or layer on a surface provided as bedding or housing for an animal

Subvalvular aortic stenosis – a narrowing of the area beneath the aortic valve, causing a degree of obstruction or blockage of blood flow through the heart

Syndactulous – two or more digits partially or completely united

Syringomyelia – a cyst within the spinal cord

Systemic lupus erythematosus – autoimmune disease in which the immune system attacks healthy skin, joints, kidneys, brain, or other organs

Swimmer puppies – a puppy that is unable to stand at three weeks of age as a result of weak muscles in the rear limbs; often paddles feet as if swimming; most recover with time; can be an inherited condition

Tail-tip necrosis – from beating an infected tail against objects, increasing the injury to the point where the skin become more brittle and in later stages the skin dies and sluffs off; amputation of the infected tip may be required if the infection cannot be healed

Thrombocytopenia – relative decrease of thrombocytes (platelets) in the blood

Thyroid disease – any dysfunction of the thyroid gland located at the base of the neck; most common types are hyper or hypothyroidism, cancer, goiter, nodules, or inflammation

Ticked – small, isolated areas of black or colored hairs on white ground

Tonsillitis – inflammation of the tonsils which are located at the back of the throat

Torsions – a twist of the stomach, intestines, or spleen

Torticollis – the head becomes persistently turned to one side, often with painful muscle spasms

Toxemia – pre-eclampsia characterized by abrupt hypertension (high blood pressure), albuminuria (leakage of large amounts of the protein albumin into the urine), and edema (swelling) in a pregnant animal

Toxoplasmosis – infected with the Toxoplasma gondii parasite

Trace – a dark stripe down the back of the Pug

Tracheal collapse – condition characterized by incomplete formation or weakening of the cartilaginous rings of the trachea resulting in flattening of the trachea

Trichiasis – abnormally positioned eyelashes that touch the cornea or conjunctiva

Trichomoniasis – sexually transmitted infection caused by a protozoan parasite

Tri-colored – three-colors: white, black, and tan.

Tufts – hair growing from the ears or between toes

Tyzzer's disease – acute epizootic bacterial disease caused by the spore-forming bacterium *Clostridium piliforme*, characterized by necrotic lesions on the liver; most commonly affected include young, stressed animals in laboratory environments

Umbilical hernia – congenital malformation of the umbilicus

Undershot jaw – see Brachygnathism

Unique juvenile-onset polyarthritis syndrome – immune-mediated inflammatory disease of the moveable joints, such as shoulder, knee, etc.

Urolithiasis – formation of stony concretions or uroliths in the bladder or urinary tract

Uroliths – stony concretions that form in the urinary system (bladder is most common in animals) in varying sizes and numbers, usually secondary to infection, due to dietary influences and genetics

Urologic diseases – congenital or acquired dysfunction of the urinary system (kidney, ureters, bladder, urethra)

Urticarial pigmentosa – rare disease caused by excessive numbers of mast cells (part of the immune system) in the skin that produce hives or lesions on the skin when irritated

Uveodermatological syndrome – autoimmune condition directed against melanin of neural crest (skin, hair, anterior uvea) and neuro-ectodermal (retina) origin; ocular signs begin with bilateral uveitis later progressing to retinal detachment, posterior synechiae, secondary glaucoma, and cataracts; the skin and hair abnormalities include leukoderma and leukotrichia involving the eyelids, nasal planum, lips, scrotum, vulva and footpads; pruritis and lymphadenopathy are common

Vaccination – a biological preparation that provides active acquired immunity to a particular disease

Vaginal hyperplasia – proliferation of the vaginal mucosa resulting in a mass protruding from the vulva as a result of estrogenic stimulation; usually occurs during or just prior to a dog being in heat

Vaginitis – inflammation of the vagina; symptoms include discharge, itching, and pain

Vascular disease – includes any condition affecting the circulatory system

Vasectomized – surgical removal of all or part of the vas deferens as a means of sterilization

Ventricular septal defects – defect in the ventricular septum, the wall dividing the left and right ventricles of the heart

Vestibular disease – sudden, non-progress disturbance of balance, common in older dogs; signs include loss of balance, disorientation, head tilt, nystagmus; can be caused from middle or inner ear infections, trauma, tumors, hypothyroidism, or drugs toxic to the ear

Vibrissae – hair that is longer and thicker, often called whiskers, that have large follicles containing blood-filled sinus tissues and by have an identifiable representation in the somatosensory cortex, meaning that they are used as a sensory organ

Vitiligo periocular leukotrichia – rare condition in cats producing "cobweb" or "snowflake" effect in the hair coat; white spots begin to appear on the coat becoming more extensive with age ultimately going completely white or left with diminishing isolated patches of color

Von Willebrand's Disease (vWD) – common inherited blood-clotting disorder in dogs from an insufficient amount of a plasma protein known as von Willebrand factor (vWF) that helps blood clot

Waardenburg syndrome – rare genetic disorder characterized by varying degrees of deafness

Walrus syndrome – fetal puppies develop edema in various organs, cavities, or tissues of the body, with swelling so large that they cannot enter the birth canal and must be taken via cesarean section; often are stillborn or die shortly after birth; it can affect one puppy or the entire litter

Waltzing disease – cavy disease producing a palsy and deafness; animals often run wildly in circles

Wheaten – pale yellow or fawn colored

Whelping – the birthing process

White dog shaker syndrome – disorder causing a dog's entire body to shake, also known as *idiopathic* cerebellitis, which is an inflammation of the cerebellum; can affect any colored dog

Whip – a long, thin tapering tail, as seen in the Siamese

Whiskers – the long bristles protruding from the cat's face

Widow's peak – used to describe a hairline that comes to a point on the forehead

Withers – the highest point of the shoulders, immediately behind the neck

Wobbler's syndrome (Cervical spondylomyelopathy (CSM)) – disease of the cervical spine commonly seen in large and giant-breed dogs characterized by compression of the spinal cord leading to neurological signs and/or neck pain; symptoms include a characteristic wobbly gait

Wry mouth – once side of the mouth grows faster than the other causing difficulties eating and grasping as the teeth do not line up in the normal occlusion

Yersinia pestis – gram-negative, rod-shaped coccobacillus, facultative anaerobic bacterium responsible for three major plagues: pneumonic, septicemic, and bubonic

Zinc deficiency – results from reduced dietary intake, inadequate absorption of, increased loss of, or increased use of zinc, which plays an essential role in biochemical pathways; has numerous manifestations, the most common of which are increased rates of diarrhea, pneumonia, and malaria

Zoonotic – disease passed from animals to humans

APPENDIX

Size Comparions Charts

- Herding Group
- Hound Group
- Non-Sporting Group
- Sporting Group
- Non-Sporting Group
- Working Group
- Working Group
- Toy Group

Herding Group

2.01 Australian Cattle Dog

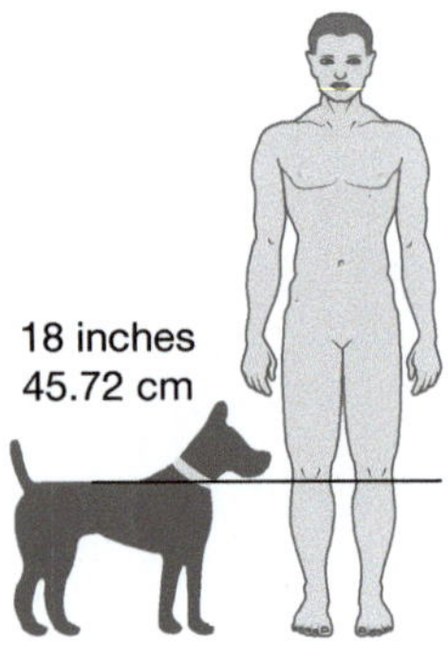

2.04 Australian Cattle Dog

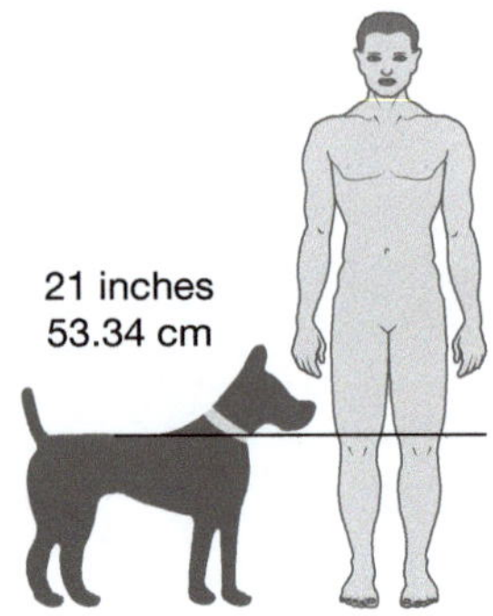

2.02 Australian Shepherd

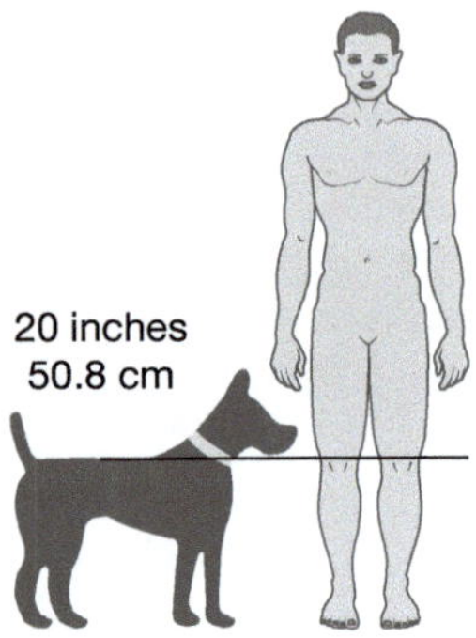

2.05 Border Collie

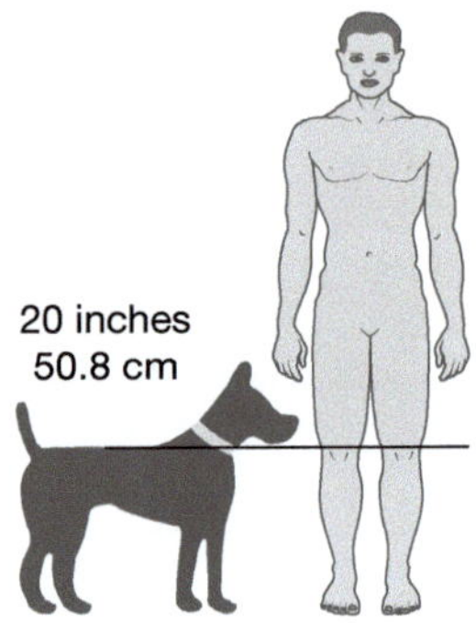

2.03 Belgian Malinois, Tervuren, and Shepherd

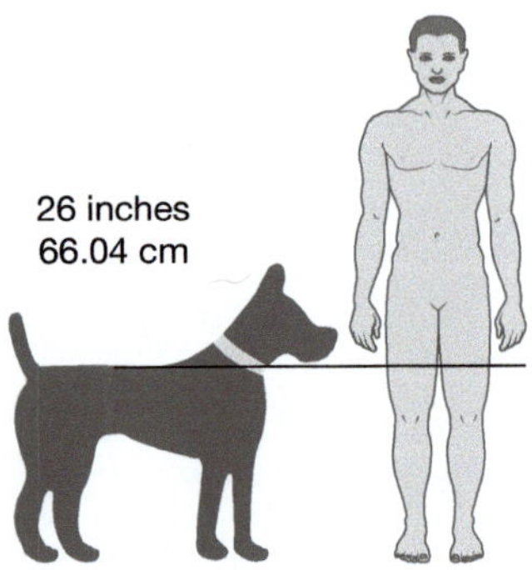

2.06 Bouvier des Flandres

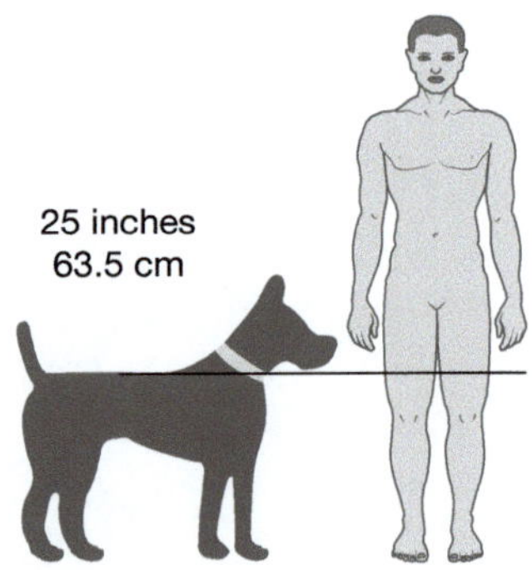

Herding Group

2.07 Briard

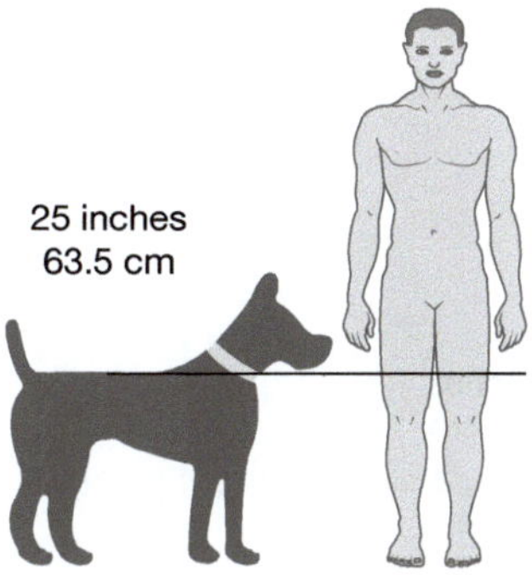

2.10 Old English Sheepdog

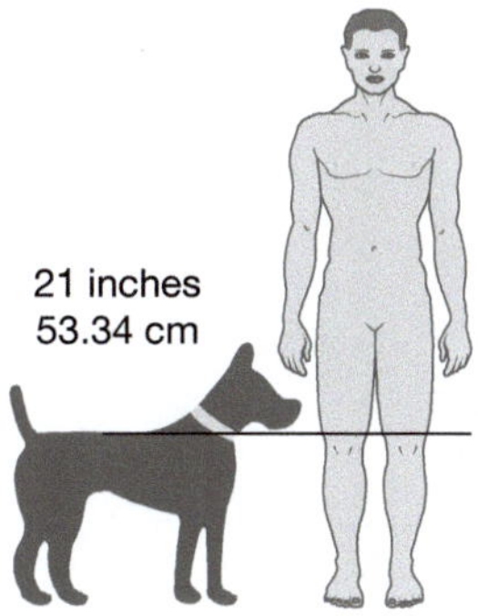

2.08 Collie

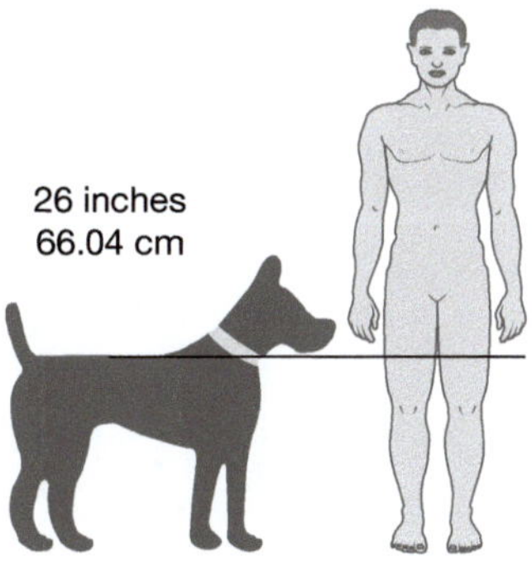

2.11 Puli

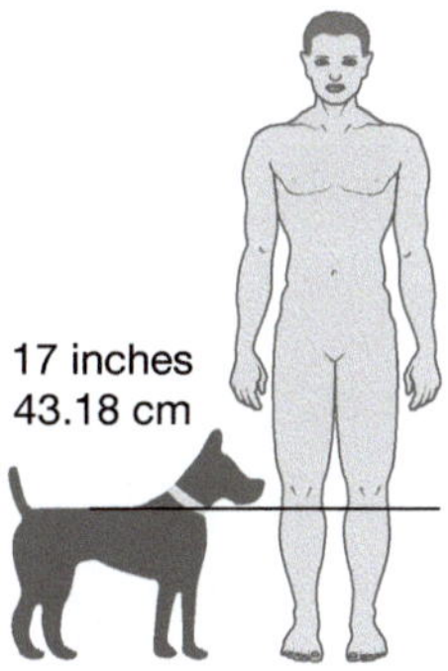

2.09 German Shepherd

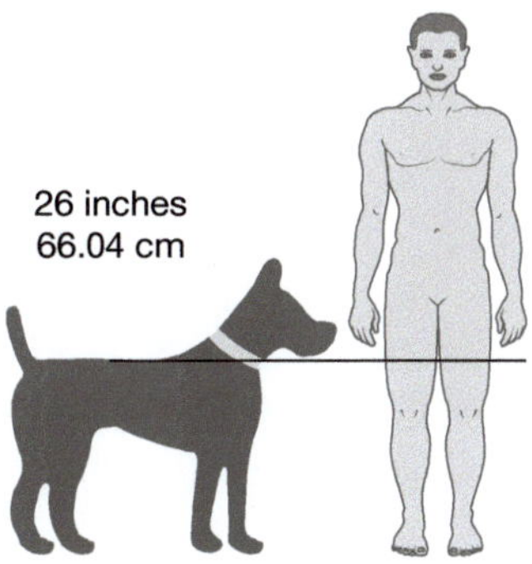

2.12 Shetland Sheepdog

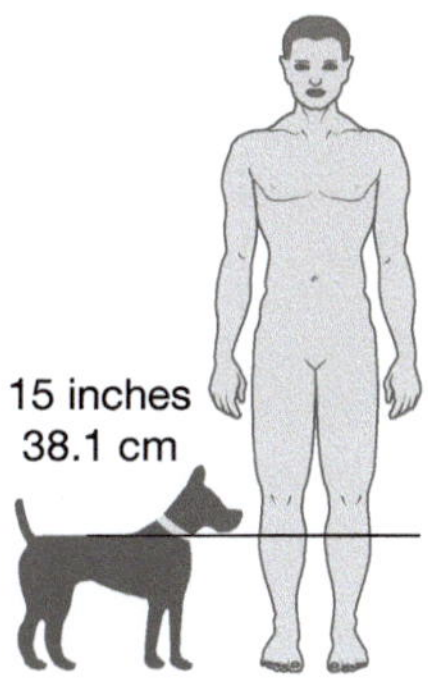

Herding Group

2.13 Welsh Corgi Cardigan

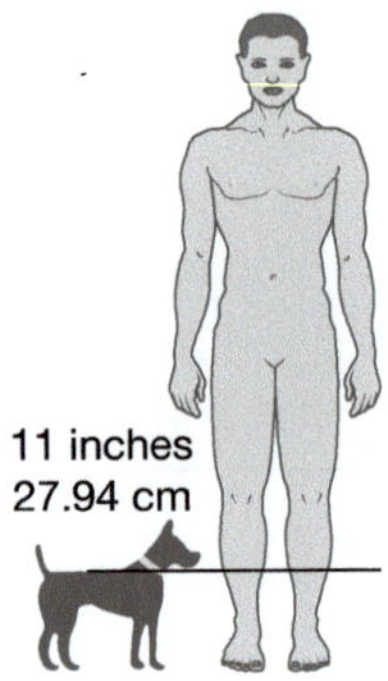

2.14 Welsh Corgi Pembroke

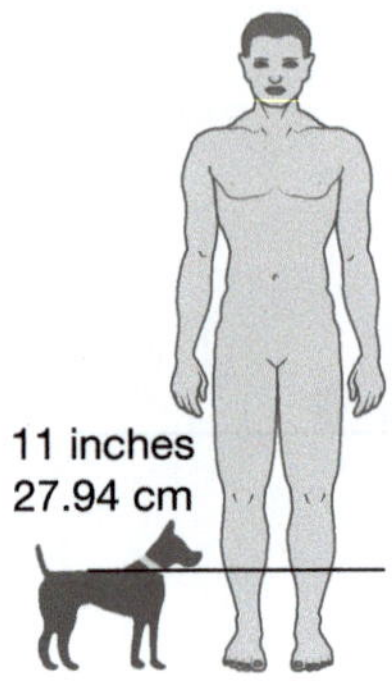

Hound Group

3.00 Afghan Hound

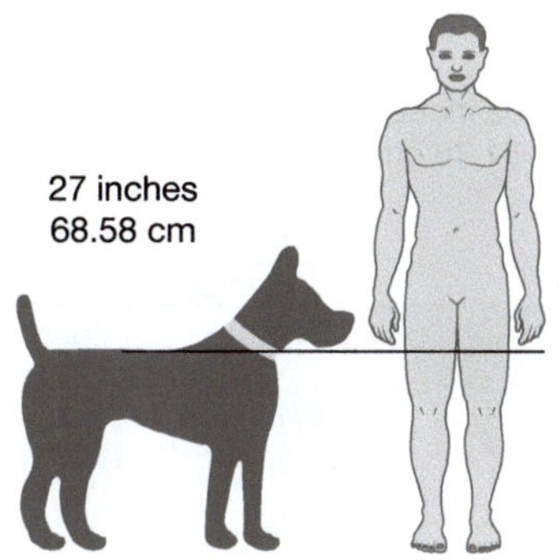

3.03 Greyhound

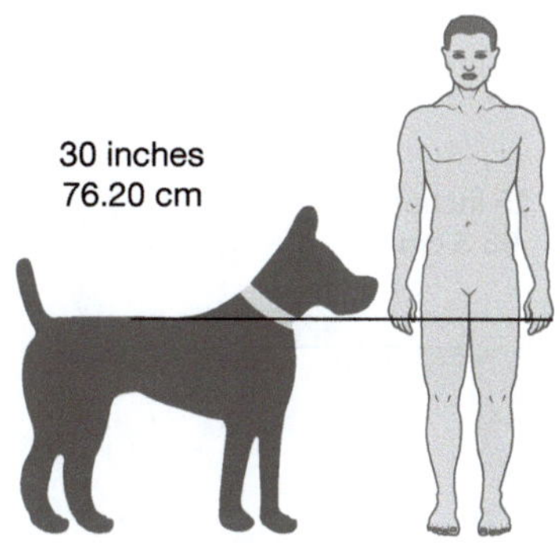

3.01 Basenji

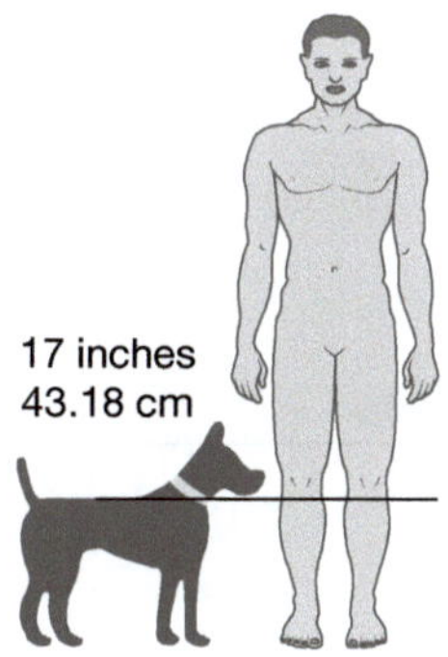

3.04 Ibizan Hound

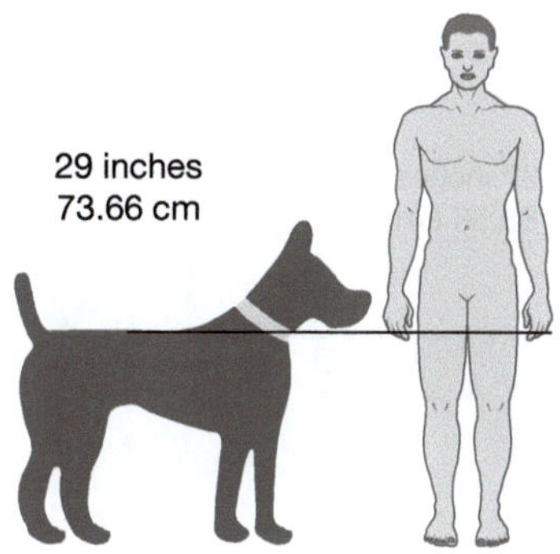

3.02 Borzoi

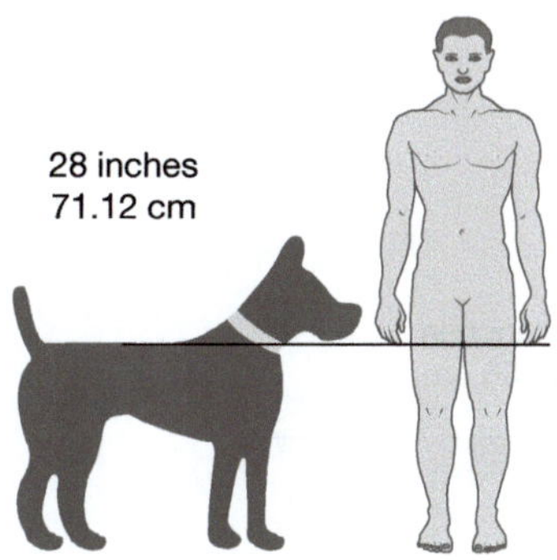

3.05 Irish Wolfhound

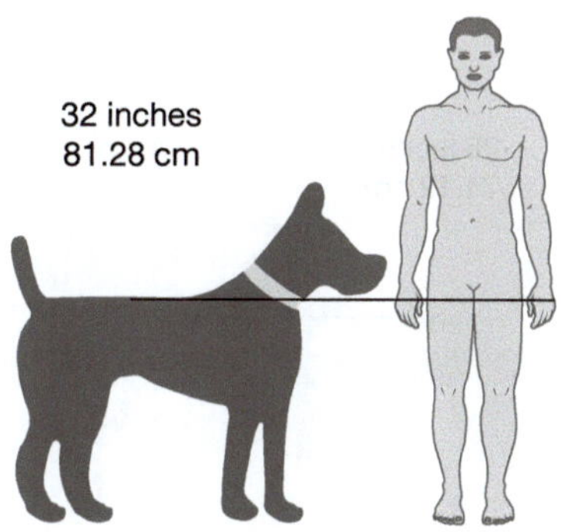

Hound Group

3.06 Pharaoh Hound

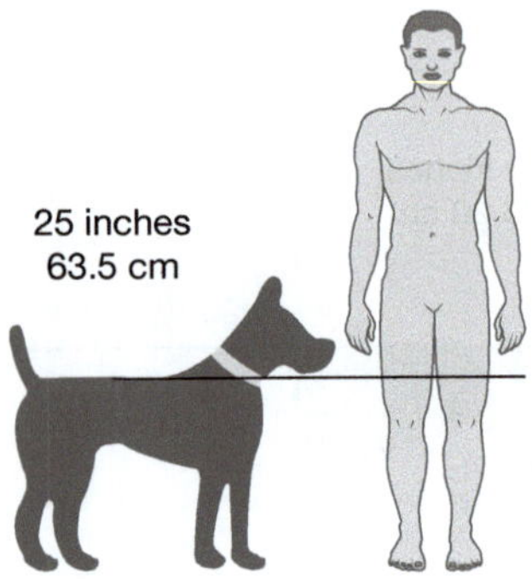

3.07 Rhodesian Ridgeback

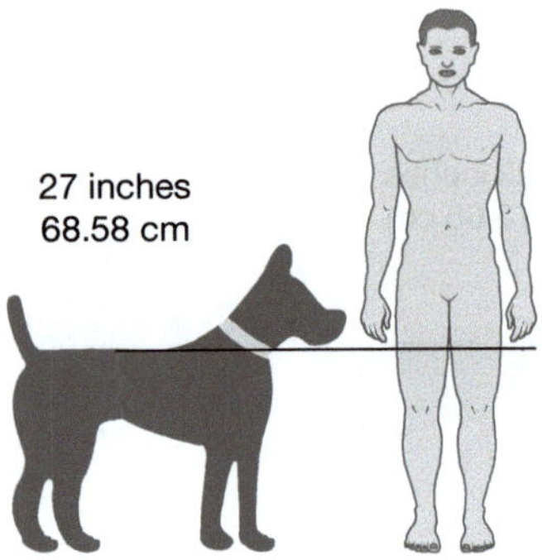

3.08 Saluki

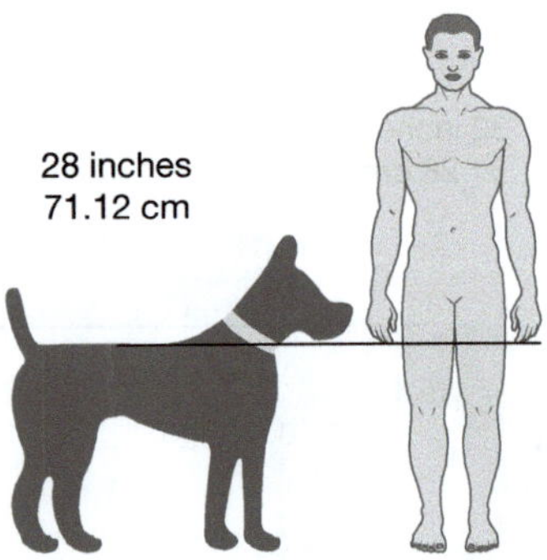

3.09 Scottish Deerhound

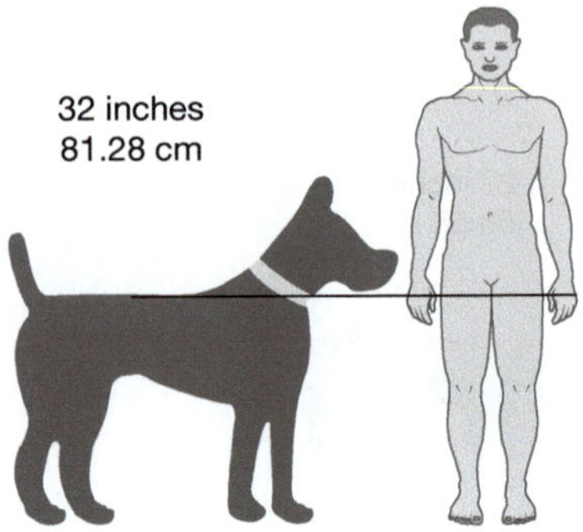

3.10 Whippet

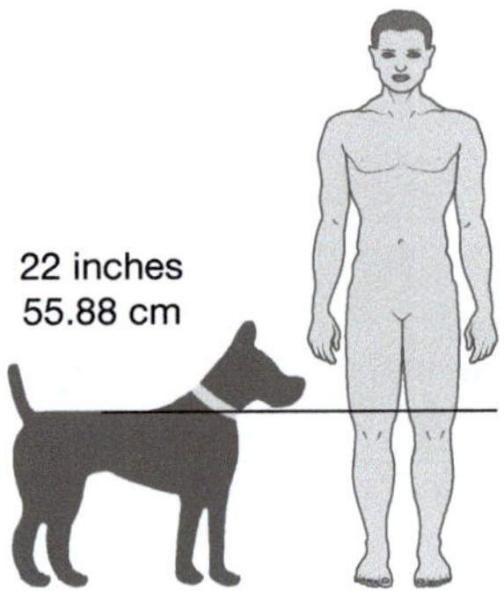

3.11 Basset Hound

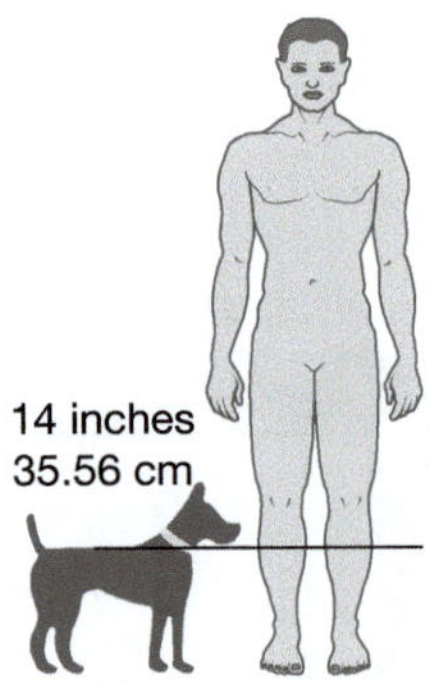

Hound Group

3.12 Beagle

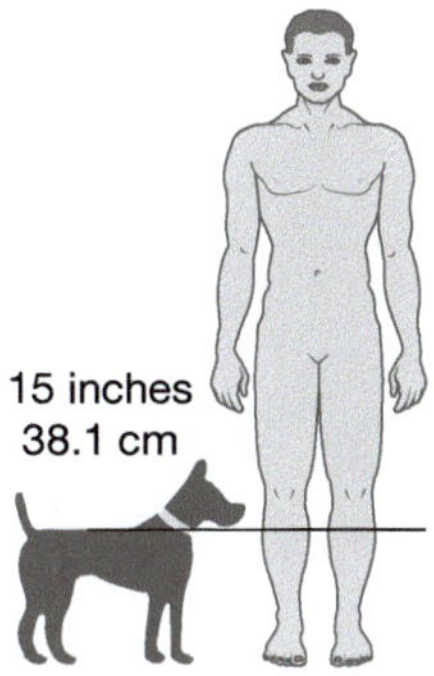

3.15 American Foxhound

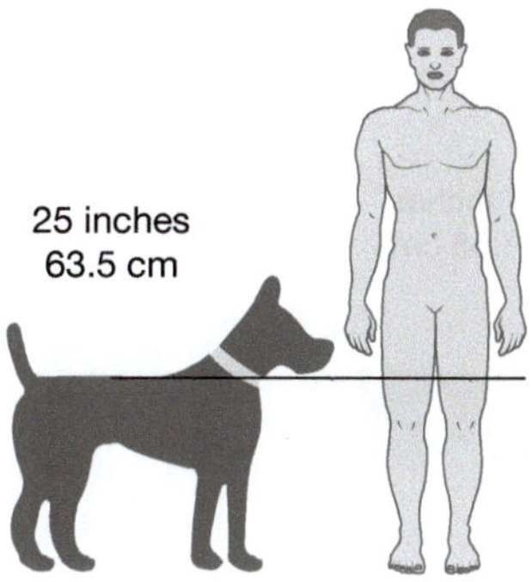

3.13 Bloodhound

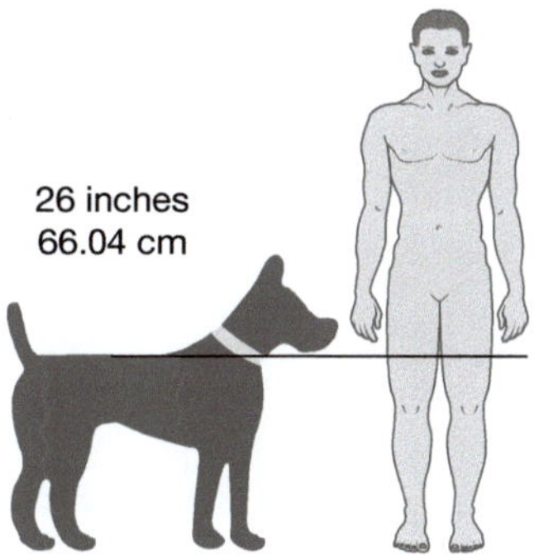

3.16 Norwegian Elkhound

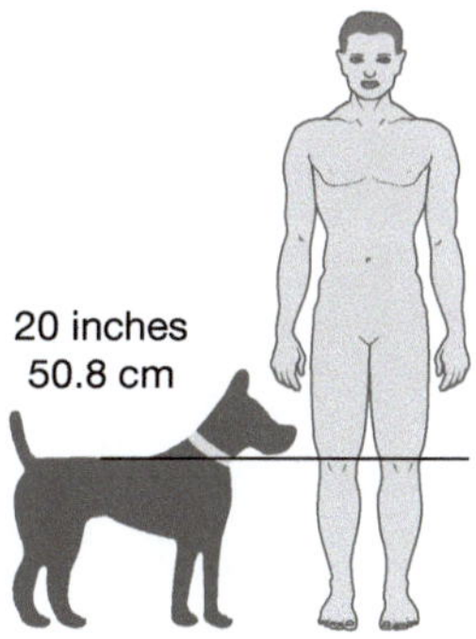

3.14 Dachshunds

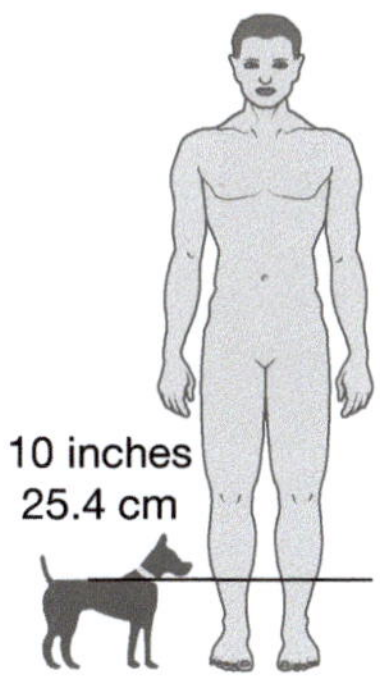

Hound Group

3.17 Otter Hound

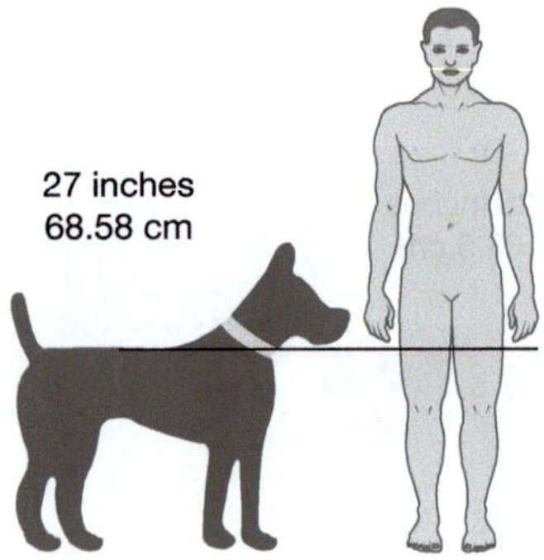

3.18 Petits Bassets Griffon

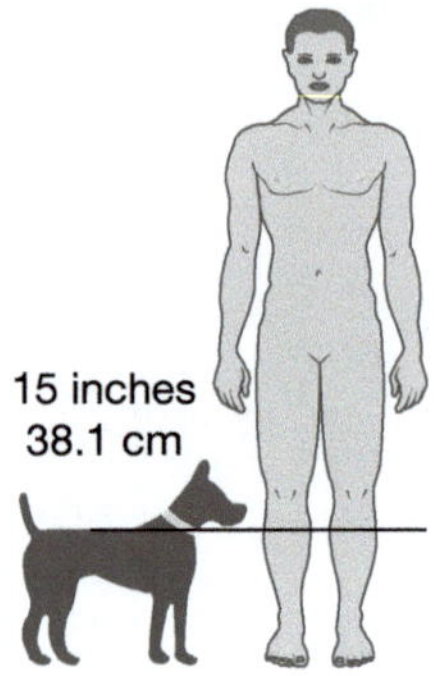

Non-Sporting Group

4.00 American Eskimo Dog

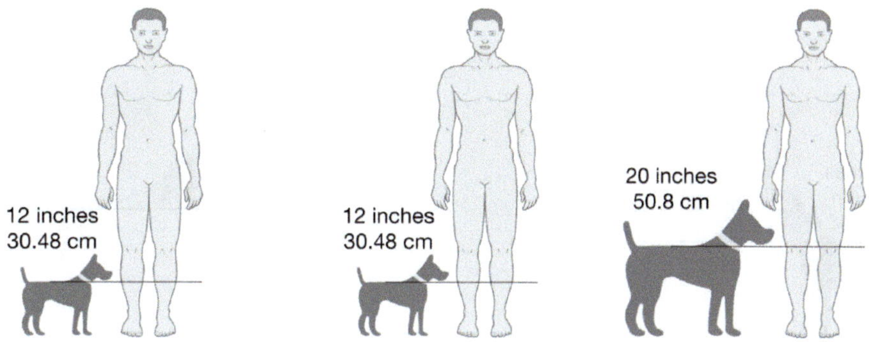

4.01 Bichon Frise

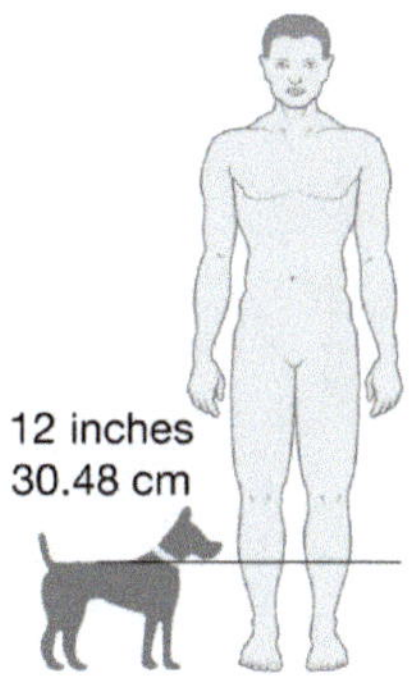

4.03 Bulldog

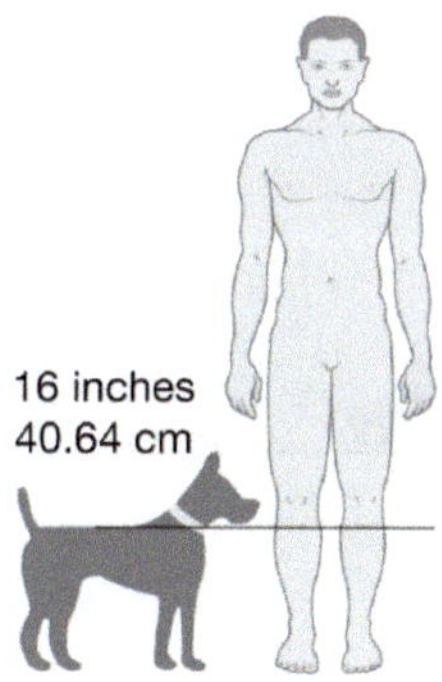

4.02 Boston Terrier

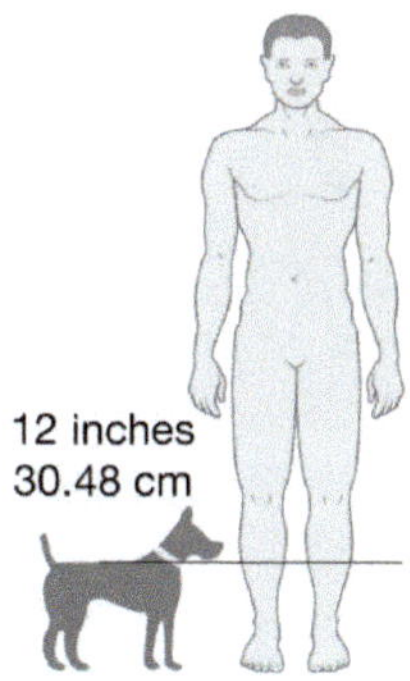

4.04 Chinese Shar-Pei

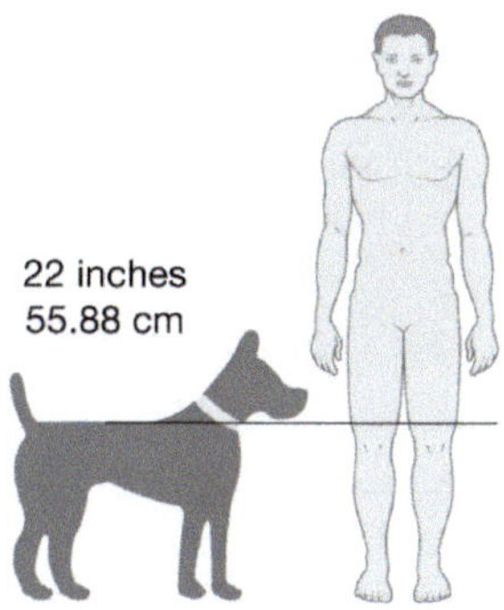

Non-Sporting Group

4.05 Chow Chow

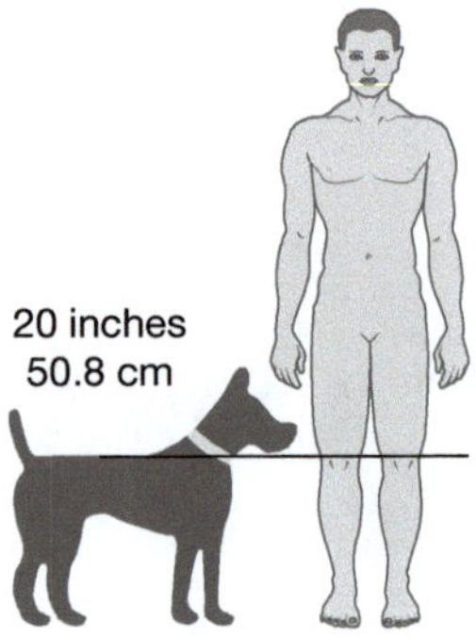

4.06 Dalmatian

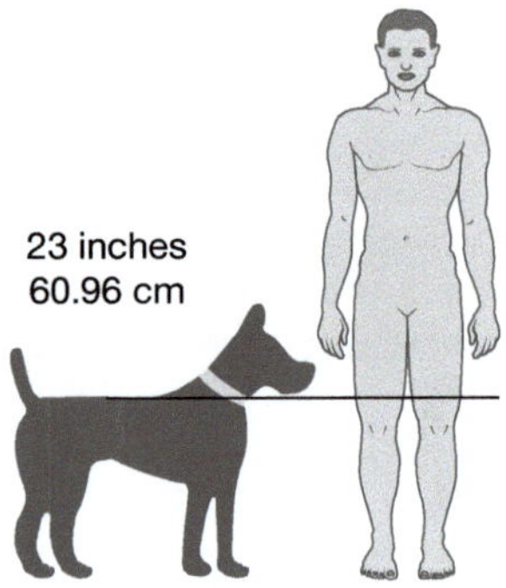

4.07 French Bulldog

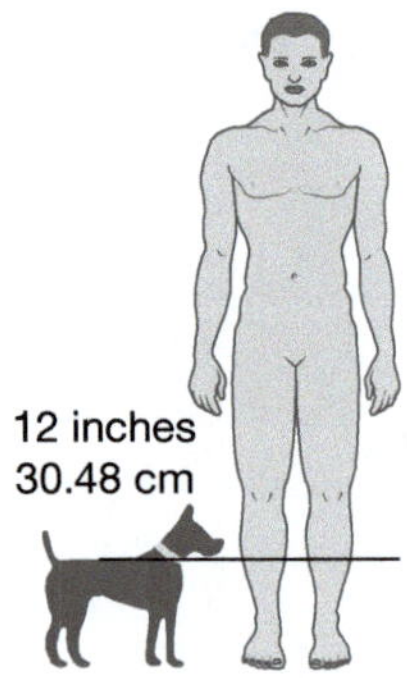

4.08 Keeshond

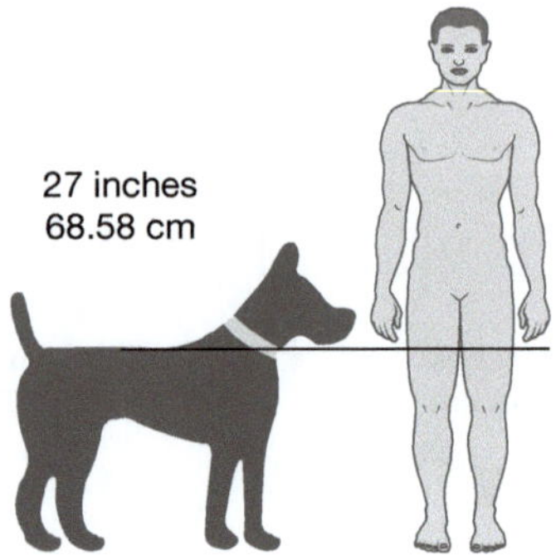

4.09 Lhasa Apso

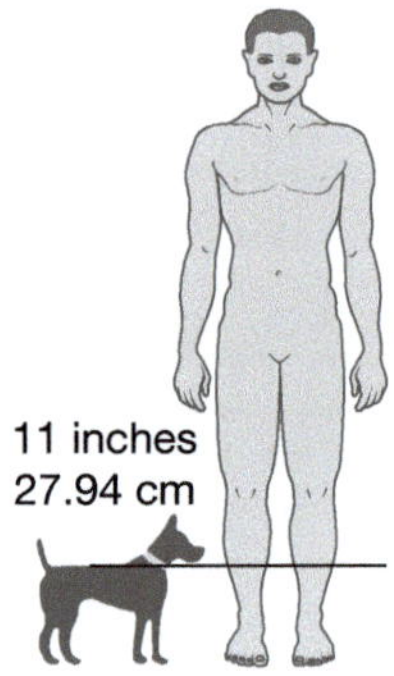

Non-Sporting Group

4.10 Poodle Standard

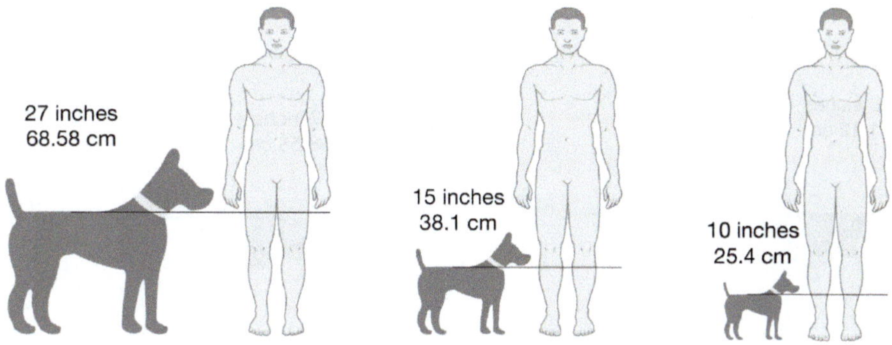

4.11 Schipperkes

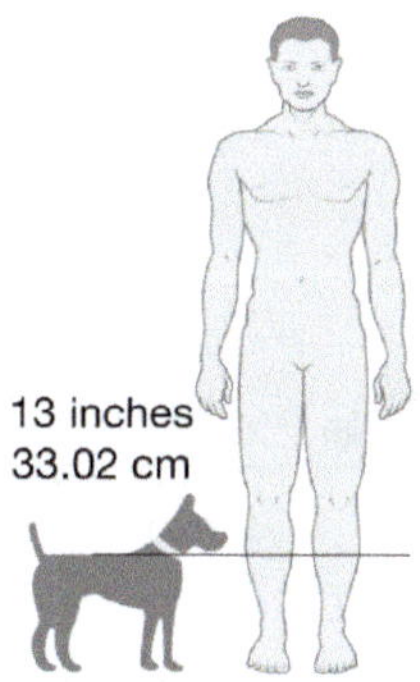

4.13 Tibetan Spaniel

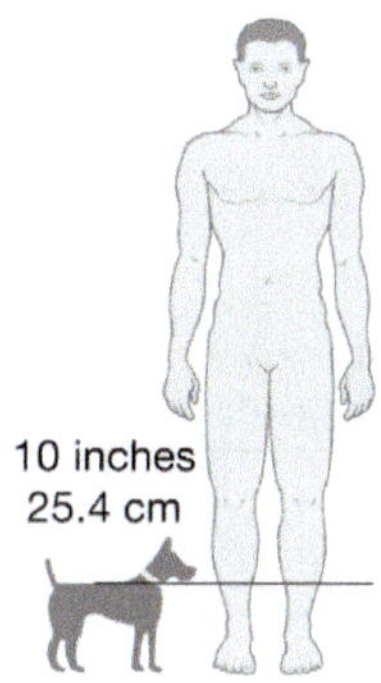

4.12 Shiba Inu

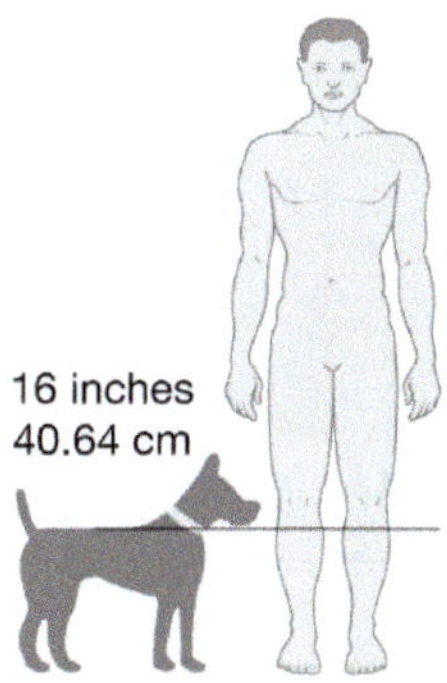

4.14 Tibetan Terrier

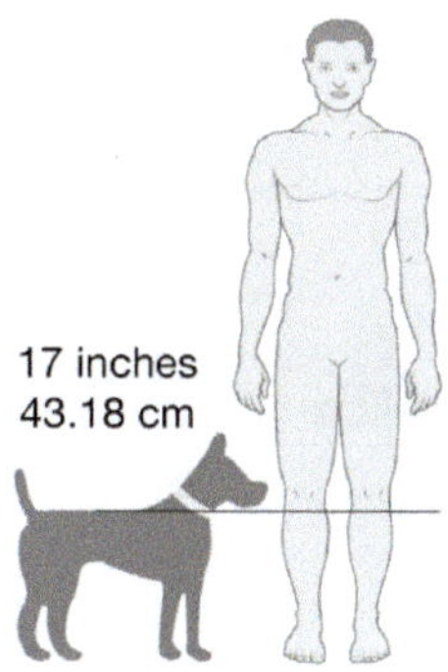

Sporting Group

5.00 English Pointer

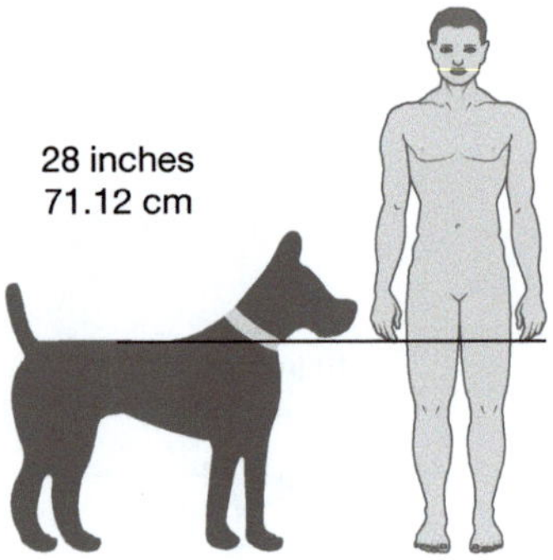

5.03 Weimaraner

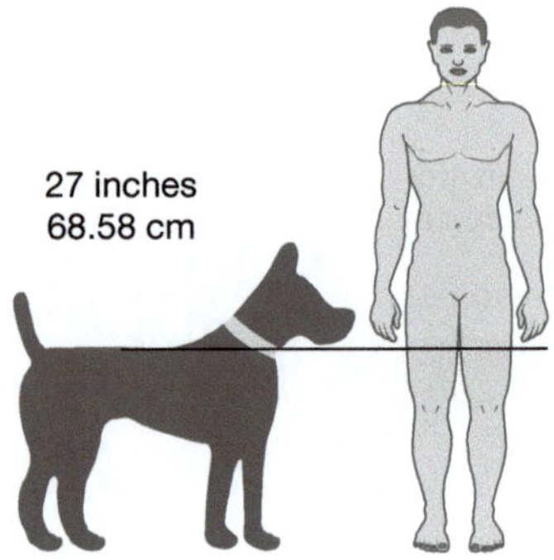

5.01 German Shorthaired Pointer

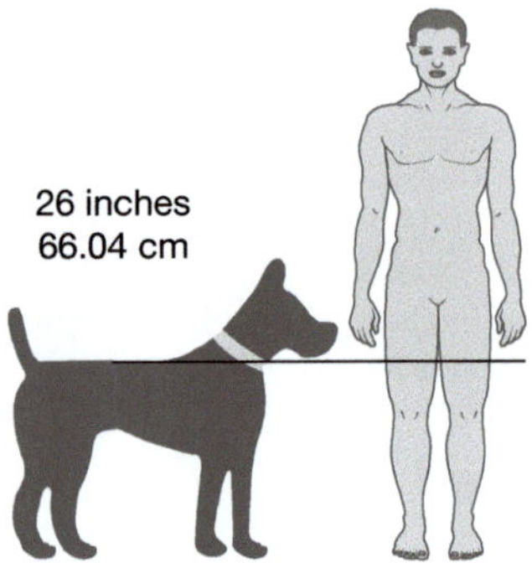

5.04 Wirehaired Pointing Griffon

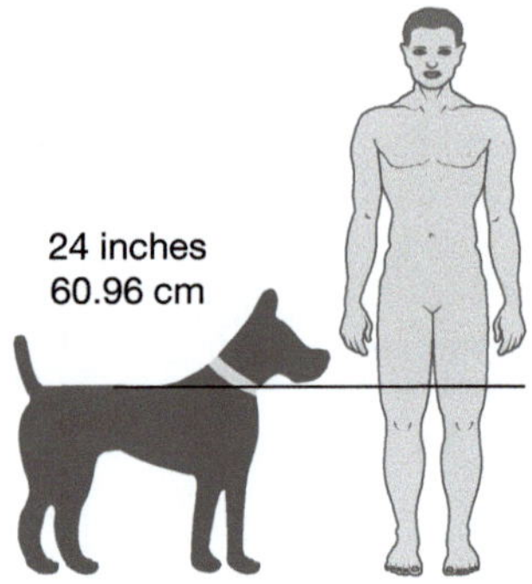

5.02 German Wirehaired Pointer

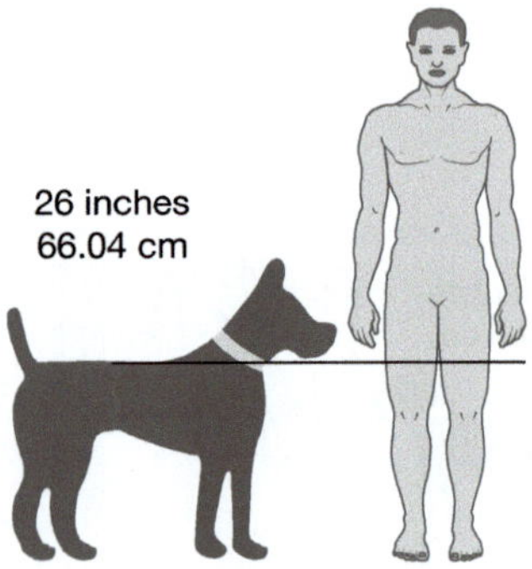

5.05 Vizsla

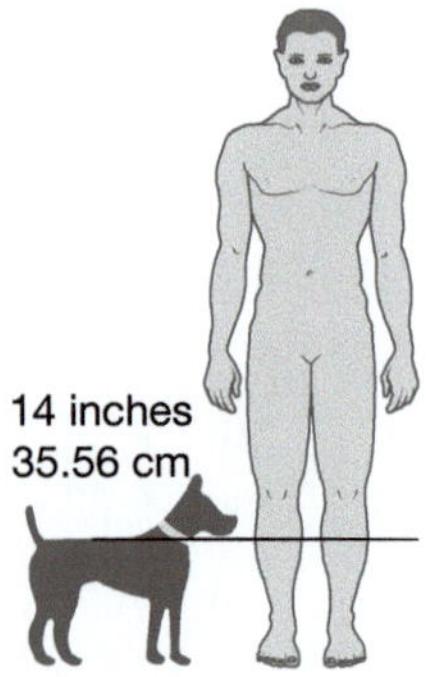

Sporting Group

5.06 Chesapeake Bay Retriever

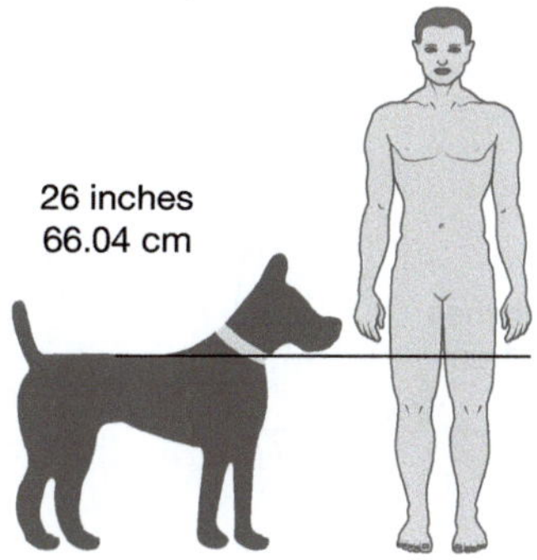

5.09 Golden Retriever

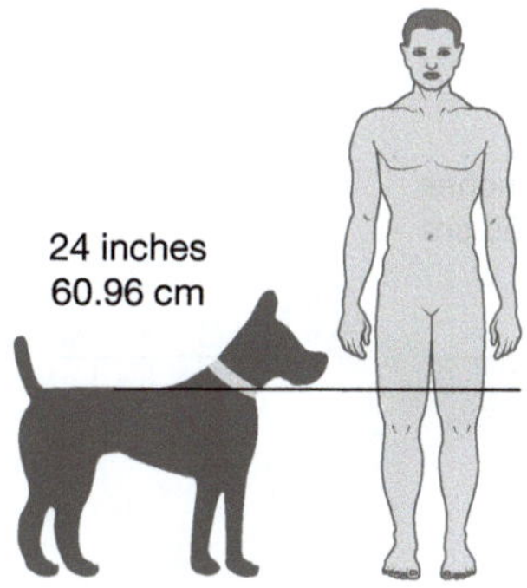

5.07 Curly-Coated Retriever

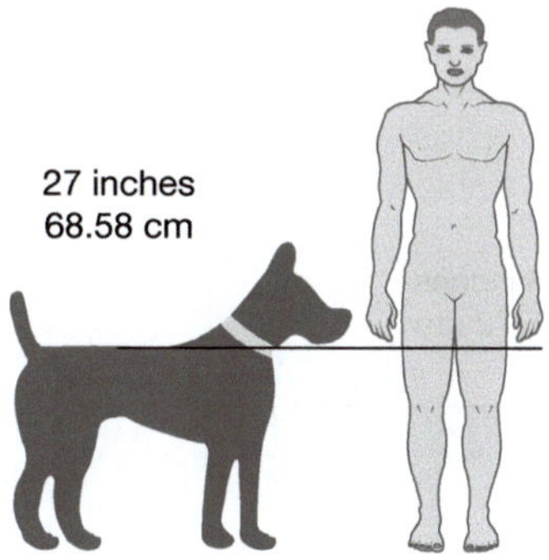

5.10 Labrador Retriever

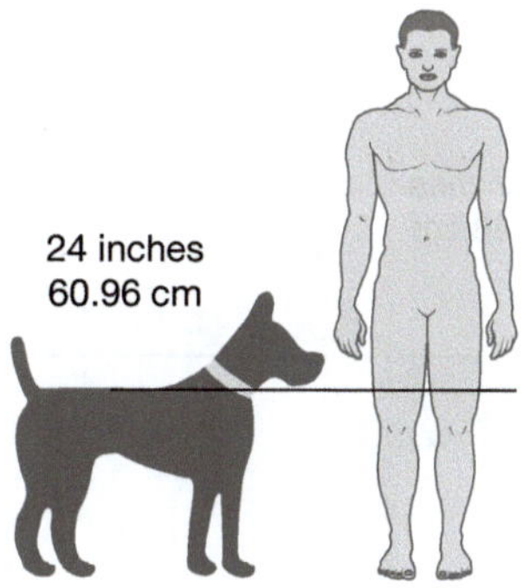

5.08 Flat-Coated Retriever

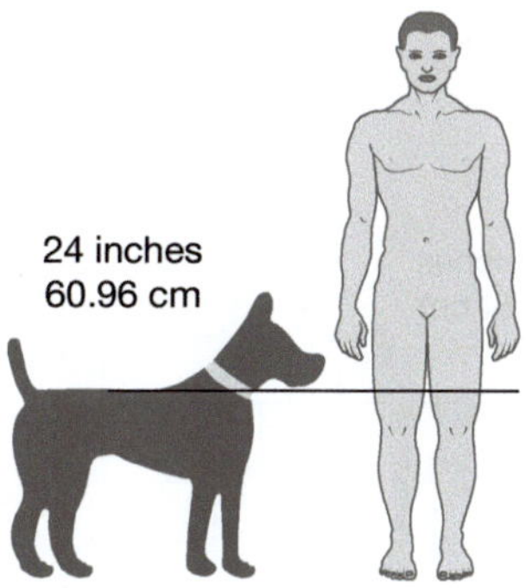

5.11 Nova Scotia Duck Tolling Retriever

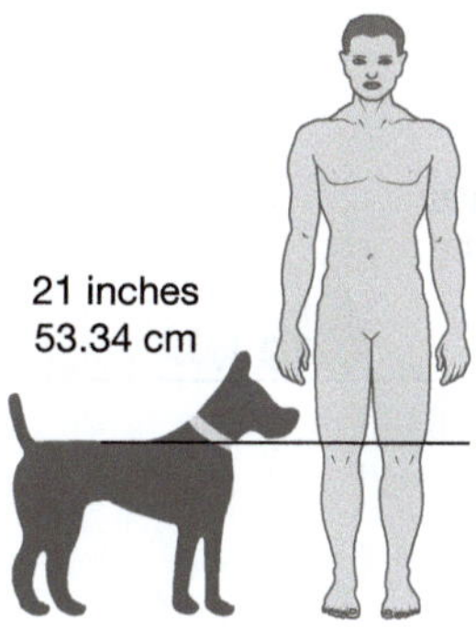

Sporting Group

5.12 English Setter

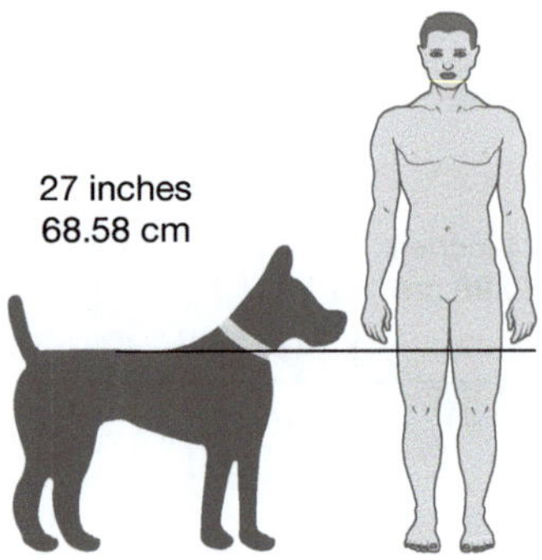

5.13 Gordon Setter

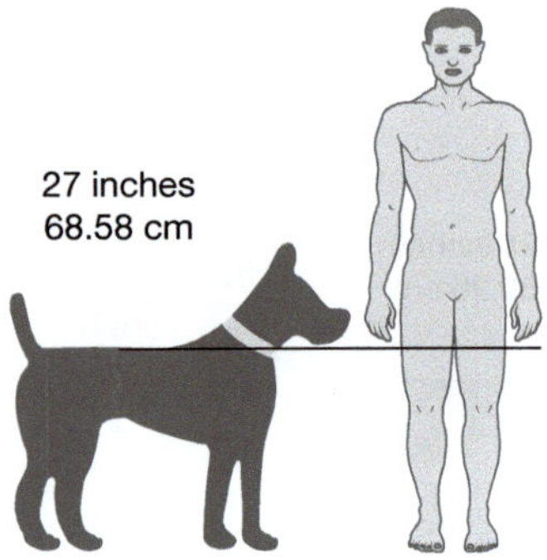

5.14 Irish Setter

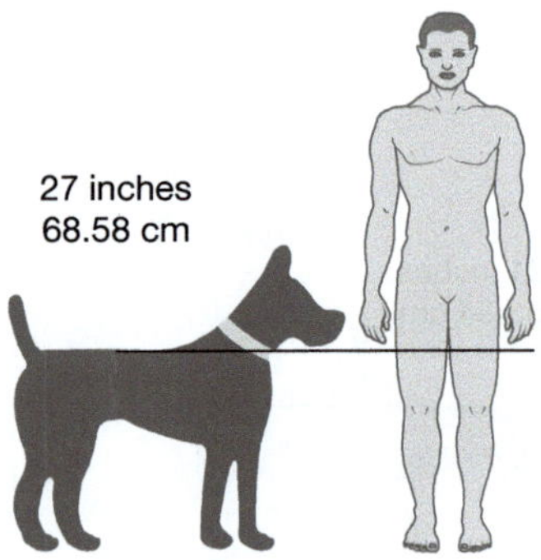

5.15 American Cocker Spaniel

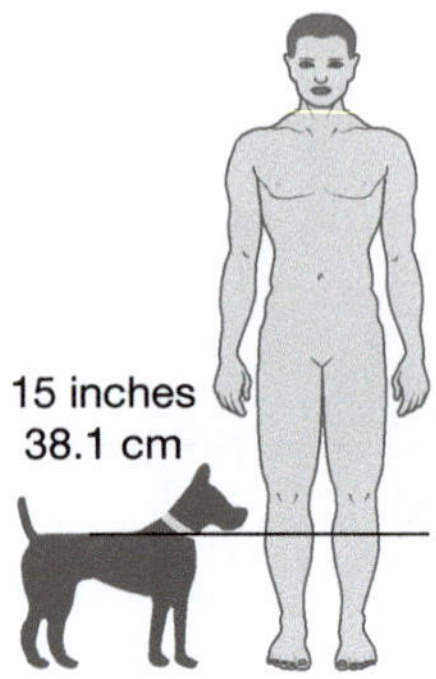

5.16 American Water Spaniel

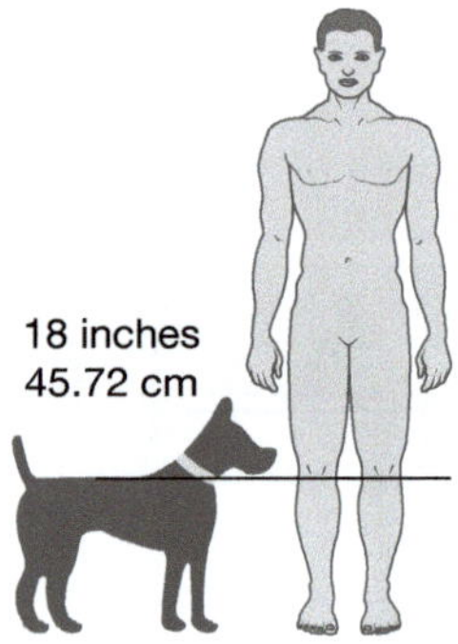

5.17 Brittany

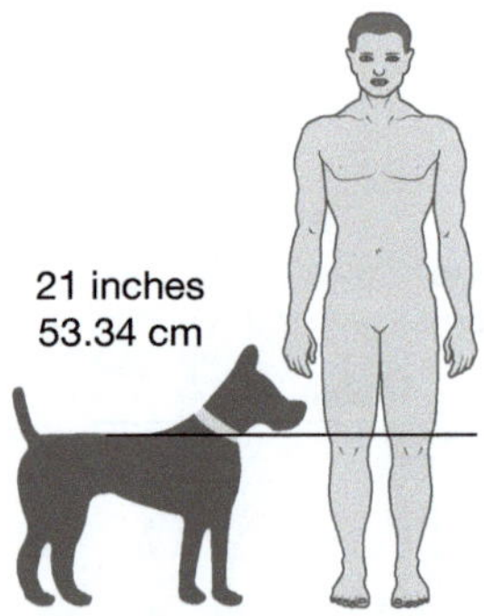

Sporting Group

5.18 Clumber Spaniel

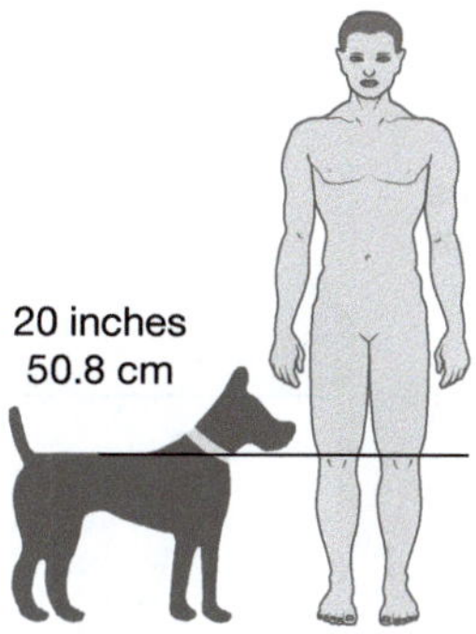

5.20 English Springer Spaniel

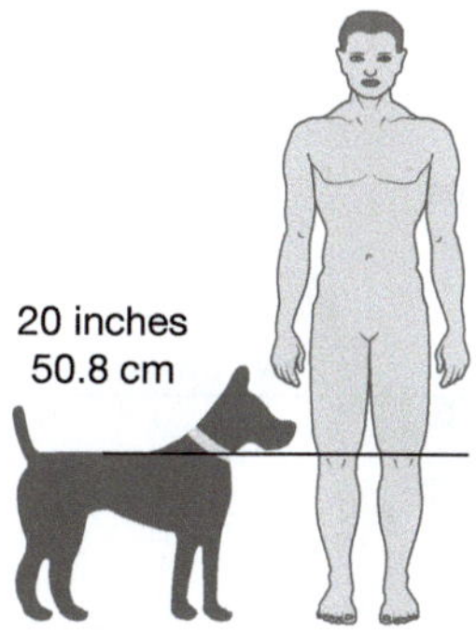

5.19 English Cocker Spaniel

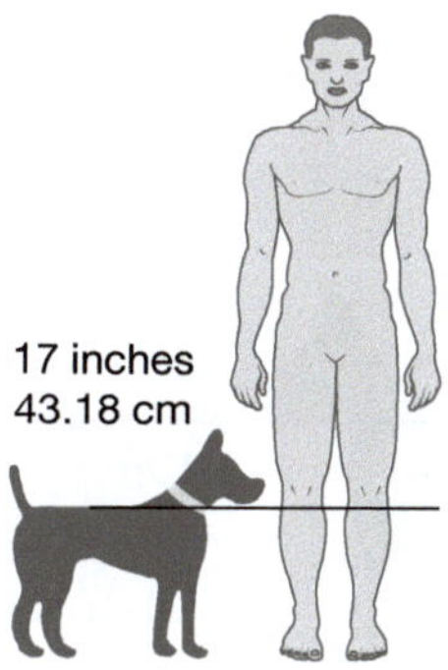

Working Group

6.00 Alaskan Malamute

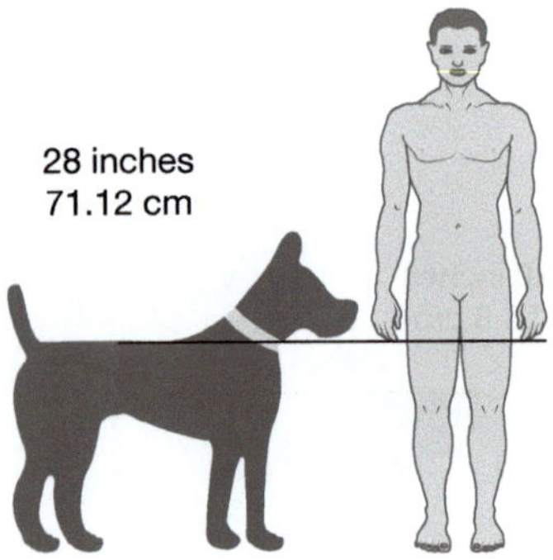

6.01 Samoyed

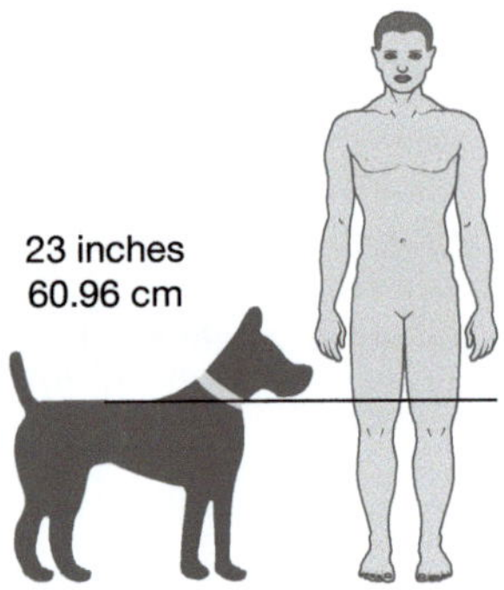

6.02 Siberian Husky

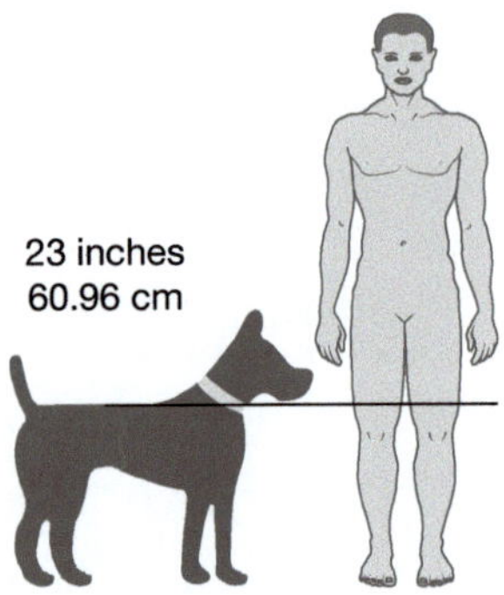

6.03 Akita

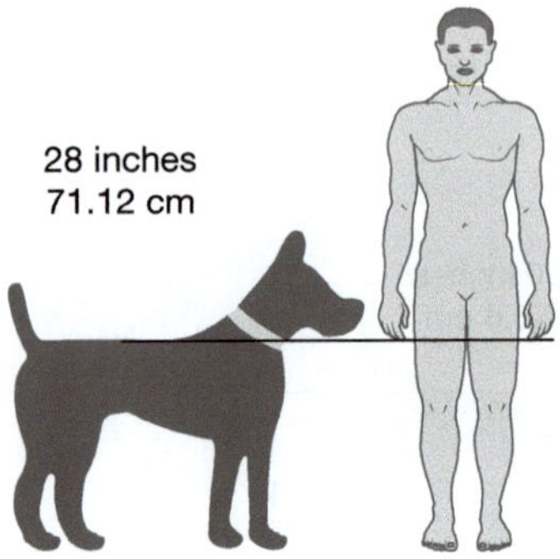

6.04 Bernese Mountain Dog

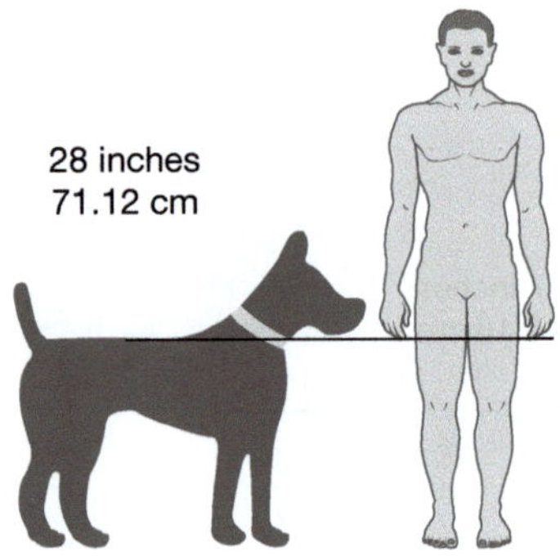

6.05 Boxer

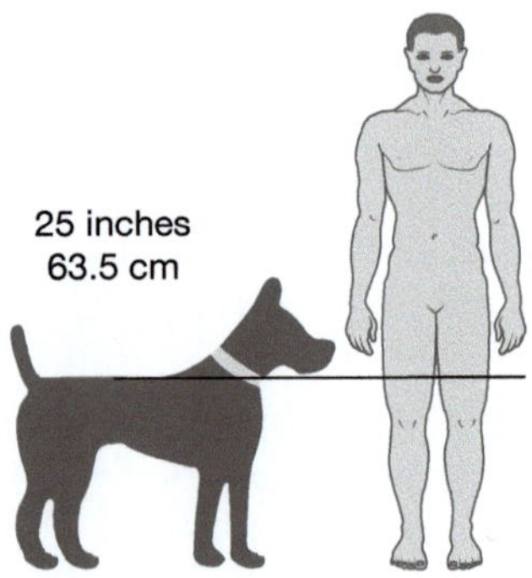

Working Group

6.06 Bullmastiff

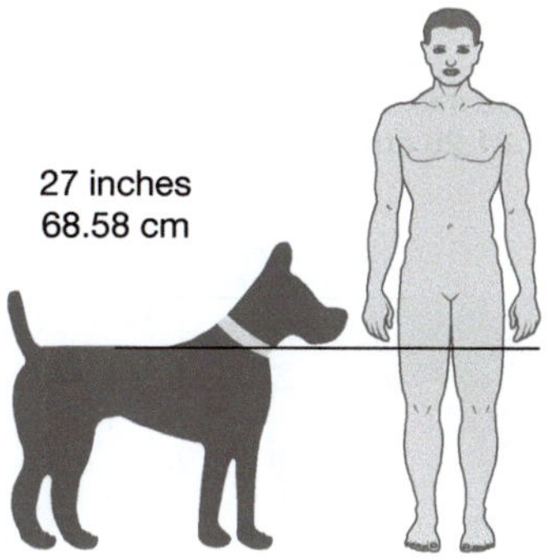

6.09 Giant Schnauzer

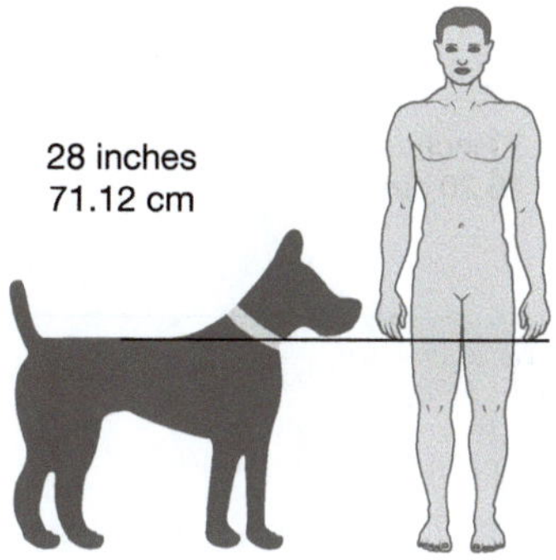

6.07 Doberman Pinscher

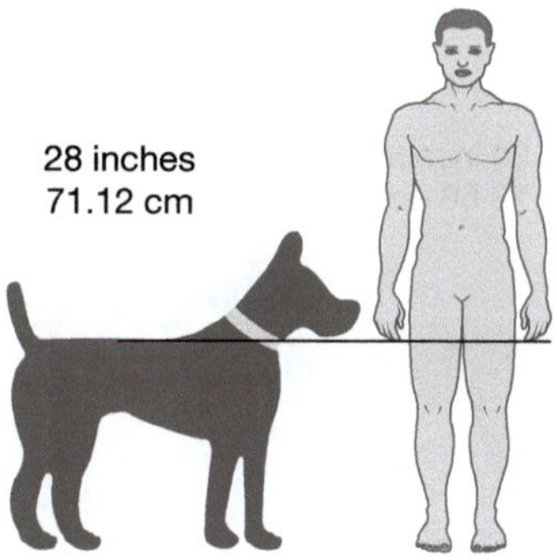

6.10 Great Dane

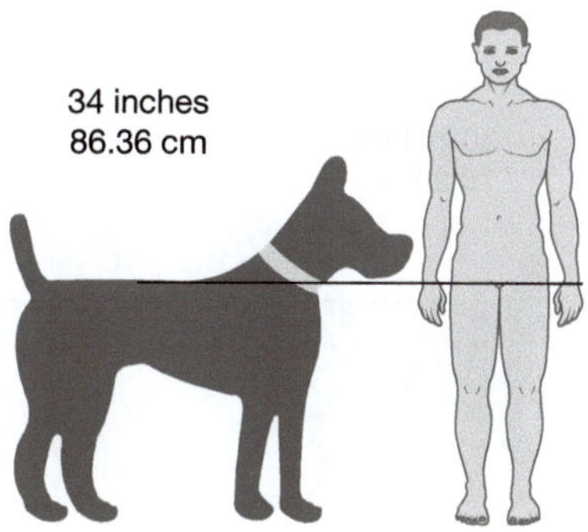

6.08 Dogue de Bordeaux

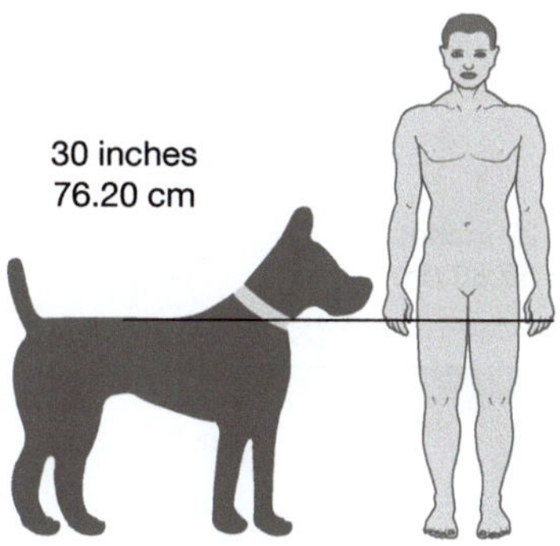

6.11 Komondor

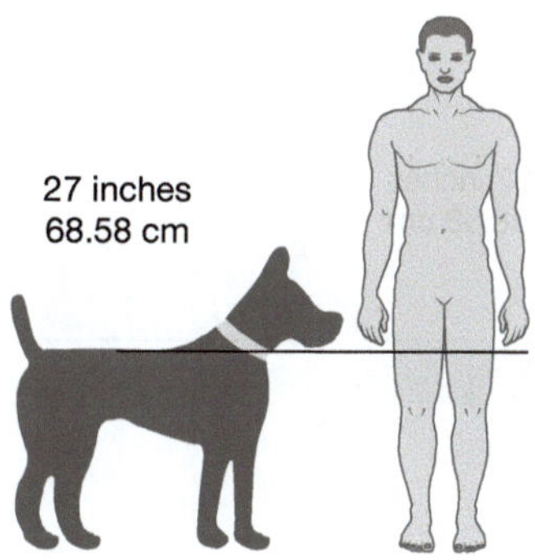

Working Group

6.12 Kuvaszk

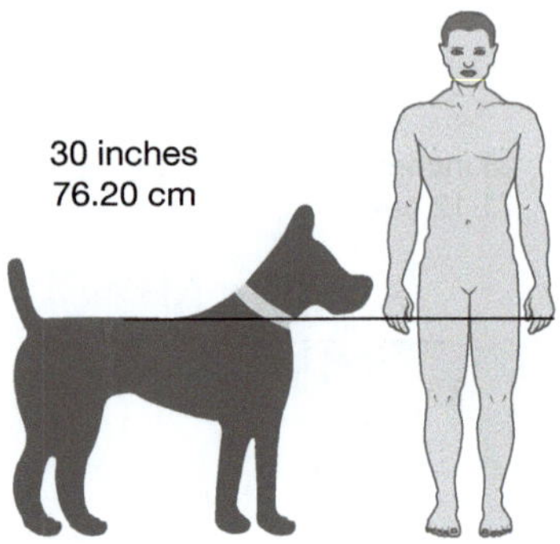

6.15 Great Pyrenees

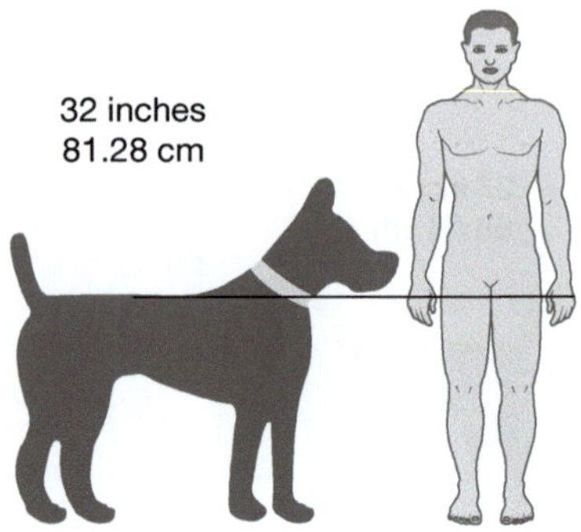

6.13 Mastiff

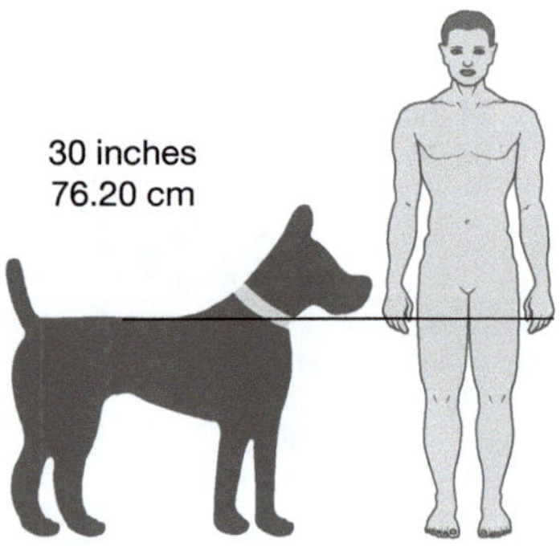

6.16 Newfoundland

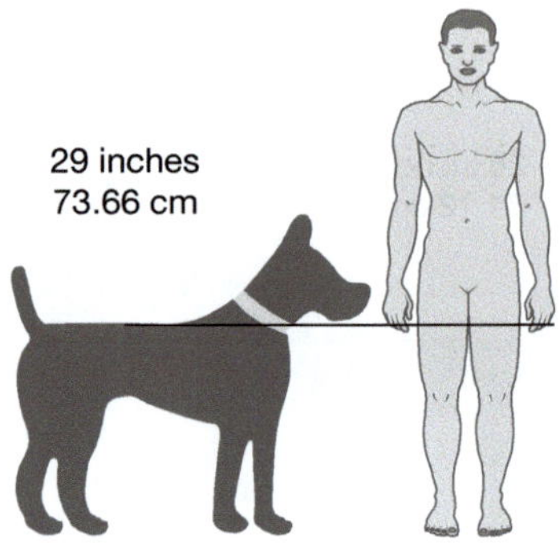

6.14 Rottweiler

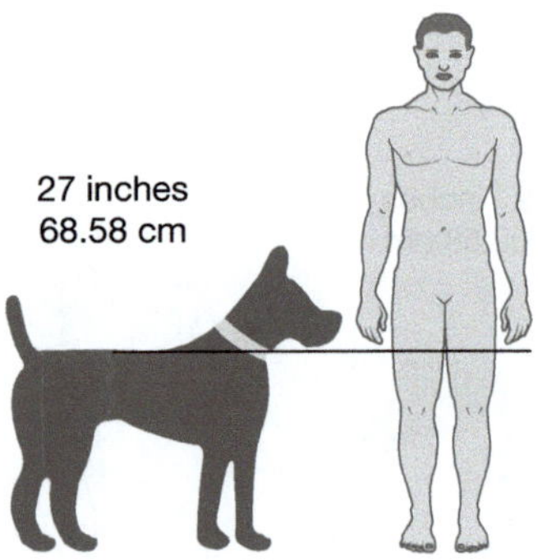

6.17 Portuguese Water Dog

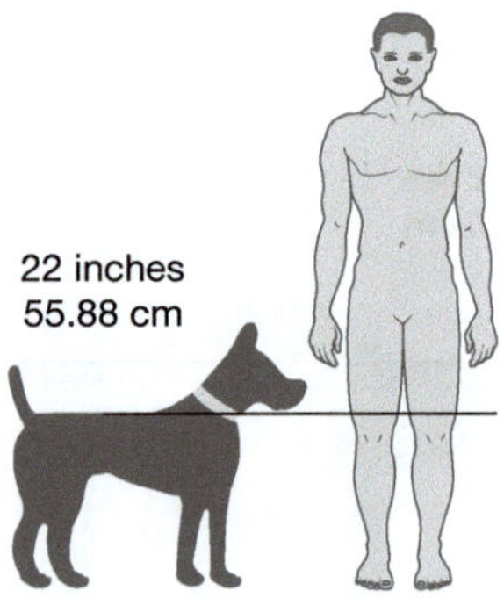

Working Group

6.18 Saint Bernard

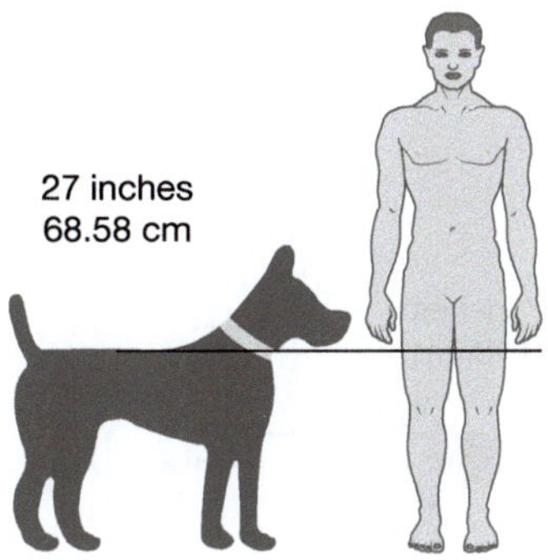

Terrier Group

7.00 Airedale Terrier

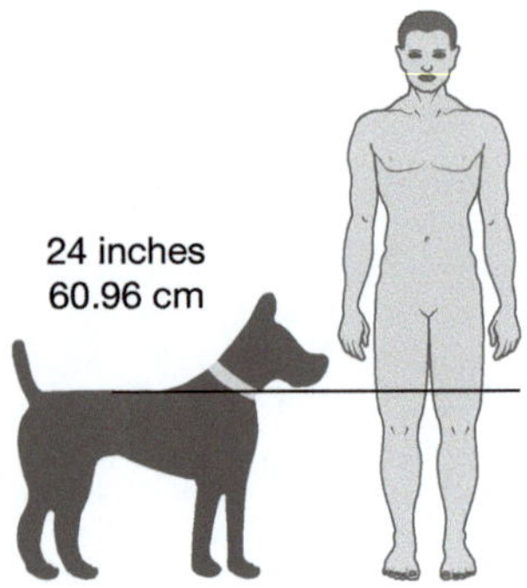

7.03 Bedlington Terrier

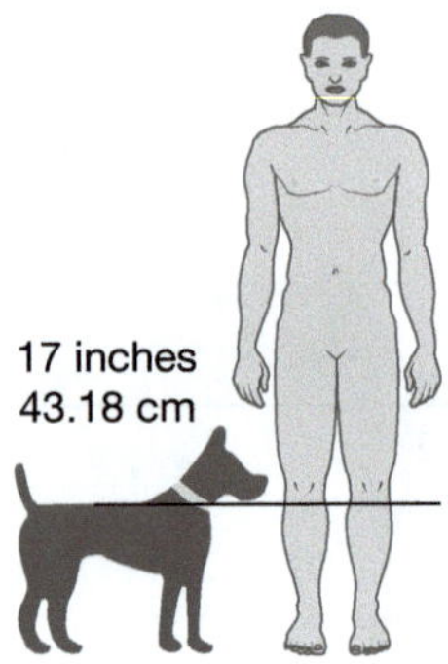

7.01 American Staffordshire Terrier

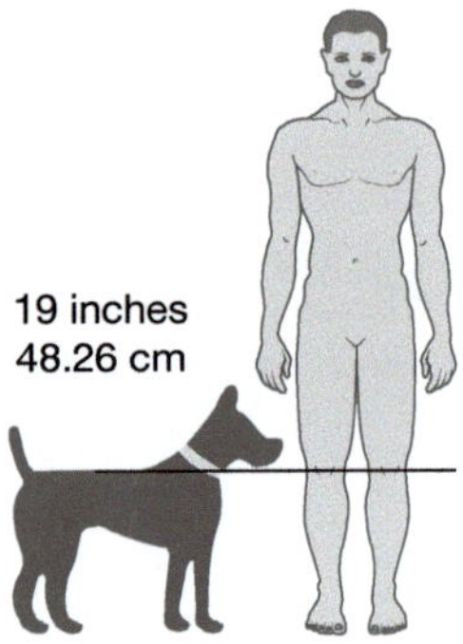

7.04 Border Terrier

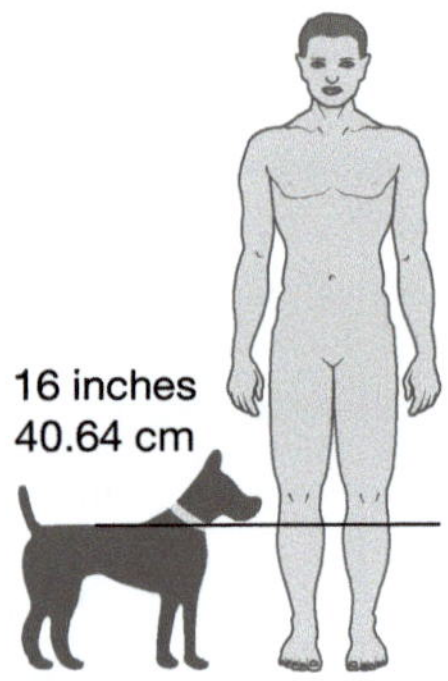

7.02 Australian Terrier

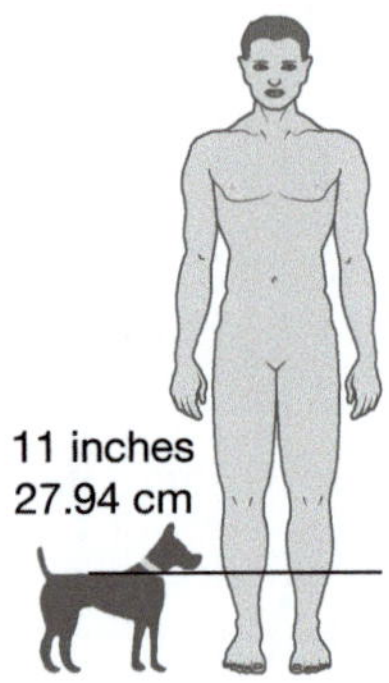

7.05 Bull Terrier

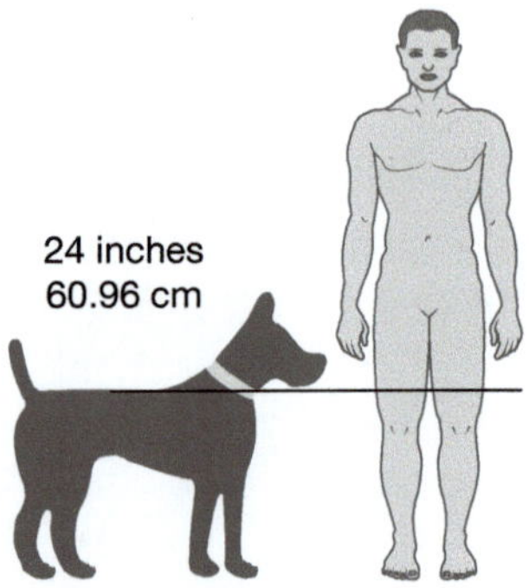

Terrier Group

7.06 Cairn Terrier

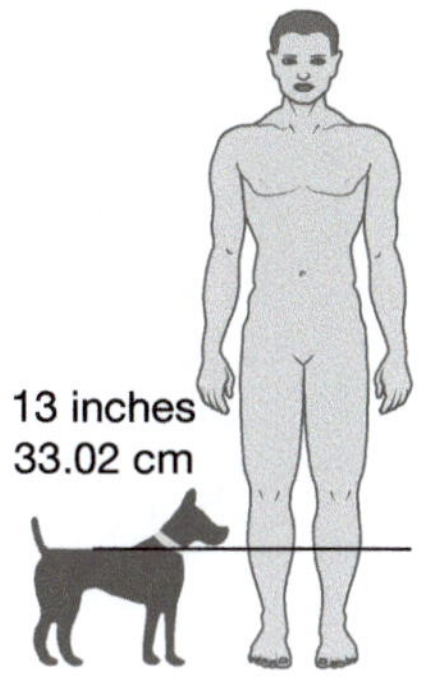

7.07 Dandie Dinmont

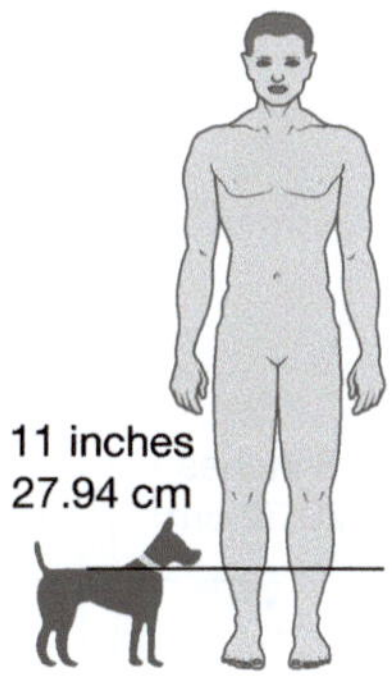

7.08 Fox Terriers

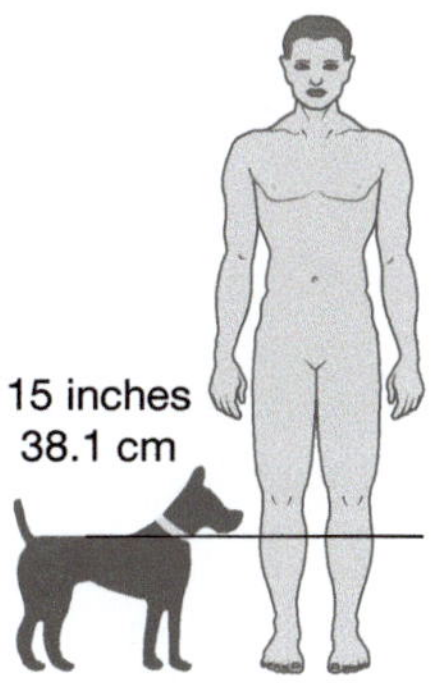

7.09 Irish Terrier

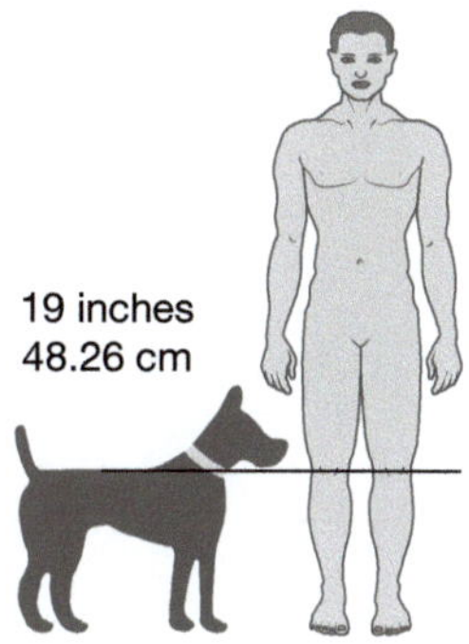

7.10 Kerry Blue Terrier

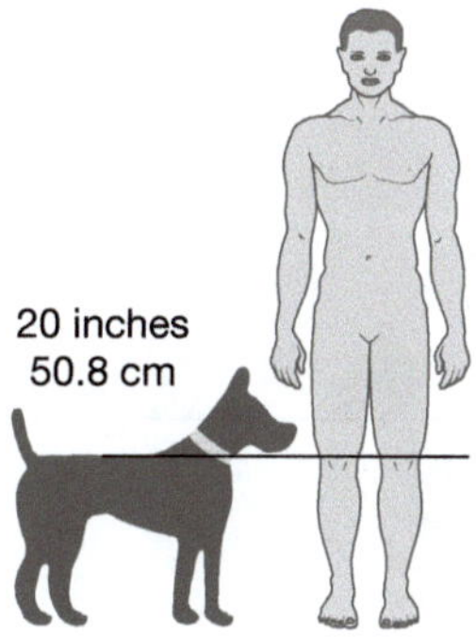

7.11 Norfolk Terriers

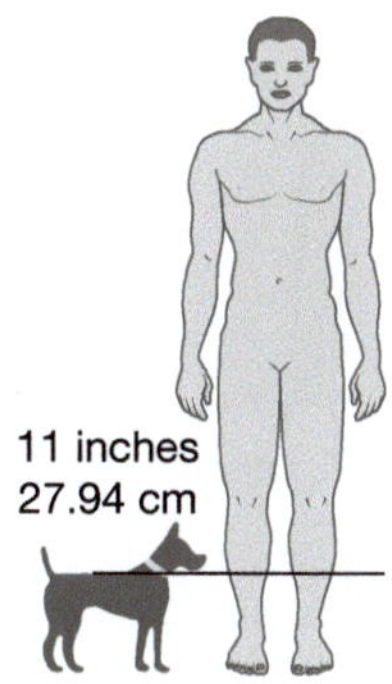

Terrier Group

7.12 Norwich Terrier

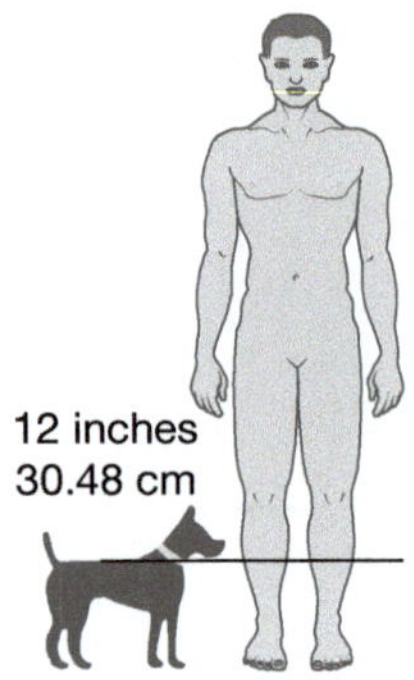

7.13 Rat Terrier

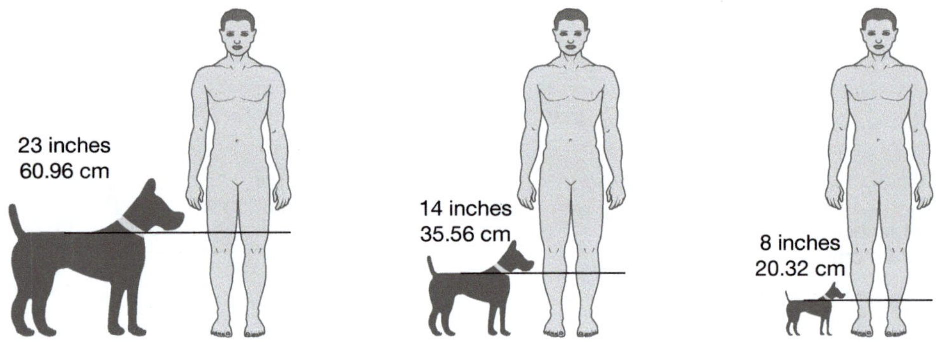

7.14 Jack Russell

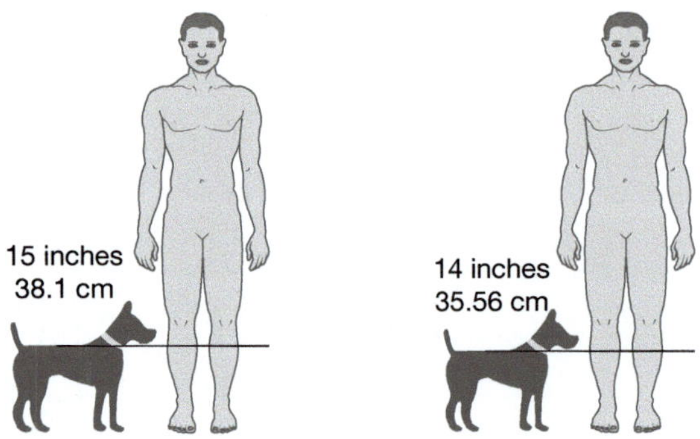

Terrier Group

7.15 Schnauzer

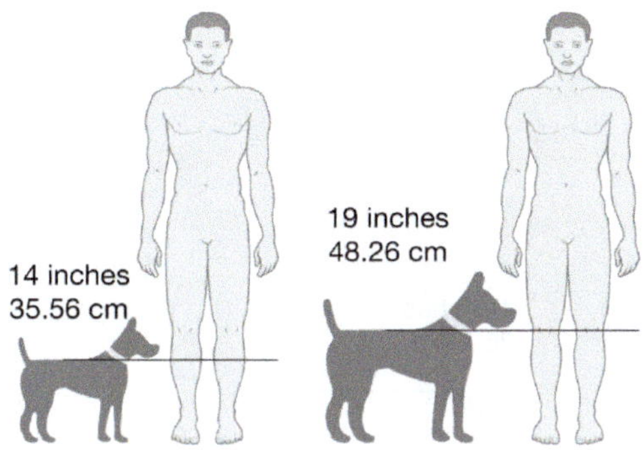

7.16 Scottish Terrier

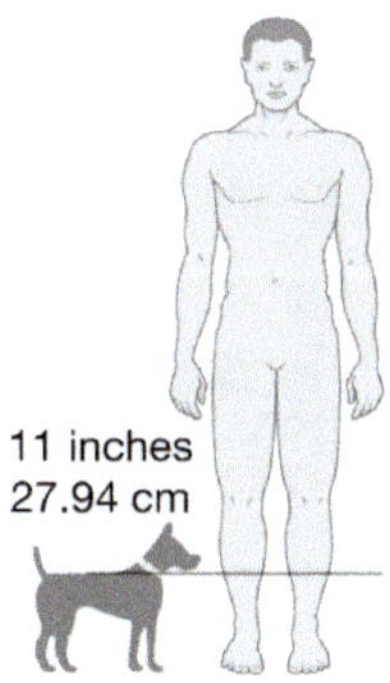

7.17 Soft Coated Wheaten Terrier

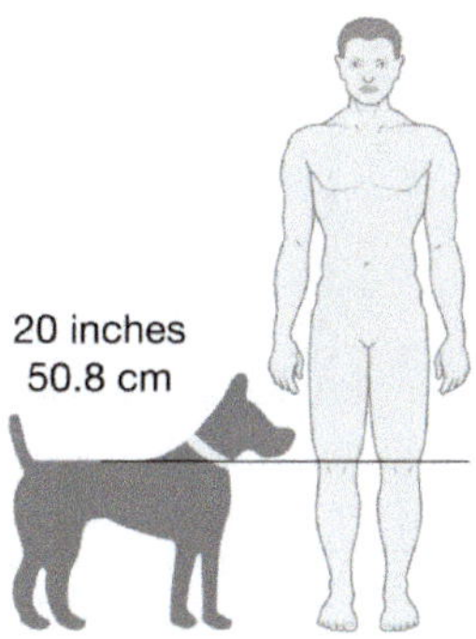

7.18 Staffordshire Bull Terrier

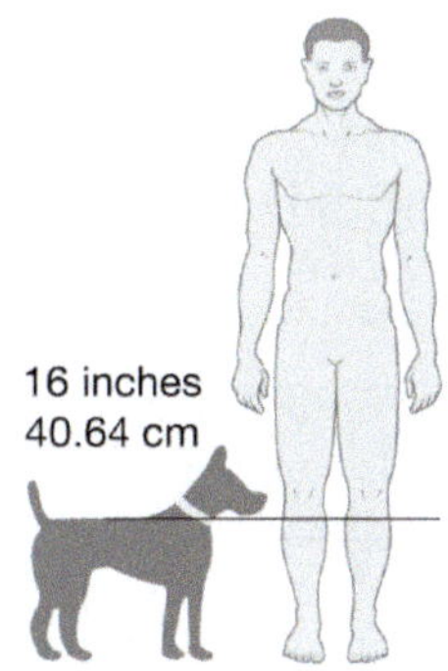

7.19 Welsh Terrier

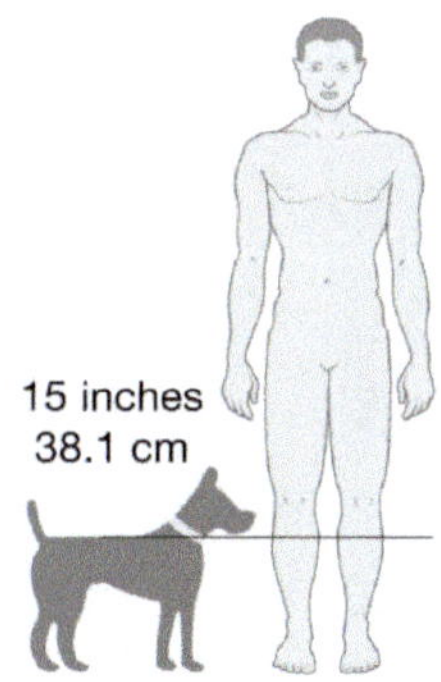

7.20 West Highland White Terrier

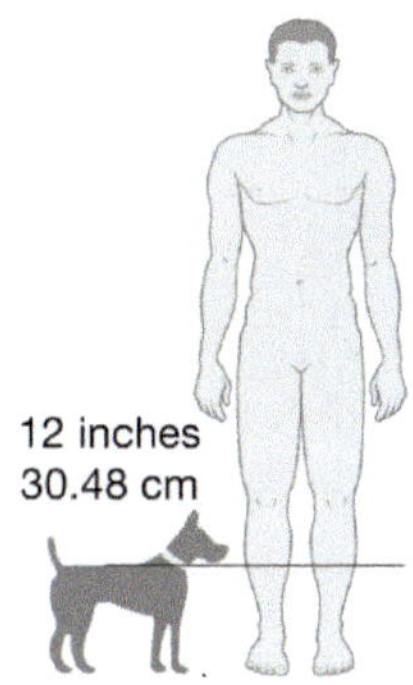

Toy Group

8.00 Affenpinscher

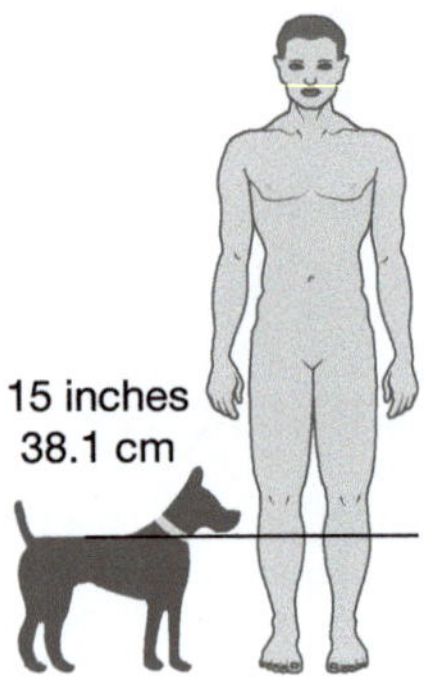

8.01 Brussels Griffon

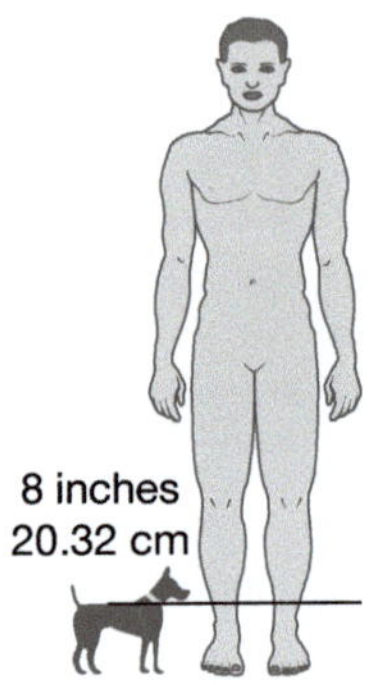

8.02 Cavalier King Charles Spaniel

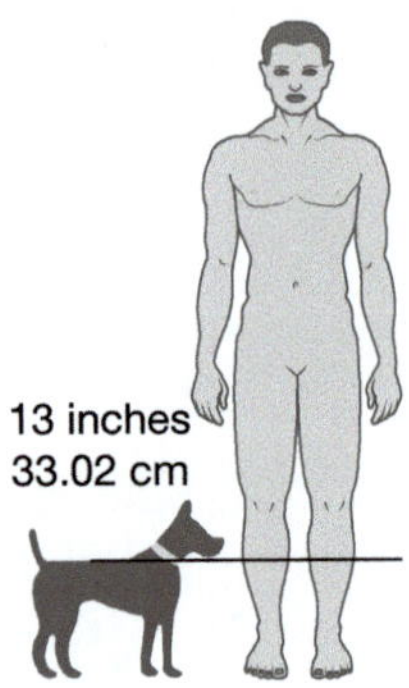

8.03 Chihuahua

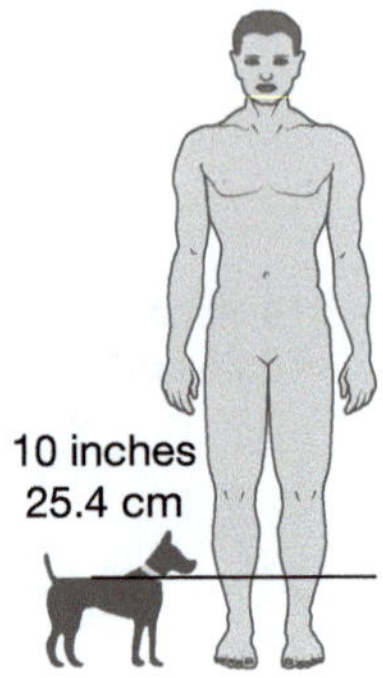

8.04 Chinese Crested

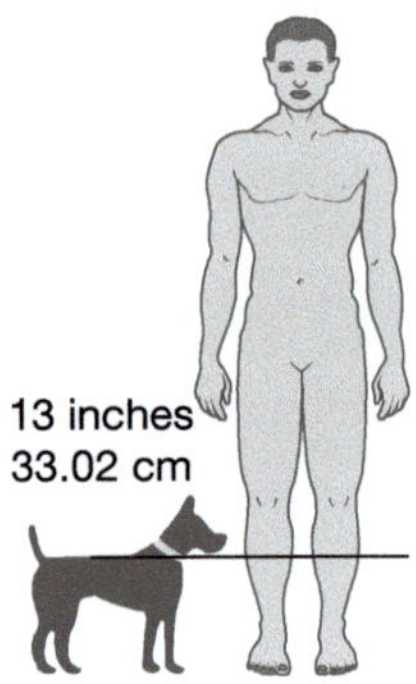

8.05 English Toy Spaniel

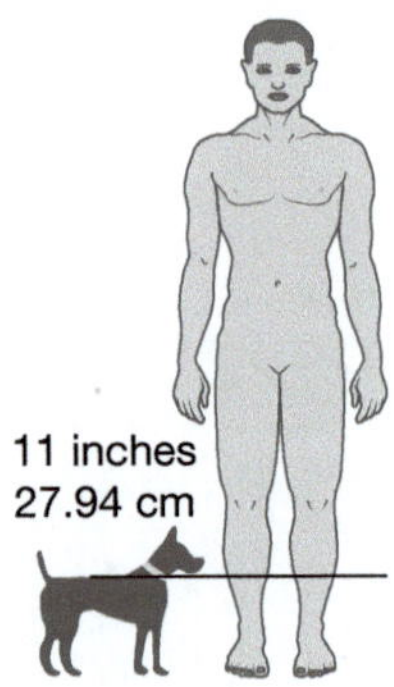

Toy Group

8.06 Havanese

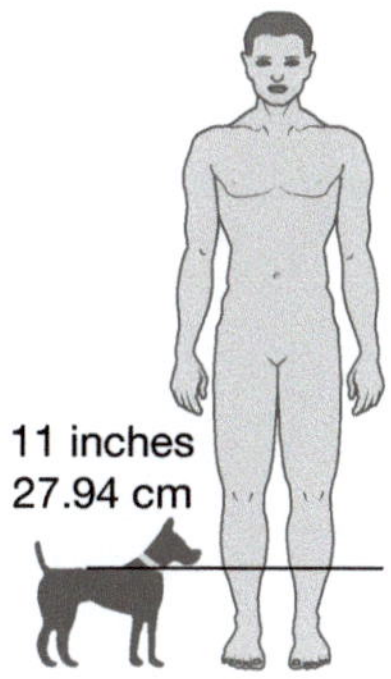

8.07 Italian Greyhound

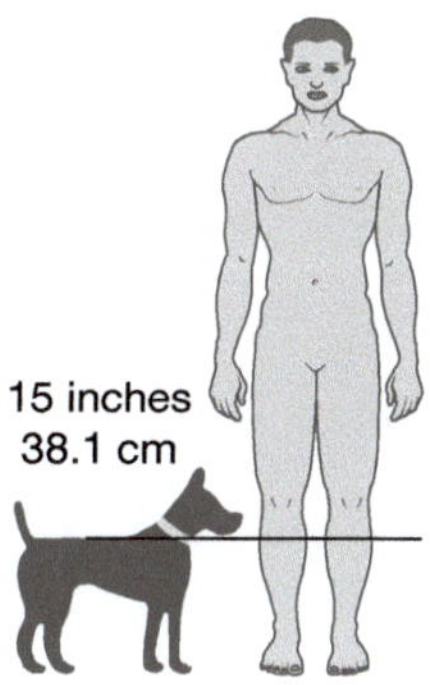

8.08 Japanese Chin

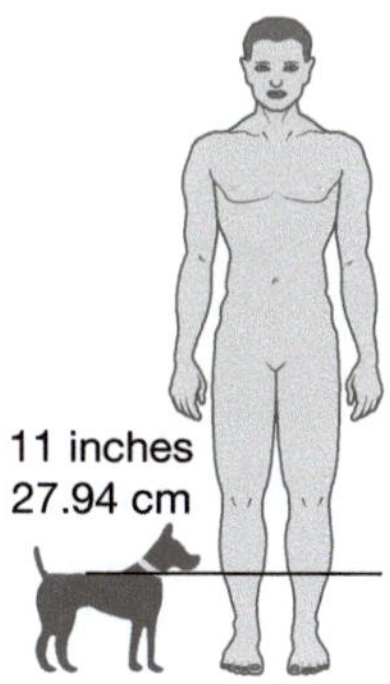

8.09 Maltese

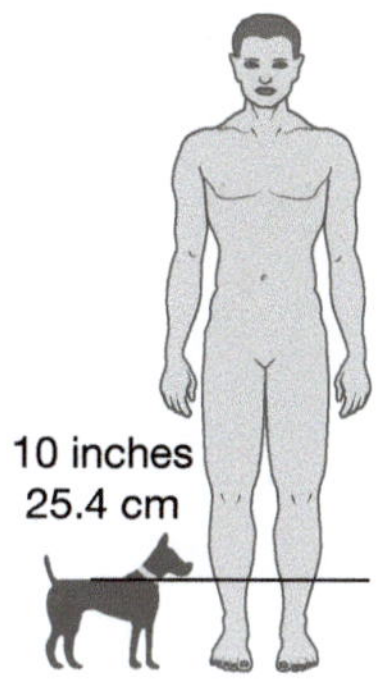

8.10 Manchester Terrier

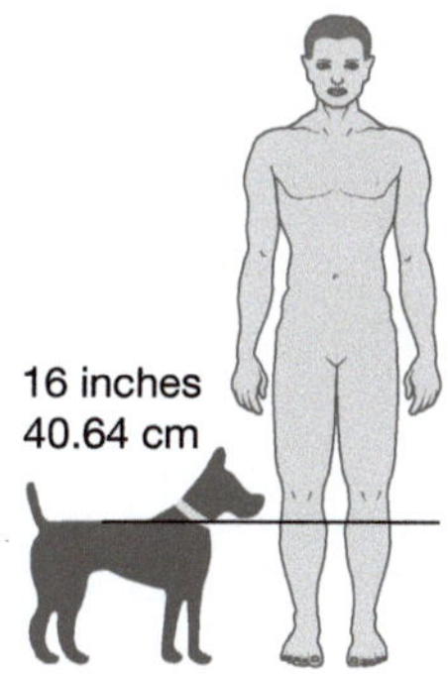

8.11 Miniature Pinscher

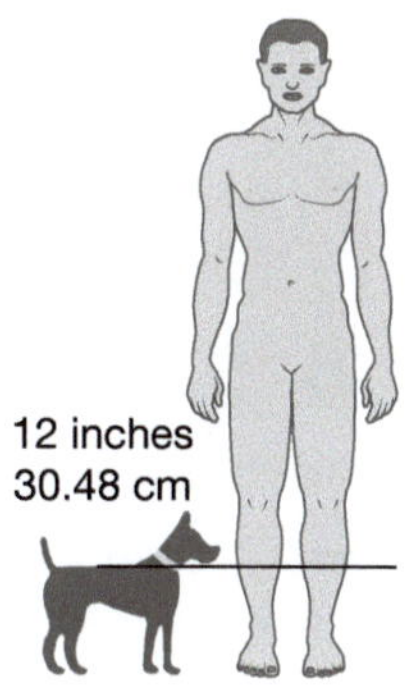

Toy Group

8.12 Papillon

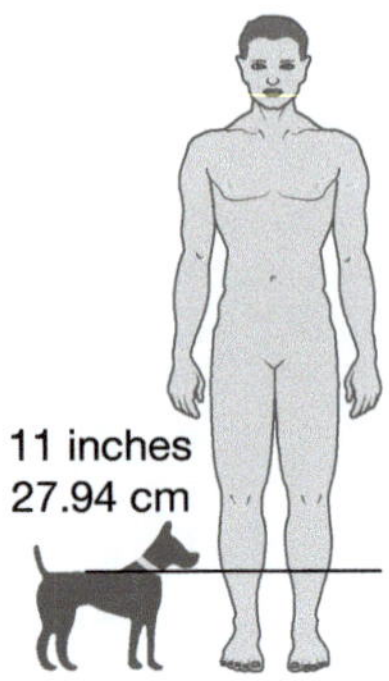

8.15 Pug

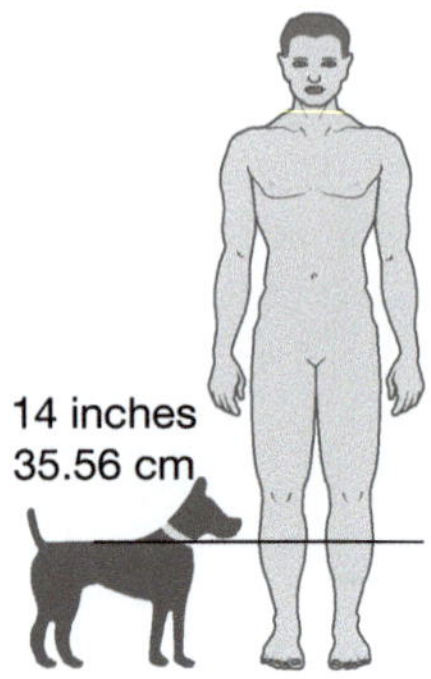

8.13 Pekingese

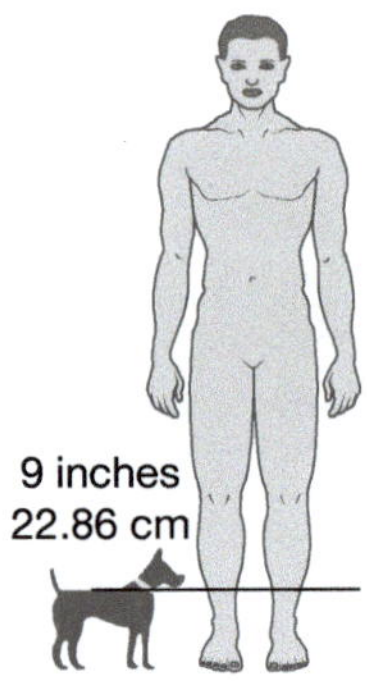

8.16 Shih Tzu

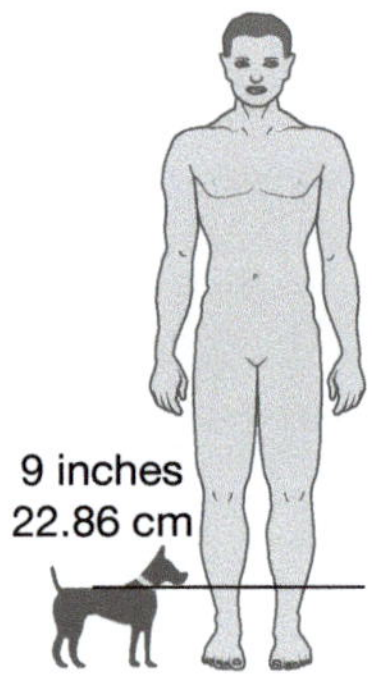

8.14 Pomeranian

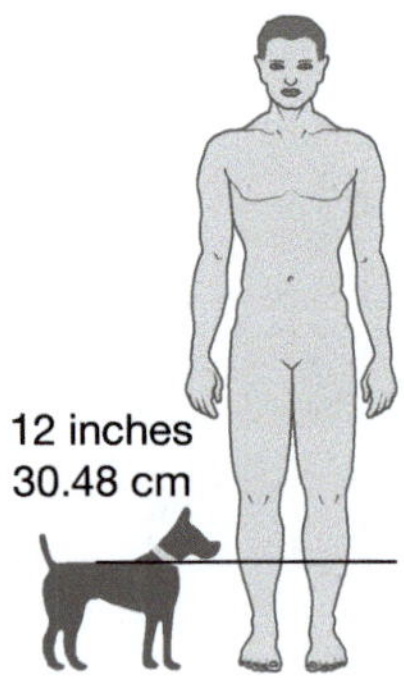

8.17 Silky Terrier

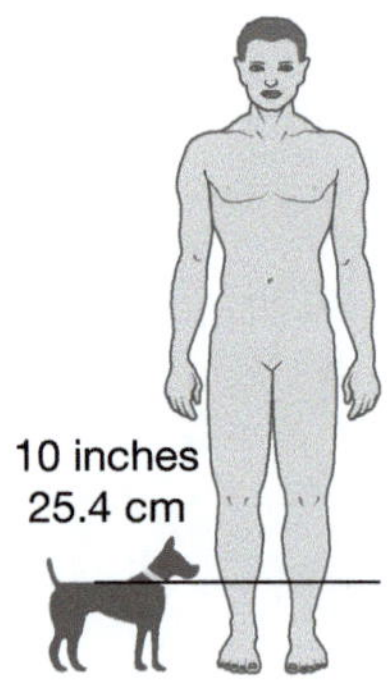

Toy Group

8.18 Yorkshire Terrier

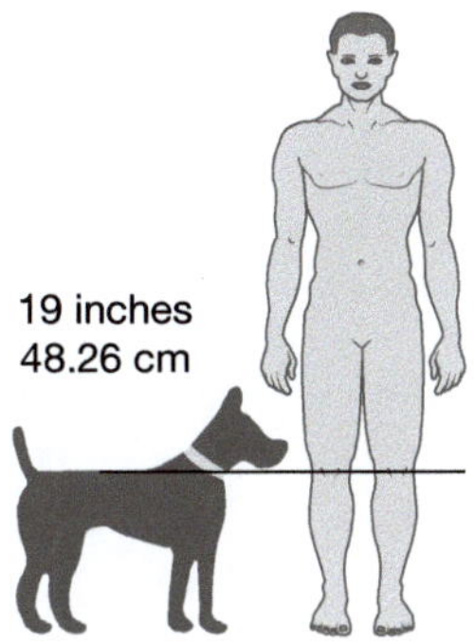

Fig. 1.01: Copyright © 2009 by Jean-Pierre Dagneaux, (CC BY-SA 3.0) at http://commons.wikimedia.org/wiki/File:Boston_Terrier_male.jpg.

Fig. 1.02: Jonny Hedberg / Copyright in the Public Domain.

Fig. 1.03: Copyright © 2010 by mjk23, (CC BY 2.0) at https://commons.wikimedia.org/wiki/File:Pointer_-_world_dog_show_2010.jpg.

Fig. 1.04: Copyright © 2008 by EGILEO, (CC BY 2.0) at http://commons.wikimedia.org/wiki/File:Bou.jpg.

Fig. 1.05: Copyright © 2007 by Pleple2000, (CC BY-SA 3.0) at https://commons.wikimedia.org/wiki/File:Terier_irlandzki_g%C5%82owa_pl.jpg.

Fig. 1.06: Copyright © 2007 by original author: copperfeesh; edited by Pharaoh Hound, (CC BY-SA 2.0) at http://commons.wikimedia.org/wiki/File:Basenji_pair_edit.jpg .

Fig. 1.07: Copyright © 2007 by pato garza, (CC BY 2.0) at http://commons.wikimedia.org/wiki/File:03_Dobermann_sitstay.jpg.

Fig. 1.08: BlueIllusions / Copyright in the Public Domain.

Fig. 1.09: Copyright © 2006 by SuperFantastic, (CC BY 2.0) at http://commons.wikimedia.org/wiki/File:Bloodhound_portrait.jpg.

Fig. 1.10: Copyright © 2006 by Pamela Carls, (CC BY-SA 2.0) at http://commons.wikimedia.org/wiki/File:Siberian_Husky_blue_eyes_Flickr.jpg.

Fig. 1.11: Copyright © 2008 by Winky, (CC BY 2.0) at http://commons.wikimedia.org/wiki/File:Anakin_Whippet.jpg.

Fig. 1.12: Copyright © 2008 by Picamik, (CC BY-SA 3.0) at http://commons.wikimedia.org/wiki/File:Canaille02.jpg.

Fig. 1.13: Copyright © 2009 by Pharaoh Hound, (CC BY-SA 3.0) at http://commons.wikimedia.org/wiki/File:Brittany_Spaniel_standing.jpg.

Fig. 1.14: Copyright © 2008 by TreeSong Aussies, (CC BY-SA 3.0) at https://commons.wikimedia.org/wiki/File:Moulderright.JPG.

Fig. 1.15: Copyright © 2005 by Partnerhund.com, (CC BY 2.0) at http://commons.wikimedia.org/wiki/File:Siberian_Husky_red_Flickr.jpg.

Fig. 1.16: Copyright © 2007 by Pleple2000, (CC BY-SA 3.0) at https://commons.wikimedia.org/wiki/File:Racib%C3%B3rz_2007_082.jpg.

Fig. 1.17: Copyright © 2009 by Pharaoh Hound, (CC BY-SA 3.0) at http://commons.wikimedia.org/wiki/File:Brittany_Spaniel_standing.jpg.

Fig. 1.18: Copyright © 2013 by State Farm, (CC BY 2.0) at https://commons.wikimedia.org/wiki/File:01_Wire_Fox_terrier.jpg.

Fig. 1.19: Chrizweatley / Copyright in the Public Domain.

Fig. 1.20: Copyright © 2005 by Grzegorz Polak, (CC BY-SA 3.0) at http://commons.wikimedia.org/wiki/File:Seter_irlandzki.jpg.

Fig. 1.21: Copyright © 2006 by Pleple2000, (CC BY-SA 3.0) at http://commons.wikimedia.org/wiki/File:Sharpei_ogon_867.jpg.

Fig. 1.22: Copyright © 2006 by Wolftrouble, (CC BY-SA 2.0) at http://commons.wikimedia.org/wiki/File:German_Shepherd_Dog_standing.jpg.

Fig. 1.23: Copyright © 2006 by Pleple2000, (CC BY-SA3.0)athttp://commons.wikimedia.org/wiki/File:Chart_perski_chor%C4%85giew_6464.jpg, 1 .

Fig. 1.24: Copyright © 2005 by Scott Feldstein, (CC BY-SA 2.0) at http://commons.wikimedia.org/wiki/File:Greyhound_brindle_standing.jpg.

Fig. 1.25: Franza1984 / Copyright in the Public Domain.

Fig. 1.26: Copyright © 2008 by TreeSong Aussies, (CC BY-SA 3.0) at https://commons.wikimedia.org/wiki/File:Moulderright.JPG.

Fig. 1.27: Copyright © 2005 by Scott Feldstein, (CC BY-SA 2.0) at http://commons.wikimedia.org/wiki/File:Greyhound_brindle_standing.jpg.

Fig. 1.28: Copyright © 2007 by Radovan Rohovsky, (CC BY-SA 3.0) at http://commons.wikimedia.org/wiki/File:Westgort_Anticipation_17_months.JPG.

Fig. 1.29: Copyright © 2003 by Tirwhan, (CC BY-SA 3.0) at http://commons.wikimedia.org/wiki/File:Irish_Wolfhound_Sam.jpg.
Fig. 1.30: Copyright © 2006 by Viborg~commonswiki, (CC BY-SA 3.0) at http://commons.wikimedia.org/wiki/File:Dogge_Odin.jpg.
Fig. 1.31: Copyright © 2006 by Trisha Shears (Ltshears), (CC BY-SA 3.0) at http://commons.wikimedia.org/wiki/File:AmericanFoxhound2.jpg.
Fig. 1.32: Copyright © 2003 by Sannse, (CC BY-SA 3.0) at http://commons.wikimedia.org/wiki/File:Border_Collie_600.jpg.
Fig. 1.33: Copyright © 2013 by Томасина, (CC BY-SA 3.0) at http://commons.wikimedia.org/wiki/File:American_Cocker_in_Tallinn_2.JPG.
Fig. 1.34: Copyright © 2003 by Sannse, (CC BY-SA 3.0) at http://commons.wikimedia.org/wiki/File:Borzoi_600.jpg.
Fig. 1.35: Copyright © 2004 by Ellen Levy Finch (Elf), (CC BY-SA 3.0) at https://commons.wikimedia.org/wiki/File:GermanWirehrPtr1_wb.jpg.
Fig. 1.36: Copyright © 2002 by Christian Madden, (CC BY-SA 2.0) at http://commons.wikimedia.org/wiki/File:Rough_Collie_KC.jpg.
Fig. 1.37: Copyright © 2006 by Psiara, (CC BY-SA 3.0) at https://commons.wikimedia.org/wiki/File:Miniature_Schnauzer_02.jpg.
Fig. 1.38: Copyright © 2006 by Eva Holderegger Walser, (CC BY-SA 2.5) at https://commons.wikimedia.org/wiki/File:Beide.jpg.
Fig. 1.39: Copyright © 2004 by Ellen Levy Finch (Elf), (CC BY-SA 3.0) at http://commons.wikimedia.org/wiki/File:GermanShorthPtr_wb.jpg.
Fig. 1.40: Copyright © 2009 by Kerstin othello, (CC BY-SA 3.0) at https://commons.wikimedia.org/wiki/File:Othello.jpg.
Fig. 1.41: Copyright © 2009 by RORLVR1, (CC BY-SA 3.0) at https://commons.wikimedia.org/wiki/File:Australian_Grand_Champion_Dancestar_Zero_O_Seven_%22Bondy%22.jpg.
Fig. 1.42: Copyright © 2007 by Radovan Rohovsky, (CC BY-SA 3.0) at http://commons.wikimedia.org/wiki/File:Westgort_Anticipation_17_months.JPG.
Fig. 1.43: Copyright © 2008 by EGILEO, (CC BY 2.0) at http://commons.wikimedia.org/wiki/File:Bou.jpg.
Fig. 1.44: Copyright © 2013 by Rangajk95, (CC BY-SA 3.0) at http://commons.wikimedia.org/wiki/File:Beaglier.JPG,.
Fig. 1.45: Copyright © 2007 by Bryon Realey, (CC BY 2.0) at https://commons.wikimedia.org/wiki/File:Vizsla_Rosie_water_harness.jpg.
Fig. 1.46: Copyright © 2013 by Томасина, (CC BY-SA 3.0) at https://commons.wikimedia.org/wiki/File:Irish_Setter_in_Tallinn_2.JPG.
Fig. 1.47: Copyright © 2013 by Томасина, (CC BY-SA 3.0) at https://commons.wikimedia.org/wiki/File:Irish_Setter_in_Tallinn_2.JPG.
Fig. 1.48: Copyright © 2007 by pato garza, (CC BY 2.0) at http://commons.wikimedia.org/wiki/File:03_Dobermann_sitstay.jpg.
Fig. 1.49: Copyright © 2006 by Erica Steer (SaNtINa/kIKs), (CC BY-SA 2.0) at http://commons.wikimedia.org/wiki/File:German_Shepherd_Dog_with_disc.jpg.
Fig. 1.50: Copyright © 2007 by Lilly M, (CC BY-SA 3.0) at http://commons.wikimedia.org/wiki/File:Keeshond_002_U.jpg.
Fig. 1.51: Copyright © 2003 by Sannse, (CC BY-SA 3.0) at http://commons.wikimedia.org/wiki/File:Pug_600.jpg.
Fig. 1.52: Dmitry Guskov / Copyright in the Public Domain.
Fig. 1.53: Copyright © 2007 by Nikki68, (CC BY-SA 3.0) at http://commons.wikimedia.org/wiki/File:Komondor_delvin.jpg.
Fig. 1.54: Sally Eller / Copyright in the Public Domain.
Fig. 1.55: Copyright © 2011 by Lhasaapso, (CC BY-SA 3.0) at http://commons.wikimedia.org/wiki/File:Lhasaapso.jpg.
Fig. 1.56: Copyright © 2007 by Radovan Rohovsky, (CC BY-SA 3.0) at http://commons.wikimedia.org/wiki/File:Westgort_Anticipation_17_months.JPG.
Fig. 1.57: Copyright © by Alexandr Osipow, (CC BY-SA 3.0) at http://commons.wikimedia.org/wiki/File:Wire-haired_Dachshund_Profil.jpg.
Fig 2.00: Copyright © 2006 by Eva Holderegger Walser, (CC BY-SA 2.5) at http://commons.wikimedia.org/wiki/File:Beide.jpg.
Fig. 2.01: Copyright © 2002 by Mdk572, (CC BY-SA 3.0) at http://commons.wikimedia.org/wiki/File:Australian_Cattle_Dog_puppy.jpg.
Fig. 2.02: Copyright © 2008 by TreeSong Aussies, (CC BY-SA 3.0) at https://commons.wikimedia.org/wiki/File:Moulderright.JPG.

Fig. 2.03: Copyright © 2003 by Sannse, (CC BY-SA 3.0) at https://commons.wikimedia.org/wiki/File:Australian_Shepherd_600.jpg.

Fig. 2.04: Copyright © 2007 by D. Williams, (CC BY 2.0) at http://commons.wikimedia.org/wiki/File:Belgian_Malinois_01.jpg.

Fig. 2.04: Copyright © 2003 by Sannse, (CC BY-SA 3.0) at http://commons.wikimedia.org/wiki/File:Belgian_Groenendael_600.jpg.

Fig. 2.04: Wickerl / Copyright in the Public Domain.

Fig. 2.05: Copyright © 2005 by Steffen Heinz, (CC BY-SA 3.0) at http://commons.wikimedia.org/wiki/File:Female_Malinois_2005-01-29.jpg.

Fig. 2.06: Copyright © 2007 by Ron Armstrong, (CC BY 2.0) at http://commons.wikimedia.org/wiki/File:Belgian_Tervuren_agility_A-frame.jpg.

Fig. 2.07: Copyright © 2008 by Kij, (CC BY-SA 4.0) at http://commons.wikimedia.org/wiki/File:MEN_Belgian_Groenendael.jpg.

Fig. 2.08: Copyright © 2003 by Sannse, (CC BY-SA 3.0) at http://commons.wikimedia.org/wiki/File:Bearded_Collie_600.jpg.

Fig. 2.09: Copyright © 2004 by Zwoenitzer, (CC BY-SA 3.0) at http://commons.wikimedia.org/wiki/File:Bearded_Collie3.jpg.

Fig. 2.10: Copyright © 2003 by Sannse, (CC BY-SA 3.0) at http://commons.wikimedia.org/wiki/File:Border_Collie_600.jpg.

Fig. 2.11: Copyright © 2006 by C. MacMillan, (CC BY-SA 3.0) at http://commons.wikimedia.org/wiki/File:BC_eye.jpg.

Fig. 2.12: Johanna363 / Copyright in the Public Domain.

Fig. 2.13: Copyright © by Roswitha Casimir, (CC BY-SA 3.0) at http://commons.wikimedia.org/wiki/File:Bouvier_des_Flandres.jpg.

Fig. 2.14: Copyright © 2006 by Ejgouda, (CC BY-SA 3.0) at http://commons.wikimedia.org/wiki/File:Fiero.JPG.

Fig. 2.15: Copyright © 2008 by Ducroccobo, (CC BY-SA 3.0) at http://commons.wikimedia.org/wiki/File:Flevobriards.jpeg .

Fig. 2.16: Copyright © 2003 by Sannse, (CC BY-SA 3.0) at http://commons.wikimedia.org/wiki/File:Rough_Collie_600.jpg.

Fig. 2.16: Copyright © 2004 by Ellen Levy Finch (Elf), (CC BY-SA 3.0) at http://commons.wikimedia.org/wiki/File:SmoothCollieTri2_wb.jpg.

Fig. 2.17: Copyright © 2002 by Christian Madden, (CC BY-SA 2.0) at http://commons.wikimedia.org/wiki/File:Rough_Collie_KC.jpg.

Fig. 2.18: Copyright © 2006 by Erica Steer (SaNtINa/kIKs), (CC BY-SA 2.0) at http://commons.wikimedia.org/wiki/File:German_Shepherd_Dog_with_disc.jpg.

Fig. 2.19: Odium~commonswiki / Copyright in the Public Domain.

Fig. 2.20: Copyright © 2005 by Mona, (CC BY-SA 3.0) at http://commons.wikimedia.org/wiki/File:Bobtail.JPG.

Fig. 2.21: Copyright © 2008 by Brigitte Hervé, (CC BY-SA 3.0) at http://commons.wikimedia.org/wiki/File:Chiot_bobtail.jpg.

Fig. 2.22: Copyright © 2007 by Puli1989, (CC BY-SA 3.0) at http://commons.wikimedia.org/wiki/File:Csalfa_Sommer_07_10_A.jpg.

Fig. 2.23: Copyright © 2003 by Sannse, (CC BY-SA 3.0) at https://commons.wikimedia.org/wiki/File:Puli_600.jpg.

Fig. 2.24: Copyright © 2007 by Chris, (CC BY 2.0) at http://commons.wikimedia.org/wiki/File:Shetland_Sheepdog_and_Rough_Collie.jpg

Fig. 2.25: Copyright © 2004 by Ellen Levy Finch (Elf), (CC BY-SA 3.0) at http://commons.wikimedia.org/wiki/File:ShetlandShpdogBlue2_wb.jpg.

Fig. 2.26: Copyright © 2003 by Sannse, (CC BY-SA 3.0) at http://commons.wikimedia.org/wiki/File:Cardigan_Welsh_Corgi_600.jpg.

Fig. 2.27: Copyright © 2008 by Pmuths1956, (CC BY-SA 3.0) at http://commons.wikimedia.org/wiki/File:Welchcorgipembroke.JPG.

Fig. 3.00: Hagon1 / Copyright in the Public Domain.

Fig. 3.01: Copyright © 2009 by Dagur Brynjólfsson, (CC BY 2.0) at http://commons.wikimedia.org/wiki/File:Hundas%C3%BDning_2009_053.jpg.

Fig. 3.02: Copyright © 2009 by fugzu, (CC BY 2.0) at http://commons.wikimedia.org/wiki/File:Basenji_Profile.jpg.

Fig. 3.03: Copyright © 2007 by original author: copperfeesh; edited by Pharaoh Hound, (CC BY-SA 2.0) at http://commons.wikimedia.org/wiki/File:Basenji_pair_edit.jpg.

Fig. 3.04: Copyright © 2003 by Sannse, (CC BY-SA 3.0) at http://commons.wikimedia.org/wiki/File:Borzoi_600.jpg.

Fig. 3.05: Copyright © 2006 by Taru T Torpström, (CC BY-SA 3.0) at http://commons.wikimedia.org/wiki/File:Borzoi_female.jpg.

Fig. 3.06: Copyright © 2005 by Scott Feldstein, (CC BY-SA 2.0) at http://commons.wikimedia.org/wiki/File:Greyhound_brindle_standing.jpg.

Fig. 3.07: Copyright © 2005 by Lilly M, (CC BY-SA 3.0) at http://commons.wikimedia.org/wiki/File:Greyhound_894.jpg,.

Fig. 3.09: Copyright © 2005 by ToB, (CC BY-SA 3.0) at https://de.wikipedia.org/wiki/Datei:Podenco_Ibicenco.jpg.

Fig. 3.09: Copyright © 2003 by Tirwhan, (CC BY-SA 3.0) at http://commons.wikimedia.org/wiki/File:Irish_Wolfhound_Sam.jpg.

Fig. 3.10: Copyright © 2007 by A. Montino, (CC BY-SA 3.0) at http://commons.wikimedia.org/wiki/File:IW-Othello.jpg.

Fig. 3.12: Copyright © 2013 by Томасина, (CC BY-SA 3.0) at https://commons.wikimedia.org/wiki/File:Pharaoh_Hound_in_Tallinn_3.JPG.

Fig. 3.13: Copyright © 2006 by Jan Eduard, (CC BY-SA 3.0) at https://en.wikipedia.org/wiki/File:Kelb-tal_Fenek_(head).jpg.

Fig. 3.13: Copyright © 2003 by Sannse, (CC BY-SA 3.0) at http://commons.wikimedia.org/wiki/File:Rhodesian_Ridgeback_600.jpg.

Fig. 3.14: Copyright © 2006 by Caronna, (CC BY-SA 2.5) at http://commons.wikimedia.org/wiki/File:Ridge.jpg.

Fig. 3.15: Copyright © 2003 by Sannse, (CC BY-SA 3.0) at http://commons.wikimedia.org/wiki/File:Saluki_600.jpg.

Fig. 3.16: Jonny Hedberg / Copyright in the Public Domain.

Fig. 3.18: Copyright © 2006 by Pleple2000, (CC BY-SA 3.0) at https://commons.wikimedia.org/wiki/File:Deerhound_305.jpg.

Fig. 3.19: Copyright © 2007 by gailf548, (CC BY 2.0) at https://commons.wikimedia.org/wiki/File:Father_and_Son.jpg.

Fig. 3.19: Copyright © 2004 by Ellen Levy Finch (Elf), (CC BY-SA 3.0) at http://commons.wikimedia.org/wiki/File:WhippetWhiteSaddled_wb.jpg.

Fig. 3.20: Copyright © 2008 by Winky, (CC BY 2.0) at http://commons.wikimedia.org/wiki/File:Anakin_Whippet.jpg.

Fig. 3.21: Copyright © 2006 by Bonnie van den Born, (CC BY-SA 3.0) at http://commons.wikimedia.org/wiki/File:Bvdb-bassethound2.jpg,.

Fig. 3.22: Copyright © 2008 by Anderson Nascimento, (CC BY 2.0) at http://commons.wikimedia.org/wiki/File:Cara_de_quem_caiu_do_caminh%C3%A3o....jpg.

Fig. 3.23: BlueIllusions / Copyright in the Public Domain.

Fig. 3.24: Copyright © 2011 by Taz80 / SEDIRI Eddy, (CC BY-SA 3.0) at http://commons.wikimedia.org/wiki/File:Beagle_Faraon.JPG.

Fig. 3.25: Copyright © 2006 by Pleple2000, (CC BY-SA 3.0) at http://commons.wikimedia.org/wiki/File:Bloodhund_767.jpg.

Fig. 3.26: Copyright © 2006 by Pleple2000, (CC BY-SA 3.0) at http://commons.wikimedia.org/wiki/File:Bloodhund_rf54.jpg.

Fig. 3.27: Copyright © 2006 by Igor Bredikhin, (CC BY 3.0) at https://commons.wikimedia.org/wiki/File:Short-haired-Dachshund.jpg.

Fig. 3.28: Copyright © by Alexandr Osipow, (CC BY-SA 3.0) at http://commons.wikimedia.org/wiki/File:Wire-haired_Dachshund_Profil.jpg.

Fig. 3.29: Copyright © 2013 by Томасина, (CC BY-SA 3.0) at http://commons.wikimedia.org/wiki/File:Dachshund_standard_longhair.JPG.

Fig. 3.30: Copyright © 2006 by Alexandr Belyaev.

Fig. 3.31: Copyright © 2007 by Ltshears, (CC BY-SA 3.0) at https://commons.wikimedia.org/wiki/File:AmericanFoxhound2.jpg.

Fig. 3.32: Dmitry Guskov / Copyright in the Public Domain.

Fig. 3.33: Machinecha~commonswiki / Copyright in the Public Domain.

Fig. 3.34: Copyright © 2013 by Sagaciousphil, (CC BY-SA 3.0) at http://commons.wikimedia.org/wiki/File:JillyCrufts-cropped2.JPG.

Fig. 3.35: Copyright © 2013 by Sagaciousphil, (CC BY-SA 3.0) at http://commons.wikimedia.org/wiki/File:JillyCrufts-cropped2.JPG.

Fig. 4.00: Copyright © 2004 by Robert Southworth, (CC BY-SA 3.0) at http://commons.wikimedia.org/wiki/File:American_Eskimo_Dog.jpg.

Fig. 4.01: Copyright © by StarSaber, (CC BY-SA 3.0) at http://commons.wikimedia.org/wiki/File:American_Eskimo_Dog_(Puppy).jpg.

Fig. 4.02: Copyright © 2007 by Akaporn Bhothisuwan, (CC BY 2.0) at http://commons.wikimedia.org/wiki/File:Bichon_Frise_red_background.jpg.

Fig. 4.03: Copyright © 2007 by Lucy, (CC BY 3.0) at http://commons.wikimedia.org/wiki/File:BichonFrise8Weeks.jpg.

Fig. 4.04: Copyright © 2004 by Ellen Levy Finch (Elf), (CC BY-SA 3.0) at http://commons.wikimedia.org/wiki/File:BostonTerrierBrindleStand_w.jpg.

Fig. 4.05: Copyright © 2009 by Jean-Pierre Dagneaux, (CC BY-SA 3.0) at http://commons.wikimedia.org/wiki/File:Boston_Terrier_male.jpg.
Fig. 4.06: Copyright © 2007 by Pleple2000, (CC BY-SA 3.0) at https://commons.wikimedia.org/wiki/File:Racib%C3%B3rz_2007_082.jpg.
Fig. 4.07: Asmadeus / Copyright in the Public Domain.
Fig. 4.08: GingerPeis / Copyright in the Public Domain.
Fig. 4.09: Copyright © by Ropompin, (CC BY-SA 3.0) at http://commons.wikimedia.org/wiki/File:Sharpei_2014-07-04_03-50.jpg.
Fig. 4.10: Copyright © 2007 by Remigiusz Józefowicz.
Fig. 4.11: Copyright © 2008 by rabendeviaregia, (CC BY-SA 2.0) at http://commons.wikimedia.org/wiki/File:Cleopatra_Chow_Chow.jpg.
Fig. 4.12: Dalmatiner24.eu / Copyright in the Public Domain.
Fig. 4.13: Dalmatiner24.eu / Copyright in the Public Domain.
Fig. 4.14: Copyright © 2008 by EGILEO, (CC BY 2.0) at http://commons.wikimedia.org/wiki/File:Bou.jpg.
Fig. 4.15: Copyright © by Frosya a, (CC BY-SA 3.0) at https://commons.wikimedia.org/wiki/File:%D0%A4%D1%80%D0%B0%D0%BD%D1%86%D1%83%D0%B7%D1%81%D0%BA%D0%B8%D0%B5_%D0%B1%D1%83%D0%BB%D1%8C%D0%B4%D0%BE%D0%B3%D0%B8_2013-09-02_11-36.JPG.
Fig. 4.16: Copyright © 2007 by Lilly M, (CC BY-SA 3.0) at http://commons.wikimedia.org/wiki/File:Keeshond_002_U.jpg.
Fig. 4.17: Copyright © 2010 by Terri Brown, (CC BY 2.0) at http://commons.wikimedia.org/wiki/File:Little_Puppy_Keeshond.jpg.
Fig. 4.18: Copyright © 2011 by Lhasaapso, (CC BY-SA 3.0) at http://commons.wikimedia.org/wiki/File:Lhasaapso.jpg.
Fig. 4.19: Copyright © 2005 by Arnold C.
Fig. 4.20: Copyright © 2005 by Belinda, (CC BY 2.0) at http://commons.wikimedia.org/wiki/File:Silver_Miniature_Poodle_stacked.jpg.
Fig. 4.21: Sally Eller / Copyright in the Public Domain.
Fig. 4.22: Copyright © 2006 by Pleple2000, (CC BY-SA 3.0) at http://commons.wikimedia.org/wiki/File:Pudel_toy_srebrny_67.jpg .
Fig. 4.23: Copyright © 2006 by Pleple2000, (CC BY-SA 3.0) at http://commons.wikimedia.org/wiki/File:Schipperke_gismo_12.jpg.
Fig. 4.24: Copyright © 2006 by Lilly M, (CC BY-SA 3.0) at http://commons.wikimedia.org/wiki/File:G%C5%82owa_gismo_-_schipperke.jpg.
Fig. 4.25: Copyright © 2012 by Wesl90, (CC BY-SA 3.0) at https://commons.wikimedia.org/wiki/File:Taro_(black_and_tan,_reu)_-_Chiko_(rood,_reu)_-_Ichigo_(rood,_teef).jpg.
Fig. 4.26: Copyright © 2011 by 増田一 , (CC BY-SA 3.0) at http://commons.wikimedia.org/wiki/File:Siro-shiba.JPG.
Fig. 4.27: Ladykransteer / Copyright in the Public Domain.
Fig. 4.28: Copyright © 2011 by Angelica, (CC BY 2.0) at http://commons.wikimedia.org/wiki/File:Tibetan_Spaniel.jpg.
Fig. 4.29: Copyright © 2006 by Katy Frankel, (CC BY 2.0) at http://commons.wikimedia.org/wiki/File:Tibetan_Terrier_Image_001.jpg.
Fig. 4.30: Copyright © 2011 by Heisstudying, (CC BY-SA 3.0) at http://commons.wikimedia.org/wiki/File:TibetanTerrier-Dali-3.jpg.
Fig. 5.00: Copyright © 2006 by Pleple2000, (CC BY-SA 3.0) at https://commons.wikimedia.org/wiki/File:Pointerka_45to_t.jpg.
Fig. 5.01: Copyright © 2010 by mjk23, (CC BY 2.0) at https://commons.wikimedia.org/wiki/File:Pointer_-_world_dog_show_2010.jpg.
Fig. 5.02: Copyright © 2004 by Ellen Levy Finch (Elf), (CC BY-SA 3.0) at http://commons.wikimedia.org/wiki/File:GermanShorthPtr_wb.jpg.
Fig. 5.03: Copyright © 2005 by Bonnie van den Born, (CC BY-SA 2.5) at http://commons.wikimedia.org/wiki/File:Duitse_staande_korthaar_10-10-2.jpg.
Fig. 5.04: Copyright © 2004 by Ellen Levy Finch (Elf), (CC BY-SA 3.0) at https://commons.wikimedia.org/wiki/File:GermanWirehrPtr1_wb.jpg.
Fig. 5.05: Copyright © 2013 by SheltieBoy, (CC BY 2.0) at http://commons.wikimedia.org/wiki/File:AKC_German_Wirehaired_Pointer_Dog_Show_2013.jpg.
Fig. 5.06: Copyright © 2008 by Rufus46, (CC BY-SA 3.0) at https://commons.wikimedia.org/wiki/File:Weimaraner_Freika-2.jpg.
Fig. 5.07: Copyright © by Jessicadoub, (CC BY-SA 3.0) at https://commons.wikimedia.org/wiki/File:Mavpointsm.jpg.

Fig. 5.08: Copyright © 2004 by Ellen Levy Finch (Elf), (CC BY-SA 3.0) at https://commons.wikimedia.org/wiki/File:Korthalsgriffon.jpg.
Fig. 5.09: Copyright © 2006 by Pleple2000, (CC BY-SA 3.0) at http://commons.wikimedia.org/wiki/File:Gryfon_Korthalsa_i78.jpg.
Fig. 5.10: Copyright © by Steffen Heinz, (CC BY-SA 3.0) at https://commons.wikimedia.org/wiki/File:Vizsla_Dorl.JPG.
Fig. 5.11: Copyright © 2007 by Bryon Realey, (CC BY 2.0) at https://commons.wikimedia.org/wiki/File:Vizsla_Rosie_water_harness.jpg.
Fig. 5.12: Copyright © 2007 by Keith Rousseau, (CC BY-SA 3.0) at http://commons.wikimedia.org/wiki/File:Chesapeake_Bay_Retriever1.jpg.
Fig. 5.13: Copyright © 2003 by Nevilley, (CC BY-SA 3.0) at http://commons.wikimedia.org/wiki/File:Chesapeakebayretriever01-l.jpg.
Fig. 5.14: Copyright © 2003 by Sannse, (CC BY-SA 3.0) at http://commons.wikimedia.org/wiki/File:Flat-Coated_Retriever_large.jpg, .
Fig. 5.15: Copyright © 2005 by Bonnie van den Born, (CC BY-SA 2.5) at https://commons.wikimedia.org/wiki/File:Flatcoat_retriever_2.jpg.
Fig. 5.16: Copyright © 2006 by SARBAJIT SARBAJNA, (CC BY 2.0) at http://commons.wikimedia.org/wiki/File:Golden_Retriever_dark_stacked.jpg.
Fig. 5.17: Copyright © 2007 by Edward, (CC BY 2.0) at http://commons.wikimedia.org/wiki/File:Golden_Retriever_standing_Tucker.jpg.
Fig. 5.18: Chrizweatley / Copyright in the Public Domain.
Fig. 5.19: Copyright © 2009 by Skansken, (CC BY-SA 3.0) at http://commons.wikimedia.org/wiki/File:Gul_labradortik.JPG.
Fig. 5.20: Copyright © 2007 by Rob Hanson, (CC BY 2.0) at http://commons.wikimedia.org/wiki/File:Labrador_Retriever_Hershey_chocolate.jpg.
Fig. 5.21: Call119 / Copyright in the Public Domain.
Fig. 5.22: Copyright © 2006 by Malgorzata Korbel, (CC BY-SA 3.0) at http://commons.wikimedia.org/wiki/File:Nova_Scotia_Duck_Tolling_Retriever.jpg.
Fig. 5.23: Copyright © 2004 by Ellen Levy Finch (Elf), (CC BY-SA 3.0) at http://commons.wikimedia.org/wiki/File:English_setter.jpg.
Fig. 5.24: Franza1984 / Copyright in the Public Domain.
Fig. 5.25: Copyright © 2005 by Pleple2000, (CC BY-SA 3.0) at http://commons.wikimedia.org/wiki/File:Seter_szkocki.jpg.
Fig. 5.26: Copyright © 2001 by R. Arkesteyn, (CC BY 3.0) at https://commons.wikimedia.org/wiki/File:Gordon_Setter_Teef.jpg.
Fig. 5.27: Copyright © 2006 by Luis Miguel Bugallo Sánchez, (CC BY-SA 3.0) at https://commons.wikimedia.org/wiki/File:Can_Setter_dog_GFDL.jpg.
Fig. 5.28: Copyright © 2013 by Томасина, (CC BY-SA 3.0) at https://commons.wikimedia.org/wiki/File:Irish_Setter_in_Tallinn_2.JPG.
Fig. 5.29: Copyright © 2013 by Томасина, (CC BY-SA 3.0) at https://commons.wikimedia.org/wiki/File:American_Cocker_in_Tallinn_2.JPG.
Fig. 5.30: Copyright © 2011 by Matt Batchelor, (CC BY 2.0) at http://commons.wikimedia.org/wiki/File:An_American_Cocker_Spaniel.jpg.
Fig. 5.31: Copyright © 2005 by Awsguy1, (CC BY-SA 3.0) at http://commons.wikimedia.org/wiki/File:Chien_d%27eau_americain_champion_1.JPG.
Fig. 5.32: Copyright © 2007 by Awsguy1, (CC BY-SA 3.0) at http://commons.wikimedia.org/wiki/File:Chien_d%27eau_americain_adulte.JPG.
Fig. 5.33: Copyright © 2009 by Pharaoh Hound, (CC BY-SA 3.0) at http://commons.wikimedia.org/wiki/File:Brittany_Spaniel_standing.jpg.
Fig. 5.34: Copyright © 2008 by Uber Phot, (CC BY-SA 2.0) at http://commons.wikimedia.org/wiki/File:Brittany_001.jpg.
Fig. 5.35: Copyright © 2005 by Akumiszcza, (CC BY 3.0) at http://commons.wikimedia.org/wiki/File:English_Cocker_Spaniel_4.jpg.
Fig. 5.36: Copyright © 2006 by Floridan.
Fig. 5.37: Copyright © 2004 by Ellen Levy Finch (Elf), (CC BY-SA 3.0) at http://commons.wikimedia.org/wiki/File:EnglishSpringerSpan2_wb.jpg.
Fig. 5.38: Copyright © 2013 by Томасина, (CC BY-SA 3.0) at https://commons.wikimedia.org/wiki/File:English_Springer_Spaniel_in_Tallinn.JPG.
Fig. 6.00: Copyright © 2008 by SCMW, (CC BY 3.0) at http://commons.wikimedia.org/wiki/File:Alaskan_Malamute.jpg.
Fig. 6.01: Copyright © 2002 by Klaus Helgert, (CC BY-SA 3.0) at http://commons.wikimedia.org/wiki/File:Alaska_Malamut_Huendin.JPG.

Fig. 6.02: Copyright © 2003 by Barney Skiandu, (CC BY-SA 3.0) at https://commons.wikimedia.org/wiki/File:Samoyed_600.jpg.
Fig. 6.03: Copyright © 2012 by Achim Herzog, (CC BY-SA 2.0) at http://commons.wikimedia.org/wiki/File:Samoyed.jpg.
Fig. 6.04: Copyright © 2005 by Utopialand, (CC BY-SA 3.0) at http://commons.wikimedia.org/wiki/File:Siberian-husky.jpg.
Fig. 6.05: Copyright © 2006 by Kelly Cookson, (CC BY-SA 2.0) at http://commons.wikimedia.org/wiki/File:Siberian_Huskies_white_and_red.jpg.
Fig. 6.06: Copyright © 2009 by Sevenfatdogs, (CC BY 3.0) at http://commons.wikimedia.org/wiki/File:Japaneseakita.jpg.
Fig. 6.07: Copyright © 2006 by Caro108, (CC BY-SA 3.0) at http://commons.wikimedia.org/wiki/File:American_Akita_-_Mad-Max2.jpg.
Fig. 6.08: Copyright © 2009 by Kerstin othello, (CC BY-SA 3.0) at https://commons.wikimedia.org/wiki/File:Othello.jpg.
Fig. 6.09: Copyright © 2006 by Ocmey, (CC BY-SA 2.5) at http://commons.wikimedia.org/wiki/File:Bouvier_Bernois_BE.jpg.
Fig. 6.10: Copyright © n.d. by Mood210, (CC BY-SA 3.0) at http://commons.wikimedia.org/wiki/File:Rj_boxer.jpg.
Fig. 6.11: Copyright © 2004 by Ellen Levy Finch (Elf), (CC BY-SA 3.0) at http://commons.wikimedia.org/wiki/File:Boxer_(dog).jpg.
Fig. 6.12: Copyright © 2006 by Fausto Moreno, (CC BY-SA 3.0) at https://commons.wikimedia.org/wiki/File:Bullmastiff_edited.JPG.
Fig. 6.13: Copyright © 2007 by Pleple2000, (CC BY-SA 3.0) at http://commons.wikimedia.org/wiki/File:Bullmastif_756.jpg.
Fig. 6.14: Copyright © 2007 by pato garza, (CC BY 2.0) at http://commons.wikimedia.org/wiki/File:03_Dobermann_sitstay.jpg.
Fig. 6.15: Fran Hogan / Copyright in the Public Domain.
Fig. 6.16: Copyright © 2007 by SheltieBoy, (CC BY 2.0) at http://commons.wikimedia.org/wiki/File:Doberman_Fawn_001.jpg.
Fig. 6.17: StBrecht / Copyright in the Public Domain.
Fig. 6.18: Copyright © 2007 by Ann Gordon, (CC BY-SA 2.0) at http://commons.wikimedia.org/wiki/File:French_Mastiff_001.jpg.
Fig. 6.19: Copyright © by Petrucha, (CC BY-SA 3.0) at http://commons.wikimedia.org/wiki/File:Nev_Lemar_Barbara1.jpg.
Fig. 6.20: Copyright © 2006 by Ninalle, (CC BY-SA 3.0) at http://commons.wikimedia.org/wiki/File:Emppumenossa.jpg.
Fig. 6.21: Copyright © 2007 by Lilly M, (CC BY-SA 3.0) at http://commons.wikimedia.org/wiki/File:Dog_niemiecki_R_01.jpg.
Fig. 6.22: Copyright © 2006 by Viborg~commonswiki, (CC BY-SA 3.0) at http://commons.wikimedia.org/wiki/File:Dogge_Odin.jpg.
Fig. 6.23: Copyright © 2007 by Radovan Rohovsky, (CC BY-SA 3.0) at http://commons.wikimedia.org/wiki/File:Westgort_Anticipation_17_months.JPG.
Fig. 6.24: Copyright © 2006 by Pleple2000, (CC BY-SA3.0)athttp://commons.wikimedia.org/wiki/File:Mastif_angielski_pregowany_nn.jpg,.
Fig. 6.25: Copyright © 2011 by JUAN RAMON RODRIGUEZ SOSA, (CC BY-SA 2.0) at http://commons.wikimedia.org/wiki/File:02_I_Exposici%C3%B3n_Monogr%C3%A1fica_Club_Rottweiler_de_Espa%C3%B1a_-_Santa_Brigida_-_Gran_Canaria..jpg.
Fig. 6.26: Copyright © by , (CC BY-SA 3.0) at http://commons.wikimedia.org/wiki/File:Rottweiler212.jpg.
Fig. 6.27: Copyright © 2012 by HeartSpoon, (CC BY-SA 3.0) at http://commons.wikimedia.org/wiki/File:Great_Pyrenees_Mountain_Dog.jpg.
Fig. 6.28: Copyright © 2011 by V.Beerenkotten, (CC BY-SA 3.0) at http://commons.wikimedia.org/wiki/File:Pyren%C3%A4enberghunde_vom_Beerenkotten.jpg.
Fig. 6.29: Copyright © 2006 by DanDee Shots, (CC BY 2.0) at http://commons.wikimedia.org/wiki/File:Newfoundland_dog_Smoky.jpg.
Fig. 6.30: Copyright © 2006 by Skern, (CC BY-SA 3.0) at http://commons.wikimedia.org/wiki/File:Newfoundland_brown.jpg.
Fig. 6.31: Copyright © 2006 by Silke Hollje-Schumacher, (CC BY-SA 3.0) at http://commons.wikimedia.org/wiki/File:C%C3%A3o_de_agua_Portugu%C3%AAs_2.jpg,.
Fig. 6.32: Copyright © 2014 by Schorle, (CC BY-SA 3.0) at http://commons.wikimedia.org/wiki/File:Portugiesischer_Wasserhund_wei%C3%9F.JPG.

Fig. 6.33: Copyright © 2008 by Cassie J, (CC BY 2.0) at http://commons.wikimedia.org/wiki/File:St_Bernard_Dog_001.jpg.
Fig. 6.34: VisualBeo / Copyright in the Public Domain.
Fig. 7.00: Copyright © 2006 by Lilly M, (CC BY-SA 2.5) at http://commons.wikimedia.org/wiki/File:Airedale_terrier_744.jpg.
Fig. 7.01: Copyright © 2005 by Angela Montillon, (CC BY-SA 2.0) at http://commons.wikimedia.org/wiki/File:Airedale_Terrier_Face.jpg.
Fig. 7.02: Copyright © by Tatanga 2006, (CC BY 3.0) at http://commons.wikimedia.org/wiki/File:AmStaff3.jpg.
Fig. 7.03: Copyright © 2008 by Eggman91972, (CC BY-SA 3.0) at http://commons.wikimedia.org/wiki/File:4YOAmstaff.JPG.
Fig. 7.04: Copyright © 2003 by David Owsiany, (CC BY-SA 3.0) at http://commons.wikimedia.org/wiki/File:Boutchie_apres_championnat_004.JPG.
Fig. 7.05: Copyright © 2004 by Ellen Levy Finch (Elf), (CC BY-SA 3.0) at http://commons.wikimedia.org/wiki/File:Bedlington_Terriers.jpg.
Fig. 7.06: Copyright © 1997 by Charles J Sharp, (CC BY-SA 3.0) at https://commons.wikimedia.org/wiki/File:Border-terrier-bitch.jpg.
Fig. 7.07: Copyright © 2009 by SheltieBoy, (CC BY 2.0) at https://commons.wikimedia.org/wiki/File:Border_Terrier_Fond_Du_Lac_County_Kennel_Club.jpg.
Fig. 7.08: Copyright © 2007 by Lilly M, (CC BY-SA 2.5) at https://commons.wikimedia.org/wiki/File:Bullterier_089_LM.jpg.
Fig. 7.09: Copyright © 2008 by Bruno ILHATl, (CC BY-SA 3.0) at https://commons.wikimedia.org/wiki/File:Bullterrier_femelle.jpg.
Fig. 7.10: Copyright © 2007 by Lilly M, (CC BY-SA 2.5) at http://commons.wikimedia.org/wiki/File:Bulterier_i_miniatura_LM.jpg.
Fig. 7.11: Copyright © 2005 by Ronald Müller-Hagen, (CC BY-SA 3.0) at http://commons.wikimedia.org/wiki/File:Cairn-Terrier-Garten1.jpg.
Fig. 7.12: Copyright © 2007 by John, (CC BY 2.0) at http://commons.wikimedia.org/wiki/File:Cairn_Terrier_portrait.jpg.
Fig. 7.13: Copyright © 2006 by Piotr Pietryka, (CC BY-SA 3.0) at http://commons.wikimedia.org/wiki/File:Fox_in_the_forest.JPG.
Fig. 7.14: Renáta Schwarzová / Copyright in the Public Domain.
Fig. 7.15: Copyright © 2013 by State Farm, (CC BY 2.0) at http://commons.wikimedia.org/wiki/File:01_Wire_Fox_terrier.jpg.
Fig. 7.16: Copyright © 2006 by Pleple2000, (CC BY-SA 3.0) at https://commons.wikimedia.org/wiki/File:Foksterier_szorstkowlosy_kufa_z_przodu_yy.jpg.
Fig. 7.17: Copyright © 2006 by Bree Bailey, (CC BY-SA 2.0) at http://commons.wikimedia.org/wiki/File:Rat_Terrier.jpg.
Fig. 7.18: Copyright © 2007 by Liza Lee Miller, (CC BY 2.0) at http://commons.wikimedia.org/wiki/File:Rat_Terrier_Ruby.jpg.
Fig. 7.19: Copyright © 2010 by Miyagawa, (CC BY-SA 3.0) at https://commons.wikimedia.org/wiki/File:Jack_Russell_Terrier_in_Park.jpg.
Fig. 7.20: Copyright © 2005 by Alephalpha, (CC BY-SA 3.0) at https://commons.wikimedia.org/wiki/File:05052881_PRT_braun_rau.jpg,.
Fig. 7.21: Copyright © 2006 by Psiara, (CC BY-SA 3.0) at https://commons.wikimedia.org/wiki/File:Miniature_Schnauzer_02.jpg.
Fig. 7.22: Copyright © 2010 by Madwren, (CC BY-SA 3.0) at http://commons.wikimedia.org/wiki/File:BlkStdSchnauzer1.jpg.
Fig. 7.23: Copyright in the Public Domain.
Fig. 7.24: Copyright © by Mider, (CC BY-SA 3.0) at http://commons.wikimedia.org/wiki/File:Scotish_terrier_burleska_2005.jpg.
Fig. 7.25: Copyright © 2013 by Томасина, (CC BY-SA 3.0) at http://commons.wikimedia.org/wiki/File:Scottish_Koira_2013.JPG.
Fig. 7.26: Copyright © 2003 by Sannse, (CC BY-SA 3.0) at http://commons.wikimedia.org/wiki/File:Soft_Coated_Wheaten_Terrier_600.jpg.
Fig. 7.27: Copyright © 2009 by RORLVR1, (CC BY-SA 3.0) at https://commons.wikimedia.org/wiki/File:Australian_Grand_Champion_Dancestar_Zero_O_Seven_%22Bondy%22.jpg.
Fig. 7.28: Aixeneta / Copyright in the Public Domain.
Fig. 7.29: Copyright © by Sannse, (CC BY-SA 3.0) at http://commons.wikimedia.org/wiki/File:Staffordshire_Bull_Terrier_600.jpg.
Fig. 7.30: Copyright © 2009 by Pleple2000, (CC BY-SA 3.0) at http://commons.wikimedia.org/wiki/File:Terier_walijski_suka_2009_pl.jpg.
Fig. 7.31: Copyright © by Shleiderbmx, (CC BY 3.0) at http://commons.wikimedia.org/wiki/File:Welsh_Terrier.JPG.

Fig. 7.32: Copyright © 2011 by SheltieBoy, (CC BY 2.0) at http://commons.wikimedia.org/wiki/File:A_West_Highland_White_Terrier_Earthdog_Trial.jpg.
Fig. 7.33: Copyright © 2008 by Antonio Villascusa, (CC BY 3.0) at http://commons.wikimedia.org/wiki/File:Laika01.jpg.
Fig. 8.00: Copyright © 2014 by Dan9186, (CC BY-SA 3.0) at https://commons.wikimedia.org/wiki/File:Brussels_Griffon_Rembrant_(jpg_format).jpg.
Fig. 8.01: Copyright © 2007 by Maria Michaëlsson, (CC BY-SA 3.0) at https://commons.wikimedia.org/wiki/File:0709_gucci_stolt.jpg.
Fig. 8.02: Copyright © 2009 by Andreweatock, (CC BY-SA 3.0) at https://commons.wikimedia.org/wiki/File:CarterBIS.Tiki.13.6.09.jpg.
Fig. 8.03: Copyright © 2007 by tanakawho, (CC BY 2.0) at https://commons.wikimedia.org/wiki/File:Cavalier_King_Charles_Spaniel_trio.jpg.
Fig. 8.04: Copyright © 2007 by Tatiana Borisova, (CC BY-SA 3.0) at http://commons.wikimedia.org/wiki/File:Chihuahuasmoothcoat.jpg.
Fig. 8.05: Copyright © 2009 by Vindhyana, (CC BY 3.0) at http://commons.wikimedia.org/wiki/File:Babylone_3ans1.jpg.
Fig. 8.06: Copyright © 2006 by Tommy Gildseth, (CC BY 3.0) at http://commons.wikimedia.org/wiki/File:IndyStands.jpg.
Fig. 8.07: Copyright © 2014 by Amy Church, (CC BY-SA 3.0) at https://commons.wikimedia.org/wiki/File:Powderpuff_Chinese_Crested_6.jpg.
Fig. 8.08: Copyright © 2007 by Pavel Marrero, (CC BY-SA 3.0) at https://commons.wikimedia.org/wiki/File:Bichon_habanero.jpg.
Fig. 8.09: Copyright © 2013 by Scheepers100, (CC BY-SA 3.0) at https://commons.wikimedia.org/wiki/File:Havanezer_Tailwhispers.JPG.
Fig. 8.10: Copyright © 2007 by christina, (CC BY 2.0) at http://commons.wikimedia.org/wiki/File:Italian_Greyhound_standing_gray.jpg.
Fig. 8.11: Copyright © 2012 by Vandeverre2 (Luigi Griffini), (CC BY-SA 3.0) at https://commons.wikimedia.org/wiki/File:YURI_-_Piccolo_Levriero_Italiano.jpg.
Fig. 8.12: Copyright © 2012 by Joliot, (CC BY-SA 3.0) at http://commons.wikimedia.org/wiki/File:Simsalachin_El_Manolito.jpg.
Fig. 8.13: Copyright © 2006 by LostinTexas (Alex Archambault), (CC BY 2.0) at http://commons.wikimedia.org/wiki/File:Japanese_Chin_portrait.jpg.
Fig. 8.14: Canino21 / Copyright in the Public Domain.
Fig. 8.15: Copyright © 2013 by SheltieBoy, (CC BY 2.0) at http://commons.wikimedia.org/wiki/File:01_AKC_Maltese_Dog_Show_2013.jpg.
Fig. 8.16: Copyright © 2007 by Sebastian Wallin, (CC BY 3.0) at http://commons.wikimedia.org/wiki/File:Miniature_pinscher.jpg.
Fig. 8.17: Copyright © by ArizonaLincoln, (CC BY 3.0) at https://en.wikipedia.org/wiki/File:Minipinsaz.jpg.
Fig. 8.18: Copyright © 2007 by Ron Armstrong, (CC BY 2.0) at https://commons.wikimedia.org/wiki/File:Papillon_portrait_Flickr.jpg.
Fig. 8.19: Copyright © 2009 by Gvdmoort, (CC BY-SA 3.0) at http://commons.wikimedia.org/wiki/File:Outdoor_Continental_Toy_Spaniel_Papillon.jpeg.
Fig. 8.20: Copyright © 2011 by Joliot, (CC BY-SA 3.0) at https://commons.wikimedia.org/wiki/File:Fenka_pekin%C3%A9ze.jpg.
Fig. 8.21: Copyright © 2005 by DiasSoares, (CC BY-SA 2.0) at http://commons.wikimedia.org/wiki/File:Pekingese_01061.jpg.
Fig. 8.22: Blackoranges / Copyright in the Public Domain.
Fig. 8.23: Jpkollman / Copyright in the Public Domain.
Fig. 8.24: Copyright © 2011 by Xidrep, (CC BY-SA 3.0) at http://commons.wikimedia.org/wiki/File:Mops-duke-mopszucht-vom-maegdebrunnen.jpg.
Fig. 8.25: Beckilee / Copyright in the Public Domain.
Fig. 8.26: Copyright © 2005 by Melanie Dullinger, (CC BY-SA 3.0) at https://commons.wikimedia.org/wiki/File:Shih-Tzu.JPG.
Fig. 8.27: Copyright © 2007 by Tomasz Wawak (Wawri), (CC BY-SA 2.5) at http://commons.wikimedia.org/wiki/File:Shih-tzu-Fibi.jpg.
Fig. 8.28: Copyright © 2012 by Ofsilkysdream, (CC BY-SA 3.0) at http://commons.wikimedia.org/wiki/File:Australian_Silky_Terrier_Alana_Of_Silky%27s_Dream.jpg.
Fig. 8.29: Copyright © by Ofsilkysdream, (CC BY-SA 3.0) at https://commons.wikimedia.org/wiki/File:Australian-Silky-Terrier-ofsilkysdream.JPG.

Fig. 8.30: Copyright © 2009 by Dagur Brynjólfsson, (CC BY-SA 2.0) at http://commons.wikimedia.org/wiki/File:Yorkshire_Terrier_at_show.jpg.
Fig. 8.31: Josch13 / Copyright in the Public Domain.
Fig. 9.00: Copyright © 2007 by Karin Langner-Bahmann, (CC BY-SA 3.0) at https://commons.wikimedia.org/wiki/File:Siam_lilacpoint.jpg.
Fig. 9.01: Copyright © 2004 by Martin Bahmann, (CC BY-SA 3.0) at http://commons.wikimedia.org/wiki/File:Kamee02.jpg.
Fig. 9.02: Copyright © 2004 by Jonik, (CC BY-SA 2.0) at http://commons.wikimedia.org/wiki/File:Manx_breed_cat_named_Inkku.jpg.
Fig. 9.03: Copyright © n.d. by Lengoma, (CC BY 3.0) at http://commons.wikimedia.org/wiki/File:Mainecoonbrowntabby.jpg.
Fig. 9.04: Copyright © n.d. by Lengoma, (CC BY 3.0) at http://commons.wikimedia.org/wiki/File:Mainecoonbrowntabby.jpg.
Fig. 9.05: Copyright © 2008 by Heikki Siltala, (CC BY-SA 3.0) at http://commons.wikimedia.org/wiki/File:European_shorthair_male_KM-PIROK_cat_show_Tampere_2008-10-04.JPG.
Fig. 9.06: Copyright © 2008 by Sean McGrath, (CC BY 2.0) at https://commons.wikimedia.org/wiki/File:Bengal_cat,_five_months_old_(2474931892).jpg.
Fig. 9.07 b : Copyright © 2004 by Martin Bahmann, (CC BY-SA 3.0) at http://commons.wikimedia.org/wiki/File:Kamee01.jpg.
Fig. 9.08 : Copyright © 2005 by Elizabeth Krumbach, (CC BY-SA 3.0) at http://commons.wikimedia.org/wiki/File:Smoke_egyptian_mau.jpg.
Fig. 9.09: Copyright © 2014 by Bambooo, (CC BY-SA 3.0) at http://commons.wikimedia.org/wiki/File:Persian_silver_dollface.jpg.
Fig. 9.10 a : Copyright © n.d. by Bombaycats, (CC BY-SA 3.0) at http://commons.wikimedia.org/wiki/File:Bombay_Katzen_of_Blue_Sinfonie.JPG.
Fig. 9.10 b: Copyright © 2006 by Parvati Napha Naramit, (CC BY-SA 2.5) at http://commons.wikimedia.org/wiki/File:Parvati,chat-adulte-femelle-race-korat.JPG.
Fig. 9.11 a: Ramair350 / Copyright in the Public Domain.
Fig. 9.11 b: Calicocindy / Copyright in the Public Domain.
Fig. 9.12: Copyright © 2013 by Toniposadas, (CC BY-SA 3.0) at https://en.wikipedia.org/wiki/File:Domestic-mediumhair-brownmackerel-torbie.jpg.
Fig. 9.13 a: Copyright © 2010 by Jiřina Choděrová, (CC BY-SA 3.0) at http://commons.wikimedia.org/wiki/File:Debbiewiki2.JPG.
Fig. 9.13 b: Copyright © 2007 by Heikki Siltala, (CC BY 3.0) at http://commons.wikimedia.org/wiki/File:Nipsu-Nappulan_Non-Stop_Exotic_shorthair.JPG.
Fig. 9.14 a: Copyright © 2010 by Sandrus, (CC BY 3.0) at https://commons.wikimedia.org/wiki/File:Siamese_cat_Amira_Laila_Basbousa*CZ.jpg.
Fig. 9.14 b: Copyright © 2007 by Karin Langner-Bahmann, (CC BY-SA 3.0) at https://commons.wikimedia.org/wiki/File:Siam_lilacpoint.jpg.
Fig. 9.15: Sieshelle / Copyright in the Public Domain.
Fig. 9.16: Copyright © 2009 by Gangburgondes, (CC BY 3.0) at http://commons.wikimedia.org/wiki/File:Dalaja_Doll-norvegien_ambre-blotched-tabby-blanc-neige2009b.jpg.
Fig. 9.17: Copyright © 2011 by SMcCandlish, (CC BY 3.0) at https://commons.wikimedia.org/wiki/File:Female_%27riser%27_Manx_cat_crouching.jpg.
Fig. 9.18: Zach Pharr / Copyright in the Public Domain.
Fig. 10.00: Copyright © 2005 by Brigitte von Szczawinski, (CC BY-SA 3.0) at https://commons.wikimedia.org/wiki/File:Valentino.jpg.
Fig. 10.01: Copyright © 2005 by Aufnahme von Joachim Berger-Uelsberg und Dr. Gabriele Uelsberg, (CC BY-SA 3.0) at https://commons.wikimedia.org/wiki/File:Abykitten.jpg.
Fig. 10.02: Copyright © 2008 by torbakhopper, (CC BY 2.0) at http://commons.wikimedia.org/wiki/File:American_bobtail_2.jpg.
Fig. 10.03: Copyright © 2005 by Rodrigo, (CC BY-SA 3.0) at https://commons.wikimedia.org/wiki/File:Grey_kitty.jpg.
Fig. 10.04: Copyright © 2008 by Nickolas Titkov, (CC BY-SA 2.0) at http://commons.wikimedia.org/wiki/File:ACL_Pointocurl_Fiorentina.jpg.
Fig. 10.05: Copyright © 2007 by Heikki Siltala, (CC BY 3.0) at http://commons.wikimedia.org/wiki/File:EC_Recknagel_Annakurnikoova_ACS_gs_24.JPG.
Fig. 10.06: Copyright © 2009 by Tomo, (CC BY 2.0) at http://commons.wikimedia.org/wiki/File:American_shorthair_blotched_tabby.jpg.

Fig. 10.07: Copyright © 2006 by Scott, (CC BY-SA 2.0) at http://commons.wikimedia.org/wiki/File:White_American_Shorthair.jpg.
Fig. 10.08: Copyright © 2010 by Фотограф:Анна Утехина, (CC BY-SA 3.0) at http://commons.wikimedia.org/wiki/File:Ghislaine_6028.jpg.
Fig. 10.09: Igatto / Copyright in the Public Domain.
Fig. 10.10: Copyright © 2008 by Sean McGrath, (CC BY 2.0) at https://commons.wikimedia.org/wiki/File:Bengal_cat,_five_months_old_(2474931892).jpg.
Fig. 10.11: Seduisant / Copyright in the Public Domain.
Fig. 10.12: Copyright © 2005 by D.Maillard, (CC BY-SA 2.5) at https://commons.wikimedia.org/wiki/File:Birma_seal_u_tortie_point.jpg.
Fig. 9.32: Copyright © 2007 by Claudia Zaino, (CC BY-SA 2.5) at https://commons.wikimedia.org/wiki/File:Minerva-21.jpg.
Fig. 10.14: Copyright © by Bombaycats, (CC BY-SA 3.0) at http://commons.wikimedia.org/wiki/File:Bombay_Katzen_of_Blue_Sinfonie.JPG.
Fig. 10.15: Copyright © 2009 by Bombaykatze, (CC BY-SA 3.0) at http://commons.wikimedia.org/wiki/File:Bombay_Katze_Ebene_of_Blue_Sinfonie.JPG.
Fig. 10.16: Copyright © 2008 by Chatterie des Millenovæ, (CC BY-SA 3.0) at https://commons.wikimedia.org/wiki/File:British_Longhair_-_Black_Silver_Shaded.jpg.
Fig. 10.17: Copyright © 2008 by DennisChristensen, (CC BY-SA 3.0) at http://commons.wikimedia.org/wiki/File:BlackTortieBritishShorthair.jpg.
Fig. 10.18: Copyright © by Hans-Jörg Hellwig, (CC BY-SA 3.0) at http://commons.wikimedia.org/wiki/File:Co-kbk-07-001.jpg.
Fig. 10.19: Copyright © 2011 by Sergey Lubov, (CC BY-SA 3.0) at http://commons.wikimedia.org/wiki/File:BurMau_Vega_Star,_champagne_burmese_kitten.jpg.
Fig. 10.20: Copyright © 2007 by Jgroland2, (CC BY-SA 3.0) at http://commons.wikimedia.org/wiki/File:Burmeser-Donna_Summer.JPG.
Fig. 10.21: Copyright © 2010 by Jiřina Choděrová, (CC BY-SA 3.0) at http://commons.wikimedia.org/wiki/File:Debbiewiki2.JPG.
Fig. 10.22: Copyright © 2006 by Groeni, (CC BY-SA 2.5) at http://commons.wikimedia.org/wiki/File:Olivia2.jpg.
Fig. 10.23: Copyright © 2007 by Bebopscrx, (CC BY-SA 3.0) at http://commons.wikimedia.org/wiki/File:Devon_rex.jpg.
Fig. 10.24: Copyright © 2006 by Pathfinder Linden, (CC BY 2.0) at http://commons.wikimedia.org/wiki/File:Devon_Rex_Izzy.jpg.
Fig. 10.25: Copyright © 2009 by Lil Shepherd, (CC BY 2.0) at http://commons.wikimedia.org/wiki/File:Egyptian_mau_-_Horus_2.jpg.
Fig. 10.26: Copyright © 2008 by liz west, (CC BY 2.0) at http://commons.wikimedia.org/wiki/File:Egyptian_Mau_Bronze.jpg.
Fig. 10.27: Copyright © 2007 by Heikki Siltala, (CC BY 3.0) at http://commons.wikimedia.org/wiki/File:Nipsu-Nappulan_Non-Stop_Exotic_shorthair.JPG.
Fig. 10.28: Copyright © 2008 by Heikki Siltala, (CC BY 3.0) at http://commons.wikimedia.org/wiki/File:Noidankattilan_Bellatrix_Exotic_shorthair.JPG.
Fig. 10.29: Copyright © 2009 by Dave Scelfo, (CC BY-SA 2.0) at http://commons.wikimedia.org/wiki/File:Havana_Brown_-_choco.jpg.
Fig. 10.30: Copyright © 2008 by Dave Scelfo, (CC BY-SA 2.0) at http://commons.wikimedia.org/wiki/File:Havana_Brown_-_brown_whiskers.jpg.
Fig. 10.31: Copyright © by ようてい, (CC BY-SA 3.0) at http://commons.wikimedia.org/wiki/File:JapaneseBobtailBlueEyedMi-ke.JPG.
Fig. 10.32: Copyright © by Chris 73, (CC BY-SA 3.0) at http://commons.wikimedia.org/wiki/File:Japanese_Bobtail_walking.JPG.
Fig. 10.33: Copyright © 2006 by Parvati Napha Naramit, (CC BY-SA 2.5) at http://commons.wikimedia.org/wiki/File:Parvati,chat-adulte-femelle-race-korat.JPG.
Fig. 10.34: Copyright © 2005 by Max Bouche, (CC BY-SA 2.5) at http://commons.wikimedia.org/wiki/File:Korat_Family.jpg.
Fig. 10.35: Copyright © 2008 by Bebopscrx, (CC BY-SA 3.0) at http://commons.wikimedia.org/wiki/File:Laperm_LH_blacktortie_white.jpg.
Fig. 10.36: Copyright © 2008 by Bebopscrx, (CC BY-SA 3.0) at http://commons.wikimedia.org/wiki/File:Laperm_blacktortie_white.jpg.
Fig. 10.37: Copyright © by Lengoma, (CC BY 3.0) at http://commons.wikimedia.org/wiki/File:Mainecoonbrowntabby.jpg.
Fig. 10.38: Harrison R. Diamond / Copyright in the Public Domain.

Fig. 10.40: Copyright © 2004 by Jonik, (CC BY-SA 2.0) at http://commons.wikimedia.org/wiki/File:Manx_breed_cat_named_Inkku.jpg.
Fig. 10.41: Sieshelle / Copyright in the Public Domain.
Fig. 10.42: Copyright © 2009 by Gangburgondes, (CC BY-SA 3.0) at http://commons.wikimedia.org/wiki/File:Dalaja_Doll-norvegien_ambre-blotched-tabby-blanc-neige2009b.jpg.
Fig. 10.43: Tom Bjornstad / Copyright in the Public Domain.
Fig. 10.44: Copyright © 2005 by Tom Bjornstad, (CC BY-SA 3.0) at https://commons.wikimedia.org/wiki/File:Ocicat_Charan-Rizzo_sitting.jpg.
Fig. 10.45: Copyright © 2011 by Daly69, (CC BY-SA 3.0) at https://commons.wikimedia.org/wiki/File:Fatale_de_la_l%C3%A9gende_d%27ali.jpg.
Fig. 10.46: Copyright © 2011 by Heikki Siltala, (CC BY 3.0) at https://commons.wikimedia.org/wiki/File:OSH_24-CH_Clamoris_Epeius_(Omar).JPG, .
Fig. 10.47: Copyright © 2006 by Lajoma, (CC BY-SA 3.0) at http://commons.wikimedia.org/wiki/File:Persialainen.jpg.
Fig. 10.48: Copyright © 2008 by Cindy See, (CC BY-SA 2.0) at http://commons.wikimedia.org/wiki/File:Chocolate_Himlayan.jpg.
Fig. 10.49: Copyright © 2014 by Bambooo, (CC BY-SA 3.0) at http://commons.wikimedia.org/wiki/File:Persian_silver_dollface.jpg.
Fig. 10.50: Bea Stankovičová / Copyright in the Public Domain.
Fig. 10.51: Copyright © by Ariele, (CC BY-SA 3.0) at http://commons.wikimedia.org/wiki/File:A_Real_Ragdoll_seal_colourpoint.jpg.
Fig. 10.52: Copyright © 2008 by Tkeiger, (CC BY 3.0) at https://commons.wikimedia.org/wiki/File:Russian_blue.jpg.
Fig. 10.53: Copyright © 2007 by ruskis, (CC BY-SA 3.0) at http://commons.wikimedia.org/wiki/File:RB_kitten.JPG.
Fig. 10.54: Copyright © 2010 by Psihopat, (CC BY-SA 3.0) at http://commons.wikimedia.org/wiki/File:Adult_Scottish_Fold.jpg.
Fig. 10.55: Copyright © 2008 by Heikki Siltala, (CC BY 3.0) at http://commons.wikimedia.org/wiki/File:Scottish_Fold_-_CFF_cat_show_Heinola_2008-05-03_IMG_7882.JPG.
Fig. 10.56: Copyright © 2005 by Jerry Pank, (CC BY-SA 2.0) at http://commons.wikimedia.org/wiki/File:Two_Siam_Seal_Point.jpg.
Fig. 10.57: Copyright © 2007 by Karin Langner-Bahmann, (CC BY-SA 3.0) at https://commons.wikimedia.org/wiki/File:Siam_lilacpoint.jpg.
Fig. 10.58: Copyright © 2014 by Phaithai Cattery, (CC BY-SA 4.0) at https://commons.wikimedia.org/wiki/File:PhaithaiChic_Thai_cat.jpg.
Fig. 10.59: Copyright © 2007 by Tgianf75, (CC BY-SA 3.0) at http://commons.wikimedia.org/wiki/File:Canadian_Sphynx_Red.jpg.
Fig. 10.60: Copyright © 2006 by PuppetMasteX , (CC BY-SA 3.0) at http://commons.wikimedia.org/wiki/File:2_Sphynx_cats_sleeping_together.jpg.
Fig. 10.61: Copyright © 2009 by Kamée, (CC BY 3.0) at http://commons.wikimedia.org/wiki/File:Tonkanesenkatze-natural-mink.jpg.
Fig. 10.62: Copyright © 2011 by Edenvane, (CC BY-SA 3.0) at https://commons.wikimedia.org/wiki/File:Chat_tonkinois_lilac_mink_(chatterie_de_l%27Esprit_d%27Edenvane).jpg.
Fig. 10.63: Copyright © 2011 by Daly69, (CC BY-SA 3.0) at https://commons.wikimedia.org/wiki/File:Angora_turc_black_tortie_smoke.jpg.
Fig. 10.64: Copyright © 2011 by Daly69, (CC BY-SA 3.0) at https://commons.wikimedia.org/wiki/File:Doudinka_d%27angoramour.jpg.
Fig. 10.65: Copyright © 2008 by Gelweo, (CC BY-SA 4.0) at http://commons.wikimedia.org/wiki/File:Female_turk_van.JPG.
Fig. 10.66: Zach Pharr / Copyright in the Public Domain.
Fig. 11.00: Steven Lek / Copyright in the Public Domain.
Fig. 11.01: V. Leche / Copyright in the Public Domain.
Fig. 11.02: Erlendaakre / Copyright in the Public Domain.
Fig. 11.03: Copyright © 2009 by Uwe Gille, (CC BY-SA 3.0) at https://commons.wikimedia.org/wiki/File:Bradygnathia-superior-rabbit.jpg.
Fig. 11.04: Copyright © 2014 by MaikS76, (CC BY-SA 3.0) at https://commons.wikimedia.org/wiki/File:Hausmauswurf_Im_Nest.jpg.
Fig. 11.05: Copyright © 2008 by Rama, (CC BY-SA 2.0 FR) at https://commons.wikimedia.org/wiki/File:Lab_mouse_mg_3140.jpg.
Fig. 11.06: Copyright © 2005 by , (CC BY-SA 2.5) at https://commons.wikimedia.org/wiki/File:Djungarian_Hamster_Pearl_White_run_wheel.jpg.
Fig. 12.00: Thirteen squared / Copyright in the Public Domain.

Fig. 12.01: Copyright © 2006 by Kjersti Holmang, (CC BY-SA 3.0) at https://commons.wikimedia.org/wiki/File:Chinchilla-Soelvmarmorert.jpg.

Fig. 12.02: Copyright © 2013 by Yabopomonofonomopo bay, (CC BY-SA 3.0) at https://commons.wikimedia.org/wiki/File:Chinchilla_lanigera_reared.jpg.

Fig. 12.03: Copyright © by Alfredo Gutiérrez, (CC BY-SA 4.0) at https://commons.wikimedia.org/wiki/File:Ferret_2008.png.

Fig. 12.04: Copyright © 2013 by Luciano Bernardi, (CC BY-SA 3.0) at https://commons.wikimedia.org/wiki/File:Coco_4074.JPG.

Fig. 12.05: Copyright © 2012 by Luciano Bernardi, (CC BY-SA 3.0) at https://commons.wikimedia.org/wiki/File:Jake_0314.jpg.

Fig. 12.06: A. Popov / Copyright in the Public Domain.

Fig. 12.07: Copyright © 2008 by Gophi, (CC BY-SA 4.0) at https://commons.wikimedia.org/wiki/File:2008-02-16_Mongolian_gerbil.jpg.

Fig. 12.08: Copyright in the Public Domain.

Fig. 12.09: Copyright © 2005 by Anna S. , (CC BY-SA 3.0) at https://commons.wikimedia.org/wiki/File:Hausmeerschweinchen.JPG.

Fig. 12.10: Copyright © 2005 by , (CC BY-SA 3.0) at https://commons.wikimedia.org/wiki/File:Blesseundbabies1.jpg.

Fig. 12.11: Copyright © 2007 by , (CC BY 2.0) at https://commons.wikimedia.org/wiki/File:Texel_guinea_pig.jpg.

Fig. 12.12: Copyright © 2007 by Just chaos, (CC BY 2.0) at https://commons.wikimedia.org/wiki/File:Teddy_guinea_pig.jpg.

Fig. 12.13: Copyright © 2007 by Christine, (CC BY 2.0) at https://commons.wikimedia.org/wiki/File:Peruvian_cavy.jpg.

Fig. 12.14: Copyright © 2006 by Adamjennison111 , (CC BY 2.5) at https://commons.wikimedia.org/wiki/File:Golden_hamster_side_1.jpg.

Copyright © 2006 Harner Blake, GNU General Public License at https://commons.wikimedia.org/wiki/File:Chomik.JPG

Fig. 12.16: Copyright © 2009 by Alex Eames, (CC BY-SA 3.0) at https://commons.wikimedia.org/wiki/File:Domspotchinese.JPG.

Fig. 12.17: Copyright © 2010 by Calle Eklund, (CC BY-SA 3.0) at https://commons.wikimedia.org/wiki/File:Erinaceus_europaeus_in_Avesta_02.jpg.

Fig. 12.18: Alex Huck / Copyright in the Public Domain.

Fig. 12.19: Copyright © 2010 by Tiia Monto, (CC BY-SA 3.0) at https://commons.wikimedia.org/wiki/File:Brown_mouse.jpg.

Fig. 12.20: Hampus Lybeck / Copyright in the Public Domain.

Fig. 12.21: Copyright © 2011 by Tiia Monto, (CC BY-SA 3.0) at https://commons.wikimedia.org/wiki/File:Pet_mice.jpg.

Fig. 12.22: Copyright © 2010 by Tiia Monto, (CC BY-SA 3.0) at https://commons.wikimedia.org/wiki/File:Pet_mouse_black_white.jpg.

Fig. 12.23: Copyright © 2007 by Stamatisclan, (CC BY-SA 3.0) at https://commons.wikimedia.org/wiki/File:Runt_and_Paxie.jpg.

Fig. 12.24: Copyright © 2008 by Heny, (CC BY-SA 3.0) at https://commons.wikimedia.org/wiki/File:Fick%C3%B3,_a_gy%C3%B6ny%C3%B6r%C5%B1_t%C3%B6rpe_rex.JPG.

Fig. 12.25: Copyright © 2013 by Bodlina, (CC BY-SA 3.0) at https://commons.wikimedia.org/wiki/File:Czech_Albino_Rabbit_2.jpg.

Fig. 12.26: Copyright © 2006 by Ross Little, (CC BY-SA 2.0) at https://commons.wikimedia.org/wiki/File:Fluffy_white_bunny_rabbit.jpg.

Fig. 12.27: Copyright © 2011 by Aramat, (CC BY-SA 2.5) at https://commons.wikimedia.org/wiki/File:Satijn_konijn_ivoor.jpg.

Fig. 12.28: Copyright © 2011 by Cliff Reppart, (CC BY-SA 2.5) at https://commons.wikimedia.org/wiki/File:English_Lop_Rabbit.jpg.

Fig. 12.29: ja / Copyright in the Public Domain.

Fig. 12.31: Copyright © 2007 by AlexK100, (CC BY-SA 2.0) at https://commons.wikimedia.org/wiki/File:Fancy_rat_blaze.jpg.

Fig. 12.32: Copyright © by Ykmyks, (CC BY-SA 3.0) at https://commons.wikimedia.org/wiki/File:Ratedumbo.jpg.

Fig. 12.33: Copyright © 2009 by 6th Happiness, (CC BY-SA 3.0) at https://commons.wikimedia.org/wiki/File:Beige_Down_Under_cc.jpg.

Fig. 12.34: Pemelet / Copyright in the Public Domain.

Fig. 12.35: Copyright © 2006 by Dawson, (CC BY-SA 2.5) at https://commons.wikimedia.org/wiki/File:Petaurus_breviceps_1.jpg.

Fig. 12.36: Copyright © 2012 by OberonNightSeer, (CC BY-SA 3.0) at https://commons.wikimedia.org/wiki/File:Sugar_Gliders_eating_Mealworms.jpg.

Fig. 12.37: Copyright © by Anke Meyring, (CC BY-SA 2.5) at https://commons.wikimedia.org/wiki/File:Sugarglider_hp.jpg.

Fig. 13.00 : Copyright in the Public Domain.
Fig. 13.01: Copyright © 2007 by Matt edmonds, (CC BY-SA 3.0) at https://commons.wikimedia.org/wiki/File:Yellow-naped_Amazon.jpg.
Fig. 13.02: Copyright © 2007 by Matt edmonds, (CC BY-SA 3.0) at https://commons.wikimedia.org/wiki/File:Yellow-naped_Amazon.jpg.
Fig. 13.02: Copyright © by Muriel Gottrop, (CC BY-SA 3.0) at https://commons.wikimedia.org/wiki/File:The_feathers_of_the_wing_the_birds.jpg.
Fig. 13.03: Copyright © 2008 by Marah09013 , (CC BY-SA 3.0) at https://commons.wikimedia.org/wiki/File:Goffins_Cockatoo_in_flight_in_harness.jpg.
Fig. 13.04: Copyright © 2009 by 3268zauber, (CC BY-SA 3.0) at https://commons.wikimedia.org/wiki/File:Fu%C3%9F_eines_Wei%C3%9Fhaubenkakadus.JPG.
Fig. 14.00 : Copyright © 2007 by L.Miguel Bugallo Sánchez, (CC BY-SA 3.0) at https://commons.wikimedia.org/wiki/File:Psittacus_erithacus_-perching_on_tray-8d.jpg.
Fig. 14.01: Copyright © 2006 by Michael Gwyther-Jones, (CC BY 2.0) at https://commons.wikimedia.org/wiki/File:Psittacus_erithacus_-captivity-6.jpg.
Fig. 14.02: Hartman3 / Copyright in the Public Domain.
Fig. 14.03: Copyright © by Remezeath, (CC BY-SA 4.0) at https://commons.wikimedia.org/wiki/File:African_grey_timneh.jpg.
Fig. 14.04: Copyright © 2007 by Peter Fuchs, (CC BY-SA 2.0) at https://commons.wikimedia.org/wiki/File:Psittacus_erithacus_timneh-parrot_on_cage.jpg.
Fig. 14.05: Copyright © 2009 by Lizzy Foulkes, (CC BY-SA 2.0) at https://commons.wikimedia.org/wiki/File:Amazona_oratrix_-Vancouver_Aquarium-8a-4c.jpg.
Fig. 14.06: Copyright © 2006 by Matthew Watts, (CC BY-SA 2.0) at https://commons.wikimedia.org/wiki/File:Blue-fronted_Amazon_(Amazona_aestiva)-head-8.jpg.
Fig. 14.07: Copyright © 2007 by Matt Edmonds, (CC BY-SA 3.0) at https://commons.wikimedia.org/wiki/File:Yellow-naped_Amazon.jpg.
Fig. 14.08: Copyright © 2009 by Jerry Tillery, (CC BY 2.0) at https://commons.wikimedia.org/wiki/File:Melopsittacus_undulatus_-Fort_Worth_Zoo-8a-4c.jpg.
Fig. 14.09: Copyright © 2010 by Jen Smith, (CC BY-SA 2.0) at https://commons.wikimedia.org/wiki/File:Melopsittacus_undulatus_-_English_Budgie_and_American_Parakeets.jpg.
Fig. 14.10: Copyright © 2006 by Workman , (CC BY-SA 3.0) at https://commons.wikimedia.org/wiki/File:Blue_Budgerigars.jpg.
Fig. 14.11: Jana (Jayleane) / Copyright in the Public Domain.
Fig. 14.12: Jana (Jayleane) / Copyright in the Public Domain.
Fig. 14.13: Copyright © 2013 by Astrophysique05, (CC BY-SA 3.0) at https://commons.wikimedia.org/wiki/File:Calopsitte_lutino.jpg.
Fig. 14.14: Copyright © 2004 by www.viajar24h.com, (CC BY 2.0) at https://commons.wikimedia.org/wiki/File:Cacatua_alba_-Bali_Bird_Park_-crest-8.jpg.
Fig. 14.15: Copyright © 2007 by Lenore Edman, (CC BY 2.0) at https://commons.wikimedia.org/wiki/File:Cacatua_galerita_-on_short_grass_-Sydney-8c.jpg.
Fig. 14.16: Copyright © by JJ Harrison, (CC BY-SA 3.0) at https://commons.wikimedia.org/wiki/File:Female_Galah_Outside_Nest.jpg.
Fig. 14.17: Copyright © 2010 by Brian Gratwicke, (CC BY 2.0) at https://commons.wikimedia.org/wiki/File:Aratinga_solstitialis_-Hamilton_Zoo_-white_flowers-8.jpg.
Fig. 14.18: Copyright © by John Graziano, (CC BY-SA 3.0) at https://commons.wikimedia.org/wiki/File:GnocchiPortrait_-_01.jpg.
Fig. 14.19: Copyright © by Eliya, (CC BY 2.0) at https://commons.wikimedia.org/wiki/File:Parrots_of_telegraph_hill.jpg.
Fig. 14.20: Copyright © by Doug Janson, (CC BY-SA 3.0) at https://commons.wikimedia.org/wiki/File:Eclectus_roratus-20030511.jpg.
Fig. 14.21: Copyright © 2011 by H. Zell, (CC BY-SA 3.0) at https://commons.wikimedia.org/wiki/File:Eclectus_roratus_01.jpg.
Fig. 14.22: Copyright © 2008 by Peter Békési, (CC BY-SA 2.0) at https://commons.wikimedia.org/wiki/File:Eclectus_roratus_(male)_-juvenile_pet_-8d.jpg.
Fig. 14.23: Copyright © 2008 by Mats Lindh, (CC BY 2.0) at https://commons.wikimedia.org/wiki/File:Rainbow_Lorikeet_(Trichoglossus_haematodus)_-drinking.jpg.
Fig. 14.24: Copyright © 2012 by Robert Howie, (CC BY-SA 2.0) at https://commons.wikimedia.org/

wiki/File:Eos_bornea_-Taronga_Zoo,_Sydney,_Australia-8a_(1).jpg.
Fig. 14.25: Copyright © 2007 by Doug Janson, (CC BY-SA 3.0) at https://commons.wikimedia.org/wiki/File:Lorius_garrulus-20070107.jpg.
Fig. 14.26: Copyright © 2007 by Peter Békési, (CC BY 2.0) at https://commons.wikimedia.org/wiki/File:Agapornis_roseicollis_-Peach-faced_Lovebird_pet_on_perch.jpg.
Fig. 14.27: Copyright © 2011 by Takashi Hososhima, (CC BY-SA 2.0) at https://commons.wikimedia.org/wiki/File:Agapornis_fischeri_-Ueno_Zoo,_Japan_-three-8a-4c.jpg.
Fig. 14.28: Copyright © 2008 by Epoulin10, (CC BY 3.0) at https://commons.wikimedia.org/wiki/File:Masked_Lovebird_(Agapornis_personata)_pet_on_cage.png.
Fig. 14.29: Copyright © 2009 by LancerEvolution, (CC BY 2.0) at https://commons.wikimedia.org/wiki/File:Ara_macao,_Ara_ararauna_and_Ara_militaris.jpg.
Fig. 14.30: Copyright © 2009 by Nori Almeida, (CC BY 2.0) at https://commons.wikimedia.org/wiki/File:Anodorhynchus_hyacinthinus_-Mato_Grosso_-Brazil-8b.jpg.
Fig. 14.31: Copyright © 2011 by , (CC BY 2.0) at https://commons.wikimedia.org/wiki/File:Ara_chloropterus_-Birmingham_Zoo,_Alabama,_USA-8a.jpg.
Fig. 14.32: Copyright © 2008 by Ujorge, (CC BY-SA 3.0) at https://commons.wikimedia.org/wiki/File:Myiopsitta_monachus_-Old_San_Juan_-Puerto_Rico.jpg.
Fig. 14.33: Copyright © 2009 by Lee, (CC BY-SA 2.0) at https://commons.wikimedia.org/wiki/File:Myiopsitta_monachus_-pet_-blue_mutation-8a.jpg.
Fig. 14.34: Copyright © by BerserkerBen, (CC BY-SA 3.0) at https://commons.wikimedia.org/wiki/File:Quaker_Parrot.jpg.
Fig. 14.35: Jeanne boleyn / Copyright in the Public Domain.
Fig. 14.36: Copyright © 2014 by RealAmnz, (CC BY-SA 3.0) at https://commons.wikimedia.org/wiki/File:Canario_photo.jpg.
Fig. 14.37: Copyright © 2007 by www.canarioscantabria.com, (CC BY 3.0) at https://commons.wikimedia.org/wiki/File:NORWICH1.jpg.
Fig. 14.38: Copyright © 2009 by Peripitus, (CC BY-SA 3.0) at https://commons.wikimedia.org/wiki/File:Taeniopygia_guttata_-_profile_-_dundee_wildlife_park.jpg.
Fig. 14.39: Copyright © 2009 by Peripitus, (CC BY-SA 3.0) at https://commons.wikimedia.org/wiki/File:Taeniopygia_guttata_-_perched_-_dundee_wildlife_park.jpg.
Fig. 14.40: Copyright © 2012 by DickDaniels, (CC BY-SA 3.0) at https://commons.wikimedia.org/wiki/File:Cut-throat_Finch_RWD7.jpg.
Fig. 14.41: Copyright © 2008 by Martin Pot, (CC BY-SA 3.0) at https://commons.wikimedia.org/wiki/File:Male_adult_Gouldian_Finch.jpg.
Fig. 14.42: Copyright © 2006 by J.M. Garg, (CC BY-SA 3.0) at https://commons.wikimedia.org/wiki/File:Blue_Rock_Pigeon_(Columba_livia)_in_Kolkata_I_IMG_9762.jpg.
Fig. 14.43: Copyright © 2007 by Iruka, (CC BY-SA 3.0) at https://commons.wikimedia.org/wiki/File:Collared_Dove_-upper_body_profile-8.jpg.
Fig. 14.44: Copyright © 2004 by Jim Gifford, (CC BY-SA 2.0) at https://commons.wikimedia.org/wiki/File:Fantail(white_self).jpg.
Fig. 14.45: Copyright © 2008 by Jim Gifford, (CC BY-SA 2.0) at https://commons.wikimedia.org/wiki/File:Ice_pigeon,_muffed(silver_chequer).jpg.
Fig. 14.46: Copyright © 2008 by Jim Gifford, (CC BY-SA 2.0) at https://commons.wikimedia.org/wiki/File:Italian_owl(ice_silver).jpg.
Fig. 14.47: Copyright © 2008 by Jim Gifford, (CC BY-SA 2.0) at https://commons.wikimedia.org/wiki/File:American_show_racer(blue_chequer).jpg.
Fig. 14.48: Copyright © 2007 by Jim Gifford, (CC BY-SA 2.0) at https://commons.wikimedia.org/wiki/File:Brunner_pouter.jpg.
Fig. 14.49: Copyright © 2007 by Jim Gifford, (CC BY-SA 2.0) at https://commons.wikimedia.org/wiki/File:English_long_faced_muffed_tumbler(black_spangle).jpg.
Fig. 14.50: Copyright © 2007 by Jim Gifford, (CC BY-SA 2.0) at https://commons.wikimedia.org/wiki/File:Ancient_(crested_magpie-marked).jpg.
Fig. 14.51: Copyright © 2008 by Graham Manning, (CC BY-SA 3.0) at https://commons.wikimedia.org/wiki/File:Carneau.jpg.

9 781516 553402

Printed by Libri Plureos GmbH in Hamburg,
Germany